THEORY OF VISCOELASTICITY

SECOND EDITION

RICHARD M. CHRISTENSEN

LAWRENCE LIVERMORE NATIONAL LABORATORY
AND STANFORD UNIVERSITY

DOVER PUBLICATIONS, INC.
MINEOLA, NEW YORK

Bibliographical Note

This Dover edition, first published in 2003, is an unabridged republication of the second edition of the work, originally published by Academic Press, Inc., New York, in 1982 under the title *Theory of Viscoelasticity: An Introduction.*

Library of Congress Cataloging-in-Publication Data

Christensen, R. M. (Richard M.)
Theory of viscoelasticity / Richard M. Christensen.—2nd ed.
p. cm.
Reprint. Originally published: New York : Academic Press, 1982.
Includes bibliographical references and index.
ISBN 0-486-42880-X (pbk.)
1. Viscoelasticity. I. Title.

TA418.2 .C48 2003
620.1'1232—dc21

2002041301

Manufactured in the United States of America
Dover Publications, Inc., 31 East 2nd Street, Mineola, N.Y. 11501

Contents

III. Thermoviscoelasticity

IV. Mechanical Properties and Approximate Transform Inversion

V. Problems of a Nontransform Type

VI. Wave Propagation

Preface to Second Edition

Viscoelasticity theory has provided a rigorous and broadly based mathematical framework from which to predict material behavior. The theory has undergone considerable development since the first edition of this book. This second edition consolidates many of these recent results.

The chapter structure of the book has been revised and enlarged to accommodate developments in three major areas: (i) approximations for practical applications, (ii) problems for which integral transform methods do not apply, and (iii) nonlinear behavior. The following new topics are treated at the level of independent, self-contained sections:

Spectrum-type representation of properties
Glass transition criterion
Heat conduction
Approximate interrelationships among properties
Approximate inversion of Laplace transform
Approximate solutions for dynamic problems
Extended correspondence principle
Crack growth modeled by local failure
Crack growth modeled by energy balance
Thermoviscoelastic stress analysis
Viscoelastic Rayleigh waves
Optimal strain history path
Nonlinear behavior of elastomers
Nonlinear acceleration waves
Viscometric flows
Nonviscometric flows
Viscoelastic lubrication

Also, the original section on nonlinear mechanical properties has been extended.

Any realistic appraisal of the linear theory reveals it to be reasonably complete and comprehensive. The status of the general nonlinear form of the theory is much less complete, even though admirable progress has been made. The present study of the nonlinear theory reflects this status. Nevertheless, the practical applications of both the linear and nonlinear theories will continue to provide opportunities for technical advancement.

I appreciate the support of my work given by the Lawrence Livermore National Laboratory in the related area of composite materials. My gratitude goes to the many people who have provided me with helpful comments on the first edition of this book. With regard to preparing this second edition, I happily acknowledge the accurate and rapid typing by Ms. Sheila Slavin. Finally, I wish to thank Kristy, Lori, and Kurt C. for help with manuscript preparation in many ways, large and small.

RICHARD CHRISTENSEN

Danville, California

Preface to First Edition

The concept of the viscoelastic behavior of materials, though old in origin, has only recently come into the prominence of widespread attention and application. A result of the recent activity in the field, caused mainly by the advent of polymers, is that a great many different aspects of the theory and the means of its application have been developed. This book is intended to integrate many of these diverse theoretical developments to provide a reasonably complete and consistent description of the linear theory of the viscoelastic behavior of materials. Also, an introductory treatment is given for the general nonlinear theory of viscoelasticity.

The approach followed here is to derive the relevant theoretical formulations from a continuum mechanics point of view, and to illustrate and discuss some of the techniques of solution of problems. The first five chapters deal with various aspects of the linear theory, under both isothermal and non-isothermal conditions, and including dynamic and quasi-static problems. The distinction between solids and fluids is drawn, and the limited applicability of the linear theory in the case of the fluids is discussed. After this considerable involvement with the linear theory, Chapter VI is concerned with a brief examination of a nonlinear theory of viscoelasticity. Separate derivations of the nonlinear theory are given for application to solids and fluids. The common characteristics of the linear and nonlinear theories are noted, as are some of the differences. The last chapter is comprised of a short study of the means of determining mechanical properties, appropriate to both the linear and nonlinear theories. Taken as a whole, the book is primarily designed to convey the theoretical characterization of the subject. However, the spirit of the formulations is strongly influenced by the desire to obtain results which ultimately are suitable for practical application. Furthermore, the inclusion of the last chapter on mechanical properties determination is intended to provide an exposure to some specific aspects of the practical application of the theory.

This book was conceived for use as a text for graduate level instruction. Some of the material presented here is based upon notes from a graduate course on the subject which I taught at the University of California, Berkeley, in 1966 and again in 1970. In this connection, it should perhaps be mentioned that several of the developments given here do not appear to be available in the literature. Whether used for instruction or reference purposes, it is presumed that this book shall be supplemented by a book on the linear theory of elasticity which employs Cartesian tensor notation. Many of the results from elasticity theory will merely be recalled here and then applied. A one semester course on this subject can be based upon the material herein. For a course on viscoelasticity shorter than one semester, Chapters I, III, and VI are basic, respectively, to the linear isothermal theory, the linear nonisothermal theory, and the nonlinear theory. Chapters II, IV, and V depend upon one or both of Chapters I and III, but may be treated independently of each other. Chapter VII relates to nearly all of the previous developments. An abbreviated treatment of the linear theory of viscoelasticity could be given at the end of a course on elasticity by covering Chapters I and II, or parts thereof.

I would like to express my appreciation to Professor P. M. Naghdi of the University of California, Berkeley, for many very helpful technical discussions over recent years, especially on the thermodynamical aspects of viscoelasticity. Also, Professor Naghdi kindly made available to me, before publication, the results of some research on a theory of thermorheologically simple materials; the presentation of Section 3.6 is partially based upon this work. Although the coverage of the book is intentionally limited to homogeneous material conditions, some of the examples and derivations were stimulated by my work on the mechanical behavior of composite polymeric materials, at Shell Development Company, to whom I am grateful. Finally, thanks are due Kristine Christensen and Clara Anderson for help with manuscript preparation and Joyce Shivel for the typing.

RICHARD CHRISTENSEN

THEORY OF VISCOELASTICITY

Chapter I Viscoelastic Stress Strain Constitutive Relations

1.1. INTRODUCTION

The general development and broad application of the linear theory of viscoelasticity is a relatively recent occurrence. In fact, the activity in this field has been primarily due to the large scale development and utilization of polymeric materials. Many of these newly developed materials exhibit mechanical response characteristics which are outside the scope of such theories of mechanical behavior as elasticity and viscosity; thus, the need for a more general theory is quite apparent.

To be more specific, the theory of elasticity may account for materials which have a capacity to store mechanical energy with no dissipation of the energy. On the other hand, a Newtonian viscous fluid in a nonhydrostatic stress state implies a capacity for dissipating energy, but none for storing it. But, then, materials which must be outside the scope of these two theories are those for which some, but not all, of the work done to deform them, can be recovered. Such materials possess a capacity to both store and dissipate mechanical energy.

A different way of characterizing these materials is through the nature of their response to a suddenly applied uniform distribution of surface tractions on a specimen. The term "suddenly applied" loading state or stress state as used in the present context does not imply rates sufficiently great to cause the excitation of a dynamic response in the specimen. An elastic material, when subjected to such a suddenly applied loading state held constant thereafter, responds instantaneously with a state of deformation which remains constant. A Newtonian viscous fluid responds to a suddenly applied state of uniform shear stress by a steady flow process. There are, however, materials for which a suddenly applied and maintained state of uniform stress induces an instantaneous deformation followed by a flow process which may or may not be limited in magnitude as time grows. A material which responds in this manner is said to exhibit both an instantaneous elasticity effect and creep characteristics. This behavior is clearly not described by either an elasticity or a viscosity theory but combines features of each.

It is instructive to consider a situation which represents a generalization of the response to a single suddenly applied change in surface tractions. Suppose a material having the instantaneous elasticity and creep characteristics described above is subjected to two nonsimultaneously applied sudden changes in uniform stress, superimposed upon each other. After the first application of stress, but before the second, the material responds in some time dependent manner which depends upon the magnitude of the first stress state. But now consider the situation that exists at an arbitrarily small interval of time after the sudden application of the second stress state. The material not only experiences the instantaneous response to the second change in surface tractions but also it experiences a continuing time dependent response due to the first applied level of stress. An elastic material would respond only to the total stress level at every instant of time. Thus, this more general type of material possesses a characteristic which can be descriptively referred to as a memory effect. That is, the material response is not only determined by the current state of stress, but is also determined by all past states of stress, and in a general sense, the material has a memory for all past states of stress. A similar situation exists if one considers the deformation as being specified, and thus, the current stress depends upon the entire past history of deformation.

It is this latter observation that materials can have a capacity for memory which will be given a simple but fundamental mathematical characterization in the next section. Thereafter a representation theorem will be used to form the linear viscoelastic stress strain constitutive relation, all under isothermal conditions. In so doing, the use of the term "memory" will be made more precise; however, at this point, it is well to take note of the fact that there are other theories of mechanical behavior of materials which have a memory of deformation but which are different in some fundamental way from what will be considered here. For example, the incremental theory of plasticity has memory effects in as much as a final state of deformation depends not only upon the final state of stress, but also upon the path in stress space traversed to reach this final state. However, the underlying difference between these two theories is that the plasticity theory is independent of the time scale involved in loading and unloading programs while the viscoelastic theory has a specific time or rate dependence.

All derivations and applications in this and the succeeding chapters assume conditions of homogeneity. In many cases the extension to certain types of inhomogeneous materials is easily obtained, while in other cases the extension is difficult if not impossible. Thus, it is necessary to consider individually the extension of methods and analyses for homogeneous viscoelastic materials to inhomogeneous materials.

Finally, note should be taken of the fact that even though most of the developments in viscoelasticity theory are recent the basic linear and isothermal field theory formulation has been available for a much longer time. While there were several early contributors, such as Maxwell, Kelvin, and Voigt, Boltzmann [1.1] in 1874 apparently supplied the first formulation of a three-dimensional theory of isotropic viscoelasticity, while Volterra [1.2] obtained comparable forms for anisotropic solids in 1909.

1.2. INTEGRAL FORM OF STRESS STRAIN CONSTITUTIVE RELATIONS, STIELTJES CONVOLUTION NOTATION

The formulation of the isothermal viscoelastic stress strain constitutive relations will now be considered. The statement of the other relevant field equations needed to form a complete theory will be deferred until the next chapter. First, the definitions of stress and strain will be briefly stated. For more detailed information reference should be made to elasticity texts, such as Sokolnikoff [1.3].

The usual Cartesian tensor notation is employed with Latin indices having the range 1, 2, 3 and repeated indices imply the summation convention. Let the coordinates of a point in the body in a fixed reference configuration be denoted by X_i, referred to rectangular Cartesian axes. In the case of solids the fixed reference configuration is taken to be the undeformed configuration. Let x_i denote the deformed configuration coordinates of the same point. The complete history of the motion of the continuum is specified by

$$x_i(\tau) = x_i(X_j, \tau), \qquad -\infty < \tau \leqslant t$$

where τ is the time variable and t denotes current time.

The components of the displacement vector are defined by

$$u_i(\tau) = x_i(\tau) - X_i .$$

A measure of the deformation is supplied by

$$\frac{\partial u_i(\tau)}{\partial X_j} = \frac{\partial x_i(\tau)}{\partial X_j} - \delta_{ij}$$

where δ_{ij} is the Kronecker symbol. Let ϵ be defined by

$$\epsilon = \sup_{\tau} | u_{i,j}(\tau)|$$

where the notation $u_{i,j} = \partial u_i/\partial X_j$ is used, $| \ |$ denotes magnitude, and sup denotes least upper bound. We say the deformation is infinitesimal at all times τ if $\epsilon \ll 1$. Furthermore, the theory of viscoelasticity which we shall develop is said to be infinitesimal if $\epsilon \ll 1$ and if the displacements of the body of continuum are small compared with all characteristic dimensions of the body. Under these conditions the infinitesimal strain tensor ϵ_{ij} is defined by

$$\epsilon_{ij}(\tau) = \tfrac{1}{2}[u_{i,j}(\tau) + u_{j,i}(\tau)]$$

where it is immaterial whether the differentiation is with respect to the x_i or X_i coordinates because the displacements are infinitesimal. Henceforth, in the infinitesimal theory, the derivatives will be written with respect to the x_i coordinates.

Stress is defined in the following way. Let δa be the area of an element in a surface which may either designate the boundary of the body, or may simply be a surface inside the body. Let **n** denote a unit vector normal to the infinitesimal surface element δa. The stress vector $\boldsymbol{\sigma}$ is defined in terms of the resultant force **G** acting on the surface element δa through

$$\sigma_i = \lim_{\delta a \to 0} \frac{G_i}{\delta a}.$$

This stress vector is defined on the side of δa corresponding to the positive direction of **n**. For each different orientation of the surface element δa, there will be a different stress vector $\boldsymbol{\sigma}$. The stress tensor σ_{ij} is defined through the transformation which relates the components of the stress vector to the orientation of the surface element; that is,

$$\sigma_i = \sigma_{ji} n_j .$$

This relationship is obtained by the application of the balance of linear momentum to a small tetrahedron. Symmetry of the stress tensor $\sigma_{ij} = \sigma_{ji}$ follows from the assumed balance of angular momentum applied to a small volume element. In the present infinitesimal theory, the area δa may be taken relative to either the reference or the deformed configuration with only negligible differences, since the displacements between these two states are of infinitesimal order. This is in contrast to the general theory presented in Chapters 8 and 9, in which none of the preceding smallness assumptions are involved.

In the present context of an infinitesimal theory, it is the constitutive relation between stress σ_{ij} and strain ϵ_{ij} which is being sought. This relationship is assumed to be linear, which is consistent with the smallness assumptions already introduced. The infinitesimal theory is said to be linear, when in addition to the linear relations already noted, the additional field equations, which are needed to comprise a complete theory, are also linear. These additional relations are stated in Chapter 2.

The hypothesis that the current value of the stress tensor depends upon the complete past history of the components of the strain tensor is formally expressed as

$$\sigma_{ij}(t) = \underset{s=0}{\overset{\infty}{\psi_{ij}}} \left(\epsilon_{kl}(t-s), \epsilon_{kl}(t) \right) \tag{1.1}$$

where $\underset{s=0}{\overset{\infty}{\psi_{ij}}}(\ )$ is a linear tensor valued functional which transforms each strain history $\epsilon_{ij}(t)$, $-\infty \leqslant t \leqslant \infty$, into a corresponding stress history $\sigma_{ij}(t)$. The functional has a parametric dependence upon the current value of strain $\epsilon_{kl}(t)$, which corresponds to the instantaneous elasticity effect mentioned in the introduction. In general, all field variables are not only functions of time, but also are functions of position x_i; but, since only local effects are being considered here, the dependence of σ_{ij} and ϵ_{ij} upon x_i is suppressed.

If the strain history $\epsilon_{ij}(t)$ is assumed to be continuous and the functional is assumed to be linear, the Riesz representation theorem [1.4] may be employed to write the functional in (1.1) as a Stieltjes integral, giving

$$\sigma_{ij}(t) = \int_0^\infty \epsilon_{kl}(t-s)\, dG_{ijkl}(s) \tag{1.2}$$

where the integrating functions $G_{ijkl}(t)$ comprise a fourth order tensor such that $G_{ijkl}(t) = 0$ for $-\infty < t < 0$ and each component is of bounded variation in every closed subinterval of $-\infty < t < \infty$. Actually, the integral in (1.2) is the convolution form of a Stieltjes integral. This form implies that the constitutive relation (1.2) is independent of any shifts in the time scale. Such behavior is known as time translation invariance, and all results in this book assume this condition. The symmetry of the stress and strain tensors imply the following relations:

$$G_{ijkl}(t) = G_{jikl}(t) = G_{ijlk}(t). \tag{1.3}$$

Now, taking $\epsilon_{ij}(t) = 0$ for $t < 0$ and assuming that $G_{ijkl}(t)$ and its first time derivative are continuous on the interval $0 \leqslant t < \infty$, then (1.2) can be written in the following form:

$$\sigma_{ij}(t) = G_{ijkl}(0)\, \epsilon_{kl}(t) + \int_0^t \epsilon_{kl}(t-s) \frac{dG_{ijkl}(s)}{ds}\, ds. \tag{1.4}$$

It is seen that (1.4) can be thought of as being obtained from (1.2) through the integration of the Dirac delta function involved in the differential of the integrating functions $G_{ijkl}(t)$ at $t = 0$. Alternatively, (1.4) can be obtained from (1.2) through the connection between Stieltjes and Riemann convolutions. Gurtin and Sternberg [1.5] recognize this latter line of reasoning to be the more rigorous and follow it in their derivation.

Another form of the stress constitutive relation can be obtained from (1.4) through the change of variable $\tau = t - s$, and an integration by parts to form

$$\sigma_{ij}(t) = \int_0^t G_{ijkl}(t-\tau) \frac{d\epsilon_{kl}(\tau)}{d\tau}\, d\tau. \tag{1.5}$$

Up until this point in the derivation $\epsilon_{kl}(t)$ has been taken to be a continuous function of time, and in fact, this was a necessary requirement for the use of the Riesz representation theorem. It would, however, be convenient to have the means of treating discontinuous strain histories. This is accomplished in Ref. [1.5] by resolving a strain history $\epsilon_{ij}(t)$ which has a step discontinuity at $t = 0$ into a uniformly convergent sequence of continuous functions with the resulting generalization of (1.5) as

$$\sigma_{ij}(t) = G_{ijkl}(t)\,\epsilon_{kl}(0) + \int_0^t G_{ijkl}(t-\tau)\,\frac{d\epsilon_{kl}(\tau)}{d\tau}\,d\tau. \tag{1.6}$$

The derivation given in Ref. [1.5] thus avoids involvement with delta functions; otherwise, (1.6) follows directly from (1.5) through the integration of the resulting delta function. Henceforth, however, in derivations given here the standard step function and delta function relations as given in Appendix A will be accepted, and the form (1.5) will be allowed for use with discontinuous strain histories. Also, although $\epsilon_{ij}(t) = 0$ for $t < 0$, in the derivation of (1.5), the lower limit in (1.5) may be changed from 0 to $-\infty$ through a shift of the time scale, as long as $\epsilon_{ij}(t) \to 0$ as $t \to -\infty$.

The stress strain relation (1.5) is one form of the general viscoelastic constitutive relations. The integrating functions $G_{ijkl}(t)$ are mechanical properties of the material and are termed relaxation functions. The determination of relaxation functions will be considered in Section 1.5 and Chapter 4. The stress strain relation (1.5) is seen to follow from nothing more than the memory hypothesis, the smoothness assumptions, and the mathematical representation theorem. It has not been necessary to appeal to physical intuition or model representations in any way. This derivation will suggest the means of proceeding in the more complicated nonisothermal case given in Chapter 3. Having obtained the stress strain relation (1.5), it now can be seen to have a very simple physical interpretation. It can be considered to be the formulation of Boltzmann's superposition principle such that the current stress is determined by the superposition of the responses to the complete spectrum of increments of strain. This point of view has been discussed extensively by Staverman and Schwarzl [1.6].

An alternative form of the stress strain relation may be obtained by reversing the roles of stress and strain in the preceding derivation in such a way that current strain is determined by the current value and past history of stress. It is then found that

$$\epsilon_{ij}(t) = \int_{-\infty}^t J_{ijkl}(t-\tau)\,\frac{d\sigma_{kl}(\tau)}{d\tau}\,d\tau \tag{1.7}$$

where

$$J_{ijkl}(t) = J_{jikl}(t) = J_{ijlk}(t) \tag{1.8}$$

and $J_{ijkl}(t) = 0$ for $-\infty < t < 0$, and $J_{ijkl}(t)$ and its first time derivative are continuous on $0 \leqslant t < \infty$. The functions $J_{ijkl}(t)$ are termed creep functions, and, as with relaxation functions, they represent mechanical properties of the material.

It will be of great practical interest to have the isotropic forms of the visco-elastic stress strain relations. Consider first the tensor of relaxation functions. The most general isotropic representation of a fourth order tensor is given by the form

$$G_{ijkl}(t) = \tfrac{1}{3}[G_2(t) - G_1(t)]\,\delta_{ij}\delta_{kl} + \tfrac{1}{2}[G_1(t)](\delta_{ik}\delta_{jl} + \delta_{il}\delta_{jk}) \tag{1.9}$$

where $G_1(t)$ and $G_2(t)$ are independent relaxation functions and δ_{ij} is the Kronecker symbol. Following a procedure parallel to that in elasticity whereby the deviatoric components of stress s_{ij} and strain e_{ij} are introduced, then (1.5), using (1.9), reduces to

$$s_{ij} = \int_{-\infty}^{t} G_1(t-\tau)\,\frac{de_{ij}(\tau)}{d\tau}\,d\tau \tag{1.10}$$

and

$$\sigma_{kk} = \int_{-\infty}^{t} G_2(t-\tau)\,\frac{d\epsilon_{kk}(\tau)}{d\tau}\,d\tau \tag{1.11}$$

where

$$s_{ij} = \sigma_{ij} - \tfrac{1}{3}\delta_{ij}\sigma_{kk}\,, \qquad s_{ii} = 0 \tag{1.12}$$

$$e_{ij} = \epsilon_{ij} - \tfrac{1}{3}\delta_{ij}\epsilon_{kk}\,, \qquad e_{ii} = 0. \tag{1.13}$$

In a similar way, the creep integral form of the stress strain relations (1.7) has the isotropic form

$$e_{ij} = \int_{-\infty}^{t} J_1(t-\tau)\,\frac{ds_{ij}(\tau)}{d\tau}\,d\tau \tag{1.14}$$

$$\epsilon_{kk} = \int_{-\infty}^{t} J_2(t-\tau)\,\frac{d\sigma_{kk}(\tau)}{d\tau}\,d\tau \tag{1.15}$$

where $J_1(t)$ and $J_2(t)$ are the two independent isotropic creep functions. It is seen that $G_1(t)$ and $J_1(t)$ are the relaxation and creep functions appropriate to states of shear, while $G_2(t)$ and $J_2(t)$ are defined relative to states of dilatation.

Obviously (1.14) and (1.15) are not relations which are independent of (1.10) and (1.11), respectively. It follows that the creep and relaxation functions $J_\alpha(t)$ and $G_\alpha(t)$ ($\alpha = 1, 2$) must be related. To display the relationship between these functions it is expedient to introduce the Laplace transformation, the complete definition of which is given in Appendix B. Let $f(t)$ be a continuous function on $0 \leqslant t < \infty$ and let it be of exponential order as $t \to \infty$. Then the Laplace transform $\bar{f}(s)$ of $f(t)$ is defined as

$$\bar{f}(s) = \int_{0}^{\infty} f(t)\,e^{-st}\,dt. \tag{1.16}$$

The Laplace transform of (1.10), (1.11), (1.14), and (1.15) gives, for $\epsilon_{ij} = 0$ for $t < 0$,

$$\begin{aligned} \bar{s}_{ij} &= s\bar{G}_1\bar{e}_{ij} \\ \bar{\sigma}_{kk} &= s\bar{G}_2\bar{\epsilon}_{kk} \\ \bar{e}_{ij} &= s\bar{J}_1\bar{s}_{ij} \\ \bar{\epsilon}_{kk} &= s\bar{J}_2\bar{\sigma}_{kk}\,. \end{aligned} \tag{1.17}$$

It follows from (1.17) that

$$\bar{J}_\alpha = (s^2\bar{G}_\alpha)^{-1}, \qquad \alpha = 1, 2. \tag{1.18}$$

The comparable elasticity relationship between moduli and compliances is $J_\alpha = G_\alpha^{-1}$. The intuitive extension of this elasticity result to viscoelasticity would suggest $J_\alpha(t) = [G_\alpha(t)]^{-1}$ which is seen from (1.18) to be incorrect. However, the initial and final value theorems of the Laplace transformation can be used to show

$$\lim_{t\to 0} J_\alpha(t) = \lim_{t\to 0}\, [G_\alpha(t)]^{-1}$$

and for solids

$$\lim_{t\to \infty} J_\alpha(t) = \lim_{t\to \infty}\, [G_\alpha(t)]^{-1}.$$

The Stieltjes convolution forms mentioned earlier in this section provide a notational convention that can be used profitably in many situations. This possibility will be discussed here and used in Sections 2.12, 5.2, and 5.3.

Following the treatment of Gurtin and Sternberg [1.5] the stress strain relation (1.2) can be written in the form

$$\sigma_{ij} = \epsilon_{kl} * dG_{ijkl} \tag{1.19}$$

where the Stieltjes convolution, $\varphi * d\psi$, of two functions $\varphi(t)$ and $\psi(t)$ is defined by

$$\varphi * d\psi = \int_{-\infty}^{t} \varphi(t-\tau)\, d\psi(\tau) \tag{1.20}$$

where $\psi(t) \to 0$ for $t \to -\infty$ and $\varphi(t)$ is continuous for $0 \leqslant t < \infty$. Under the further assumption that $\varphi(t) = 0$ for $t < 0$ then the form (1.20) can be shown to obey the commutivity relation

$$\varphi * d\psi = \psi * d\varphi. \tag{1.21}$$

Using this property, the stress strain relation (1.2) can be equivalently written as

$$\sigma_{ij} = G_{ijkl} * d\epsilon_{kl}$$

which of course symbolizes the form (1.5). In a similar manner the isotropic form of the stress strain relations can be written as

$$s_{ij} = G_1 * de_{ij} = e_{ij} * dG_1$$

and (1.22)

$$\sigma_{kk} = G_2 * d\epsilon_{kk} = \epsilon_{kk} * dG_2$$

where

$$\epsilon_{ij}(t) = 0 \qquad \text{for} \quad t < 0.$$

The associativity and distributivity properties of Stieltjes convolutions will be useful in the future applications of this notation. These properties, under the above assumptions on $\varphi(t)$ and $\psi(t)$, state the following:

$$\varphi * d(\psi * d\omega) = (\varphi * d\psi) * d\omega = \varphi * d\psi * d\omega \tag{1.23}$$

and

$$\varphi * d(\psi + \omega) = \varphi * d\psi + \varphi * d\omega. \tag{1.24}$$

The proofs of these identities are straightforward, and can be found in Ref. [1.5].

1.3. CONSEQUENCES OF FADING MEMORY AND THE DISTINCTION BETWEEN VISCOELASTIC SOLIDS AND FLUIDS

As was mentioned in Section 1.2, the stress strain relations derived there were based only upon the memory hypothesis, the smoothness assumptions, and a mathematical representation theorem. It is now of some interest to examine the consequences of a further possible physical hypothesis. Specifically we seek to determine any restrictions which may be imposed upon the stress strain relations for the memory effect being restricted to be that of a particular type. The motivation for doing this comes from the extensive theoretical work which has been done upon the type of memory effect known as fading memory. An early mathematical statement of the concept of fading memory was given by Volterra [1.7]. A more recent and more complete statement of a fading memory hypothesis has been given by Coleman and Noll in several references which are mentioned in the discussion by Coleman and Mizel [1.8]. This fading memory hypothesis has been used to great advantage in several works such as Coleman's thermodynamical formulation [1.9] of a nonlinear theory of viscoelasticity. The formal definition of this type of memory effect is more involved than is needed here for application to the infinitesimal deformation theory. Such a formal definition will be profitably used in Chapter 8 for the derivation of a nonlinear theory of viscoelasticity.

For the present purposes of a linear theory, a simple definition of fading memory of the following type will suffice. If the current value of one field variable has a linear functional type dependence upon the complete past history of another field variable, then a fading memory hypothesis implies that current value of the first variable depends more strongly upon the recent history than upon the distant history of the second variable. More specifically, the dependence of the current value of the first variable upon the values of the second variable at previous times is determined through a weighting function which must assign a continuously decreasing dependence upon past events which are continuously more distant from the current time. The use of the term "weighting function" is made clear in the following restriction of the viscoelastic stress strain relations to satisfy the fading memory hypothesis. In the stress strain relation (1.2) or (1.4), the slopes of the relaxation functions are seen to be weighting functions which characterize the extent to which the current value of each component of stress is influenced by the values of the strain at past times. For each component of the viscoelastic stress strain relations to satisfy the fading memory type behavior, it is sufficient that the magnitude of the slope of each component of the relaxation function tensor be a continuously decreasing function of time; thus,

$$\left|\frac{dG_{ijkl}(t)}{dt}\right|_{t=t_1} \leqslant \left|\frac{dG_{ijkl}(t)}{dt}\right|_{t=t_2} \qquad \text{for} \quad t_1 > t_2 > 0. \tag{1.25}$$

This fading memory hypothesis is reasonable in the sense that it would be physically unrealistic to expect materials to have a growing memory for the more distant events. In fact, all relaxation function experimental measurements have produced results in accordance with the criterion (1.25). It is a simple matter to establish the fact that the fading memory hypothesis also is satisfied by the restrictions on the components of the creep function tensor as

$$\left|\frac{dJ_{ijkl}(t)}{dt}\right|_{t=t_1} \leqslant \left|\frac{dJ_{ijkl}(t)}{dt}\right|_{t=t_2} \qquad \text{for} \quad t_1 > t_2 > 0. \tag{1.26}$$

Further restrictions upon the forms of relaxation functions are given in Section 3.3.

The other item to be discussed in this section concerns the distinction between viscoelastic solids and fluids. While it may be intuitively obvious that an elastic medium is a solid, and a viscous medium is a fluid, the situation for viscoelastic materials is less clear since it has some elements of both elastic and viscous behaviour. It is not obvious whether a viscoelastic material should be a solid or a fluid or could be either. In fact it can be either, as will be shown. In a general sense the distinction between solids and fluids is not as simple as it might at first seem to be. The rigorous definition of each, as given by Truesdell and Noll [1.10] is fairly involved and will be considered in Chapter 8. A consequence of the formal definition of a fluid is that a fluid must necessarily be isotropic, in contrast to solids. For our present purposes we will distinguish between isotropic solids and fluids in the following simple and nonrigorous physical manner. A viscoelastic fluid, when subjected to a fixed simple shear state of stress, will respond with a steady state of flow, after transient effects have died out. Also, a viscoelastic fluid, when subjected to a fixed simple shear state of deformation, will produce a stress state which will eventually decay to zero. Contrary to this, an isotropic viscoelastic solid when subjected to a simple shear state of deformation, will have a corresponding component of stress which remains nonzero as long as the state of deformation is maintained. In other words, the viscoelastic fluid has an unlimited number of undeformed configurations while the viscoelastic solid may have only one.

The implications relative to the mechanical properties of the distinction between viscoelastic solids and fluids will now be determined. In the remainder of this section it will be convenient to distinguish between the coordinates of the particles in a particular fixed reference state, and the coordinates of the particles in any state of possible deformation. This is similar to the situation outlined at the beginning of Section 1.2 where the coordinates of a particle in the reference and deformed configurations were denoted by X_i and $x_i = x_i(\tau, X_j)$, respectively. This convention is in contrast to our usual infinitesimal theory notation that x_i are the coordinates in a fixed reference state.

Consider a viscoelastic material which is subjected to a simple shear state of deformation specified by the displacement components from the fixed reference configuration as

$$u_1(x_i, t) = \hat{u}X_2 h(t), \qquad u_2 = u_3 = 0$$

where $h(t)$ is the unit step function. Using this in the strain displacement relations

$$\epsilon_{ij} = \frac{1}{2}\left(\frac{\partial u_i}{\partial X_j} + \frac{\partial u_j}{\partial X_i}\right)$$

then (1.10) gives the only nonzero relation between the components of stress and strain as

$$s_{12}(t) = [G_1(t)/2]\hat{u}$$

where it is recalled that

$$G_1(t) = 0, \qquad t < 0.$$

It follows from the definition of an isotropic viscoelastic solid that for an isotropic viscoelastic material to be a solid it is necessary and sufficient that

$$\lim_{t\to\infty} G_1(t) \to \text{nonzero constant} \qquad \text{(isotropic solid)}.$$

For a viscoelastic material to be a fluid it is necessary that

$$\lim_{t\to\infty} G_1(t) \to 0 \qquad \text{(fluid)}.$$

However, this requirement on the limiting behavior of the relaxation function in shear is not sufficient for a material to be a fluid. As we will next show, sufficiency requires that the material satisfy the steady state flow requirement.

Since a viscoelastic fluid has the property to allow flow in a steady state condition, we should be able to derive a relevant coefficient of viscosity. The existence of such a coefficient of viscosity, in steady flow, ensures that the material is a fluid. In obtaining this coefficient of viscosity, we begin by first recalling the appropriate stress constitutive relation for a Newtonian viscous fluid, given by

$$s_{ij} = 2\eta d_{ij}$$

where

$$d_{ij} = \frac{1}{2}\left[\frac{\partial \dot{x}_i(t)}{\partial x_j(t)} + \frac{\partial \dot{x}_j(t)}{\partial x_i(t)}\right] \tag{1.27}$$

and where $x_i(t)$ are the coordinates at current time t and $\dot{x}_i(t)$ are the components of velocity. Note the distinction between the reference configuration used here and the fixed reference configuration used in an infinitesimal theory. Here the reference configuration is the current configuration $x_i(t)$, while in an infinitesimal theory the reference configuration, given by coordinates X_i, is fixed and for a solid is normally taken to be the undeformed configuration. In a general (non-infinitesimal) state of deformation, the rate of deformation tensor d_{ij} given by (1.27), and the rate of the infinitesimal strain tensor,

$$\dot{\epsilon}_{ij} = \frac{1}{2}\frac{d}{dt}\left(\frac{\partial u_i}{\partial X_j} + \frac{\partial u_j}{\partial X_i}\right)$$

are completely different. But, in the special case of simple shear flow, they can be shown to be the same. This state of flow is specified by

$$\begin{aligned} x_1(t) &= \dot{\gamma} X_2 t h(t) + X_1 \\ x_2(t) &= X_2 \\ x_3(t) &= X_3 . \end{aligned} \tag{1.28}$$

To specify the state of simple shear flow in terms of displacements requires the form

$$u_i(t) = x_i(t) - X_i$$

which defines the components of displacement in terms of the coordinates. Then the flow state is equivalently defined by

$$\begin{aligned} u_1(t) &= \dot{\gamma} X_2 t h(t) \\ u_2(t) &= u_3(t) = 0. \end{aligned} \tag{1.29}$$

With these relations and (1.28) it can now be shown that $\dot{\epsilon}_{ij}$ and d_{ij} are identical. Thus, the state of simple shear flow will be used to determine the equivalent coefficient of viscosity for a viscoelastic fluid, in terms of its relaxation function.

Using the specified displacements (1.29) in (1.10) gives

$$s_{12}(t) = \frac{\dot{\gamma}}{2} \int_0^t G_1(t - \tau) \, d\tau. \tag{1.30}$$

Using the same state of deformation (1.28) in (1.27) for a Newtonian viscous fluid gives

$$s_{12} = \eta\dot{\gamma}. \tag{1.31}$$

The steady state stress will be achieved by the viscoelastic fluid at large values of time. Thus, a comparison of (1.30) and (1.31) at large values of time gives

$$\eta = \frac{1}{2} \int_0^\infty G_1(s) \, ds. \tag{1.32}$$

It follows that if a viscoelastic fluid, in a state of simple shear flow, is subjected only to steady rates of deformation, it will behave exactly as a Newtonian viscous fluid with a coefficient of viscosity given by (1.32), where the integral is assumed to exist. Actually, the use of a coefficient of viscosity defined by (1.32) not only implies a steady rate of deformation of the viscoelastic fluid, but it also implies a vanishingly small rate of deformation. This is necessary for the relation (1.30), which is based upon the infinitesimal theory, to be true at large values of time, as is implied by (1.32). The viscosity determined by (1.32) is called the zero shear rate viscosity. This nonrigorous examination of the flow characteristics of viscoelastic fluids under steady flow conditions will be superseded by the general nonlinear viscoelastic fluid treatments given in Chapters 8 and 9.

For a viscoelastic fluid the creep function in shear has the form

$$J_1(t) = \tilde{J}_1(t) + (t/2\eta)$$

where $\tilde{J}_1(t)$ approaches a finite positive asymptote for large values of time and η is the zero shear rate viscosity defined in (1.32). The derivation of this form is included as an exercise.

For polymeric materials, the distinction between solids and fluids is very simple when considered on a molecular scale. In the case of fluids, the individual long chain molecules are completely unconnected and over long time spans have unlimited mobility with respect to each other. But for solids, there are discrete chemical bonds between adjacent molecules which are called cross links and which prevent unlimited flow. However, in the case of a very lightly cross linked material, the distinction between a solid and a fluid may be almost unnoticeable insofar as the methods and techniques suggested by an infinitesimal theory are concerned.

We now have the means to distinguish between viscoelastic solids and fluids. In the later applications, both types of materials will be considered. However, it is necessary to remember that in applying stress strain relations of the type discussed in this chapter, whether they represent solids or fluids, their application must not in general violate the assumptions of infinitesimal deformation, measured with respect to a fixed reference configuration. In this sense, the distinction between solids and fluids for application in an infinitesimal deformation theory is not involved with differences in methods and techniques of solving problems; rather, the distinction is more important in assessing the appropriateness of applying the theory to particular types of problems. Thus, since the linear theory cannot be used to solve general (noninfinitesimal deformation) flow problems for viscoelastic fluids, its main usefulness in the case of fluids is in deriving the appropriate forms of the constitutive relations, which are widely determined and interpreted experimentally.

1.4. DIFFERENTIAL OPERATOR FORM OF STRESS STRAIN CONSTITUTIVE RELATIONS

The creep and relaxation integral forms of the stress strain constitutive relations are by no means the only possible forms. Two other forms will be given; the first given here involves differential operators, the second will be given in Section 1.6.

Relevant to isotropic materials, we consider the following differential operator form of a relation between the deviatoric components of stress and strain:

$$p_0 s_{ij}(t) + p_1 \frac{ds_{ij}(t)}{dt} + p_2 \frac{d^2 s_{ij}(t)}{dt^2} + \cdots = q_0 e_{ij}(t) + q_1 \frac{de_{ij}(t)}{dt} + q_2 \frac{d^2 e_{ij}(t)}{dt^2} + \cdots. \tag{1.33}$$

Or, more compactly, we write this as

$$P(D)\, s_{ij}(t) = Q(D)\, e_{ij}(t) \tag{1.34}$$

where

$$P(D) = \sum_{k=0}^{N} p_k D^k \qquad Q(D) = \sum_{k=0}^{N} q_k D^k \tag{1.35}$$

with D designating the operator d/dt. Relation (1.34) is certainly a possible relationship between stress and strain, but at this point it is not clear whether it has any significance for viscoelasticity. To investigate this possibility it is helpful to see if any connection between (1.34) and (1.10) can be established. To this end we take the Laplace transform of (1.34), using the relationships on the transforms of derivatives, then (1.34) becomes

$$P(s)\,\bar{s}_{ij}(s) - \frac{1}{s}\sum_{k=1}^{N} p_k \sum_{r=1}^{k} s^r s_{ij}^{(k-r)}(0)$$

$$= Q(s)\,\bar{e}_{ij}(s) - \frac{1}{s}\sum_{k=1}^{N} q_k \sum_{r=1}^{k} s^r e_{ij}^{(k-r)}(0) \tag{1.36}$$

where

$$P(s) = \sum_{k=0}^{N} p_k(s)^k$$
$$Q(s) = \sum_{k=0}^{N} q_k(s)^k \tag{1.37}$$

and $s_{ij}^{(k-r)}(0)$ designates the $k - r$ order derivative of $s_{ij}(t)$ evaluated at $t = 0$, with a similar relationship for $e_{ij}(t)$. The Laplace transform of (1.10) is given by the first equation in (1.17). This is seen to be equivalent to (1.36) if

$$s\bar{G}_1 = Q(s)/P(s) \tag{1.38}$$

and

$$\sum_{r=k}^{N} p_r s_{ij}^{(r-k)}(0) = \sum_{r=k}^{N} q_r e_{ij}^{(r-k)}(0), \qquad k = 1, 2, \ldots, N. \tag{1.39}$$

Equation (1.39) supplies a requirement upon the initial conditions; thus, the initial conditions upon stress and strain are not completely independent, and relations such as (1.39) must be satisfied.

Relation (1.38) gives the conditions under which the relaxation integral and differential operator forms of the deviatoric stress strain relation are equivalent. In an entirely similar manner, the independent dilatational part of the isotropic stress strain relations can be written in differential operator form and the equivalence with the relaxation integral form can be ascertained. This would then imply a relation of the type

$$L(D)\,\sigma_{kk}(t) = M(D)\,\epsilon_{kk}(t)$$

where $L(D)$ and $M(D)$ are operators similar to those in (1.35) but with independent coefficients.

1.5. RELAXATION AND CREEP CHARACTERISTICS, MECHANICAL MODELS

Two examples will be given to illustrate the connections implied by relations of the type of (1.38). For simplicity of discussion only $G_1(t)$ and $J_1(t)$ will be considered, although similar results will be understood to apply for $G_2(t)$ and $J_2(t)$.

Consider a state of simple shear with nonzero stress and strain components σ_{12} and ϵ_{12}. We take the strain as being specified in terms of a unit step function, through

$$\epsilon_{12}(t) = \epsilon_0 h(t) \qquad (1.40)$$

where ϵ_0 is the amplitude. The relaxation integral relation (1.10) subject to (1.40) becomes

$$\sigma_{12}(t) = G_1(t)\,\epsilon_0\,.$$

The resulting stress is directly related to the relaxation function. The procedure just described suggests the means of determining relaxation functions. Typical relaxation functions have a form of decaying functions of time as shown in Fig. 1.1. Such results also show the slope $dG_1(t)/dt$ to decay with time which is consistent with the fading memory hypothesis of Section 1.3. Experimentally

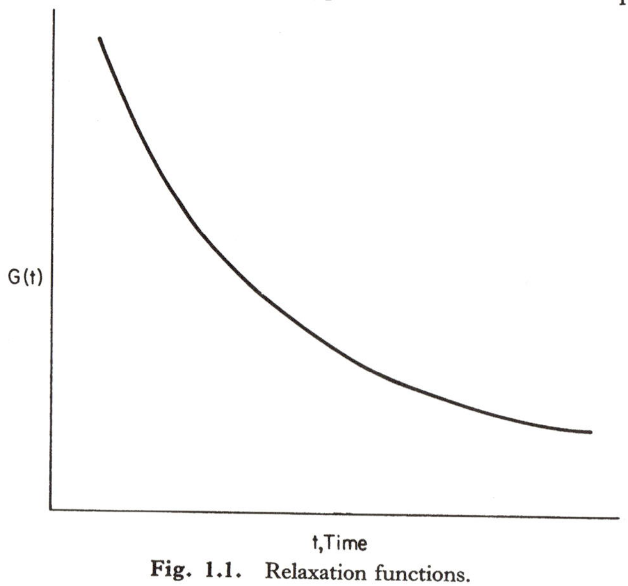

Fig. 1.1. Relaxation functions.

determined relaxation functions will be considered in Chapter 4. At this point observe that the simplest relaxation function having the fading memory characteristics of Section 1.3 would be that of the single decaying exponential

$$G_1(t) = G_0 e^{-t/t_1} h(t) \tag{1.41}$$

where G_0 is the amplitude and t_1 is a positive time constant that determines the rate of decay. Parameter t_1 is called the relaxation time. Using the Laplace transform of (1.41) in (1.38) and recalling (1.37) there results

$$\frac{G_0}{s + (1/t_1)} = \frac{\sum_{k=0}^{N} q_k(s)^k}{\sum_{k=0}^{N} p_k(s)^{k+1}}. \tag{1.42}$$

With $N = 1$, Eq. (1.42) is satisfied by

$$p_0 = \frac{1}{t_1}, \qquad p_1 = 1, \qquad q_0 = 0, \qquad q_1 = G_0. \tag{1.43}$$

The corresponding differential operator stress strain relation from (1.34) is then given by

$$\frac{1}{t_1}\sigma_{12}(t) + \frac{d\sigma_{12}(t)}{dt} = G_0 \frac{d\epsilon_{12}(t)}{dt}. \tag{1.44}$$

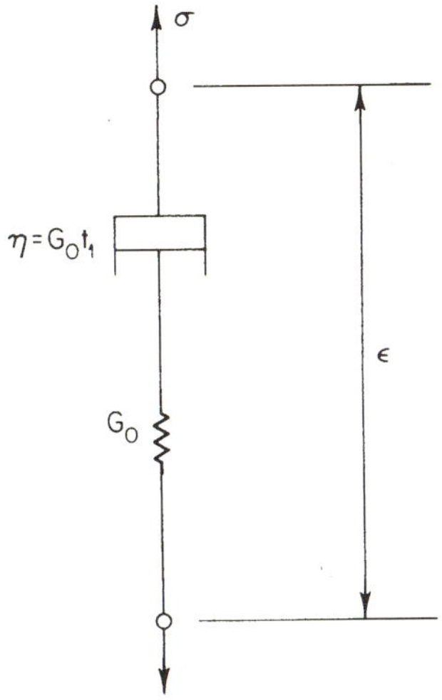

Fig. 1.2. Maxwell model.

A very simple physical interpretation of (1.44) can be devised in terms of a mechanical model involving springs and dashpots. In fact, the model shown in Fig. 1.2, with the spring constant and viscous resistance coefficient shown, is described exactly by the relation (1.44). The single term exponential relaxation function (1.41) can then be represented by the mechanical model in Fig. 1.2, which is usually referred to as a Maxwell model. It is easy to see that a relaxation function, represented by the sum of a constant plus a series of decaying exponential terms, could be interpreted in terms of a mechanical model which combines elements of the type shown in Fig. 1.2 arranged in a parallel fashion as shown in Fig. 1.3. Such a model is referred to as a generalized Maxwell model.

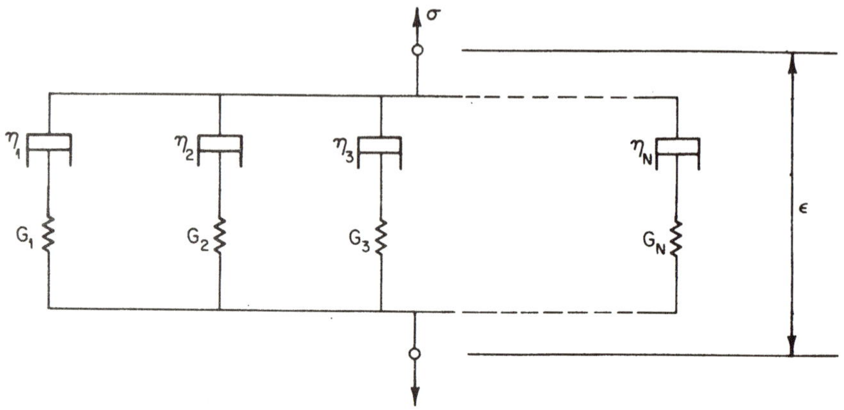

Fig. 1.3. Generalized Maxwell model.

As a second example, we again consider a state of simple shear but now we take the stress as being specified in terms of a unit step function, as

$$\sigma_{12}(t) = \sigma_0 h(t) \tag{1.45}$$

where σ_0 is the amplitude. The creep integral relation (1.14) subject to (1.45) becomes

$$\epsilon_{12}(t) = J_1(t)\,\sigma_0\,. \tag{1.46}$$

The resulting strain thus directly gives the relevant creep function, and thereby suggests the experimental means of determining creep functions. Typical creep functions are functions which increase with time but with decreasing slope, and an asymptote may or may not be approached as t becomes large depending on whether the material is a solid or a fluid. A typical creep function is shown in Fig. 1.4. One of the simplest creep functions of this type is given by

$$J_1(t) = J_0(1 - e^{-t/t_1})\,h(t) \tag{1.47}$$

where J_0 is the amplitude and t_1 is a positive time constant which determines the rate of decay of the first derivative of $J_1(t)$. Using the Laplace transform of (1.47) in (1.18) combined with (1.38) results in a form expressed by

$$\frac{1 + t_1 s}{J_0} = \frac{\sum_{k=0}^{N} q_k(s)^k}{\sum_{k=0}^{N} p_k(s)^k}\,. \tag{1.48}$$

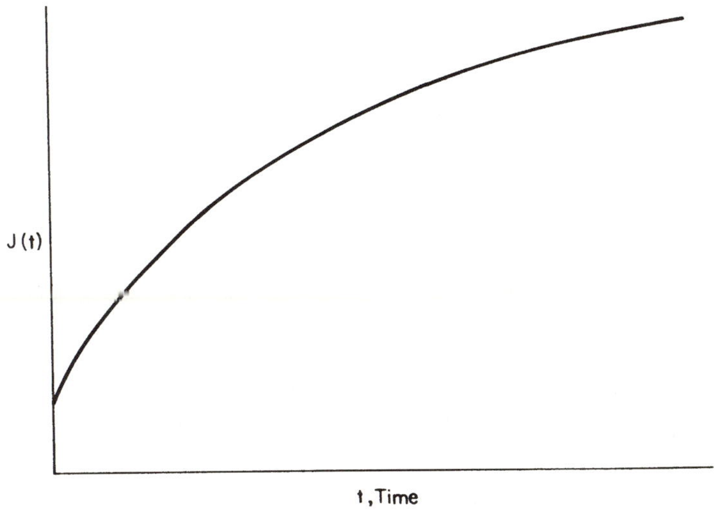

Fig. 1.4. Creep functions.

With $N = 1$, Eq. (1.48) is satisfied by

$$p_0 = J_0\,, \qquad p_1 = 0, \qquad q_0 = 1, \qquad q_1 = t_1\,. \tag{1.49}$$

The differential operator stress strain relation (1.34) which corresponds to (1.49) is then

$$J_0\sigma_{12}(t) = \epsilon_{12}(t) + t_1\,\frac{d\epsilon_{12}(t)}{dt}\,. \tag{1.50}$$

The mechanical model which corresponds to (1.50) is called a Kelvin or Voigt model, and is represented in Fig. 1.5. Creep functions, which are represented by a series of terms involving a constant minus decaying exponential terms, could be interpreted in terms of a mechanical model which combines elements of the type shown in Fig. 1.5. When these are arranged in a series order as shown in Fig. 1.6, such a model is called a generalized Kelvin model.

Mechanical model representations, when applied to the deviatoric stress strain relations, offer an obvious distinction between viscoelastic solids and fluids. The Maxwell model, of course, corresponds to a viscoelastic fluid, while the Kelvin model represents a solid. The generalized models also offer the obvious distinction between these two types of material behavior.

Although the first example given here involved the relation between a relaxation function and the differential operator form of the stress strain relations, the corresponding creep function may readily be obtained through (1.18).

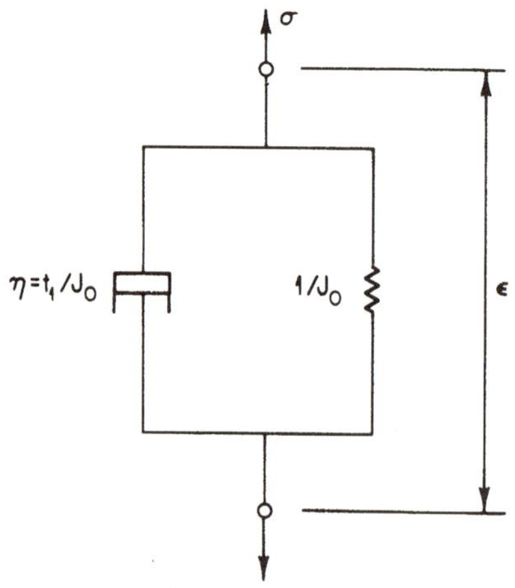

Fig. 1.5. Kelvin model.

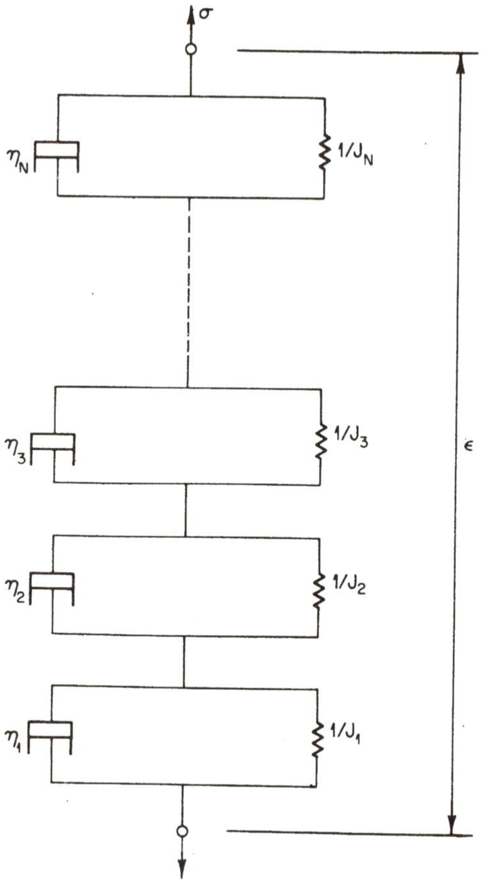

Fig. 1.6. Generalized Kelvin model.

The second example involved a particular creep function (1.47), but this is, in a sense, a degenerate case since a corresponding bounded relaxation function does not exist. However, the addition of a separate positive constant in the representation (1.47) of the creep function will ensure the existence of a corresponding relaxation function. Perhaps this is most easily understood through the consideration of the corresponding mechanical models shown in Figs. 1.2 and 1.5.

Although there are extensive studies of mechanical model representations available—see for example, Bland [1.11]—no further general reference will be made to them here. The point of view adopted here is that even though mechanical models may be of interpretive aid in certain situations, they are usually unnecessarily restrictive and they certainly are not fundamental to the development of a consistent theory of materials with memory.

1.6. STEADY STATE AND FOURIER TRANSFORMED STRESS STRAIN CONSTITUTIVE RELATIONS

There are practical situations in which viscoelastic bodies may be subjected to steady state oscillatory conditions. The previously derived stress strain relations apply in such cases, but it is reasonable to expect that special forms might arise under steady state harmonic oscillation conditions. This situation will now be examined.

We consider first the case of isotropic materials, and let the stress strain relation

$$\tilde{\sigma} = \int_{-\infty}^{t} G_\alpha(t-\tau) \frac{d\tilde{\epsilon}(\tau)}{d\tau} \, d\tau \tag{1.51}$$

designate either the deviatoric or the dilatational part of the stress strain relations, depending upon whether $\alpha = 1$ or 2. We let the strain history be specified as being a harmonic function of time according to

$$\tilde{\epsilon}(t) = \tilde{\epsilon}_0 e^{i\omega t} \tag{1.52}$$

where $\tilde{\epsilon}_0$ is the amplitude and ω is the frequency of oscillation. Before substituting (1.52) into (1.51), it is necessary to decompose $G_\alpha(t)$ into two parts

$$G_\alpha(t) = \mathring{G}_\alpha + \hat{G}_\alpha(t) \tag{1.53}$$

where

$$\hat{G}_\alpha(t) \to 0 \qquad \text{as} \quad t \to \infty.$$

Now, substituting (1.53) into (1.51) to separate it into two parts and then substituting in $\tilde{\epsilon}(t)$ from (1.52), there results

$$\tilde{\sigma}(t) = \mathring{G}_\alpha \tilde{\epsilon}_0 e^{i\omega t} + i\omega\tilde{\epsilon}_0 \int_{-\infty}^{t} \hat{G}_\alpha(t-\tau)\, e^{i\omega\tau} \, d\tau. \tag{1.54}$$

With the change of variable $t - \tau = \eta$, (1.54) can be written as

$$\tilde{\sigma}(t) = \left[\mathring{G}_\alpha + \omega \int_0^\infty \sin \omega\eta \, \hat{G}_\alpha(\eta)\, d\eta + i\omega \int_0^\infty \cos \omega\eta \, \hat{G}_\alpha(\eta)\, d\eta\right] \tilde{\epsilon}_0 e^{i\omega t}. \tag{1.55}$$

To be consistent with the steady state conditions assumed for strain history, the stress will be taken to have the same steady state form with

$$\tilde{\sigma}(t) = G_\alpha^*(i\omega)\, \tilde{\epsilon}_0 e^{i\omega t} \tag{1.56}$$

where $G_\alpha^*(i\omega)$, the complex modulus, is a complex function of frequency and is to be determined. Decomposing $G_\alpha^*(i\omega)$ into real and imaginary parts, gives

$$G_\alpha^*(i\omega) = G_\alpha'(\omega) + iG_\alpha''(\omega). \tag{1.57}$$

Combining (1.55)–(1.57) results in the forms

$$G_\alpha'(\omega) = \mathring{G}_\alpha + \omega \int_0^\infty \hat{G}_\alpha(\eta) \sin \omega\eta \, d\eta \tag{1.58}$$

$$G_\alpha''(\omega) = \omega \int_0^\infty \hat{G}_\alpha(\eta) \cos \omega\eta \, d\eta. \tag{1.59}$$

$G_\alpha'(\omega)$ and $G_\alpha''(\omega)$ are sometimes referred to as the storage and loss moduli, respectively. Also, $G_\alpha^*(i\omega)$ is sometimes referred to as the dynamic modulus, however this terminology can be misleading since $G_\alpha^*(i\omega)$ has no connection whatsoever with either the presence or neglect of inertia terms in the balance of momentum relations. Knowing the relaxation function, these relations determine the real and imaginary parts of the complex modulus $G_\alpha^*(i\omega)$ which enter the steady state form of the viscoelastic stress strain relation (1.56). It is interesting to examine the limiting case relationships implied by (1.58) and (1.59). To this end, we integrate these two relations by parts to obtain

$$G_\alpha'(\omega) = \mathring{G}_\alpha + \hat{G}_\alpha(0) + \int_0^\infty \frac{d\hat{G}_\alpha(\eta)}{d\eta} \cos \omega\eta \, d\eta \tag{1.60}$$

and

$$G_\alpha''(\omega) = -\int_0^\infty \frac{d\hat{G}_\alpha(\eta)}{d\eta} \sin \omega\eta \, d\eta. \tag{1.61}$$

At zero frequency, it follows that for exponentially decaying type relaxation functions

$$G_\alpha'(0) = \mathring{G}_\alpha = G_\alpha(t)\,|_{t\to\infty} \tag{1.62}$$

and

$$G_\alpha''(0) = 0. \tag{1.63}$$

At infinite frequency, with the change of variable $\omega\eta = \tau$, it can be shown that (1.60) and (1.61) give

$$G_\alpha'(\infty) = \mathring{G}_\alpha + \hat{G}_\alpha(0) = G_\alpha(t)\,|_{t\to 0} \tag{1.64}$$

$$G_\alpha''(\infty) = 0. \tag{1.65}$$

Equations (1.62)–(1.65) relate the limiting values of the complex modulus to those of the relaxation function.

We note from relations (1.64) and (1.65) that as the frequency of excitation becomes very large the imaginary part of the complex modulus vanishes.

This implies that, under these circumstances, the material is behaving as an elastic solid. Similarly, relations (1.62) and (1.63) reveal that as the frequency of excitation becomes very small, the material behaves either as an elastic solid or as a viscous fluid. In other words viscoelastic solids, undergoing very fast or very slow processes, respond in an elastic manner, while a viscoelastic fluid behaves elastically for very fast processes and viscously for very slow processes. These latter observations are proved rigorously in Section 1.7.

The simple form of the stress strain relation (1.56) may be alternatively written as

$$\tilde{\sigma}(t) = | G_\alpha{}^*(i\omega)| \, \tilde{\epsilon}_0 e^{i(\omega t + \varphi_\alpha)} \tag{1.66}$$

where $| G_\alpha{}^* |$ is the magnitude of $G_\alpha{}^*$ and

$$\varphi_\alpha(\omega) = \tan^{-1} [G''_\alpha(\omega)/G_\alpha{}'(\omega)]. \tag{1.67}$$

The quantity $\tan \varphi_\alpha$ is sometimes referred to as the loss tangent. The interpretation of (1.66) is physically significant. The steady state harmonic strain lags behind the stress by an amount given by the phase angle φ_α. Of course in physical applications it is either the real or imaginary parts of both stress and strain which are meaningful, rather than the full complex form.

The derivation of relations (1.66) and (1.67) suggests the experimental determination of complex moduli through observing the relationship between stress and strain in a specimen undergoing simple harmonic deformation. The details of such a procedure are considered in Chapter 4. Schematically representative curves of $G_\alpha{}'$ and G''_α versus frequency for viscoelastic solids are as shown in Fig. 1.7. It is by no means necessary that there be a single relative maximum on the curve $G''_\alpha(\omega)$ versus ω or $\log \omega$. Indeed, there may be several such local maxima, each of which is related in some way to rather complicated behavior on a molecular scale.

Equations (1.58) and (1.59) may be recognized as Fourier sine and cosine transforms, and thus may be inverted to give

$$\hat{G}_\alpha(t) = \frac{2}{\pi} \int_0^\infty \frac{G_\alpha{}'(\omega) - \mathring{G}_\alpha}{\omega} \sin \omega t \, d\omega \tag{1.68}$$

or

$$\hat{G}_\alpha(t) = \frac{2}{\pi} \int_0^\infty \frac{G''_\alpha(\omega)}{\omega} \cos \omega t \, d\omega. \tag{1.69}$$

Equation (1.68) may be written in alternate form by carrying out the integration of the second term in the integrand. This gives, using (1.53),

$$G_\alpha(t) = \frac{2}{\pi} \int_0^\infty \frac{G_\alpha{}'(\omega)}{\omega} \sin \omega t \, d\omega. \tag{1.70}$$

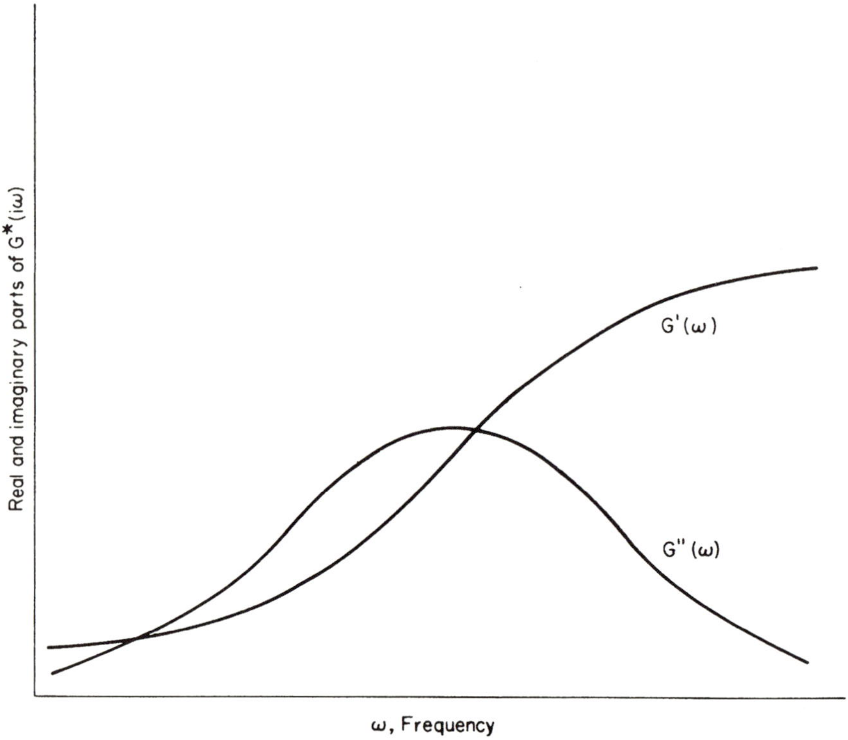

Fig. 1.7. Complex modulus.

Equations (1.68) and (1.69) afford the means of determining the relaxation functions if the complex modulus were considered to be known.

It is apparent, from (1.68) and (1.69), that the real and imaginary parts of the complex modulus (1.57) are not independent of each other, and must be interrelated in some manner. This relationship is easily found by substituting $\hat{G}_\alpha(t)$ from (1.69) into (1.58), and, with some reductions, there results

$$G_\alpha'(\omega) - \mathring{G}_\alpha = \frac{2}{\pi}\int_0^\infty \frac{G_\alpha''(\lambda)\,\omega^2}{\lambda(\omega^2 - \lambda^2)}\,d\lambda. \tag{1.71}$$

This relationship and other similar ones are given by Gross [1.12].

At this juncture a more general viewpoint will be adopted, and rather than assuming steady state harmonic conditions, the Fourier transformation of the general relaxation integral forms of the stress strain relations will be determined. This will provide a form of the stress strain relations comparable to the Laplace transformed forms (1.17). Both the Fourier and Laplace transformed stress strain relations will have applications in the following developments. The Fourier transform pair are given by

$$\bar{f}(\omega) = \int_{-\infty}^{\infty} f(t)\, e^{-i\omega t}\, dt \tag{1.72}$$

and

$$f(t) = \frac{1}{2\pi} \int_{-\infty}^{\infty} \bar{f}(\omega)\, e^{i\omega t}\, d\omega. \tag{1.73}$$

Using (1.72), the Fourier transform of the deviatoric stress strain relation (1.10) is given by

$$\bar{s}_{ij}(\omega) = \int_{-\infty}^{\infty} \left[\int_{-\infty}^{t} G_1(t-\tau) \frac{de_{ij}(\tau)}{d\tau}\, d\tau \right] e^{-i\omega t}\, dt. \tag{1.74}$$

Using the transform pair (1.72) and (1.73) along with $G_1{}^*(i\omega)$ from (1.57)–(1.59) reduces (1.74) to the simple form

$$\bar{s}_{ij}(\omega) = G_1{}^*(i\omega)\, \bar{e}_{ij}(\omega). \tag{1.75}$$

A similar procedure applied to the dilatational part of the stress strain relations (1.11) gives

$$\bar{\sigma}_{kk}(\omega) = G_2{}^*(i\omega)\, \bar{\epsilon}_{kk}(\omega). \tag{1.76}$$

The Fourier transformed stress strain relations (1.75) and (1.76) provide a compact and physically meaningful form of the viscoelastic stress strain relations. Comparable forms can be established for the anisotropic case.

1.7. ACCELERATED AND RETARDED PROCESSES

It is useful to examine the character of viscoelastic deformation in very fast and very slow processes. We let the following one-dimensional stress strain relation symbolize the relation (1.10) with only one nonzero shear component, and symbolize the dilatational relation (1.11). Thus,

$$\sigma(t) = \int_{0}^{t} G_\alpha(t-\tau) \frac{d\epsilon(\tau)}{d\tau}\, d\tau, \qquad \alpha = 1, 2. \tag{1.77}$$

For a given strain history $\epsilon(t)$, an accelerated history is defined as

$$\epsilon(\gamma t), \qquad \gamma > 1.$$

Relative to $\epsilon(t)$, a retarded history is defined as

$$\epsilon(\gamma t), \qquad \gamma < 1.$$

The effect of accelerating the strain history is the same as contracting or compressing the time scale. Accordingly, it is the stress $\sigma(t/\gamma)$ which we wish to relate to the strain history $\epsilon(\gamma t)$, $\gamma > 1$. We use (1.77) for this accelerated strain history and thereby get

$$\sigma\left(\frac{t}{\gamma}\right) = \int_0^{t/\gamma} G_\alpha\left(\frac{t}{\gamma} - \tau\right) \frac{d\epsilon(\gamma\tau)}{d\tau}\, d\tau. \tag{1.78}$$

We change the dummy variable in (1.78) such that

$$\sigma\left(\frac{t}{\gamma}\right) = \int_0^{t} G_\alpha\left(\frac{t}{\gamma} - \frac{\eta}{\gamma}\right) \frac{d\epsilon(\eta)}{d\eta}\, d\eta. \tag{1.79}$$

Now, as γ becomes very large, it is clear that

$$\sigma(t/\gamma) = G_\alpha(0)\, \epsilon(t), \qquad \gamma \to \infty. \tag{1.80}$$

Thus, for a sufficiently accelerated process, the response approaches that of an elastic material with modulus $G_\alpha(0)$ the initial value of the relaxation function. As has already been observed, discontinuous changes in strain cause a discontinuity in stress proportional to $G_\alpha(0)$, which is completely consistent with (1.80).

Relative to retarded strain histories, the stress $\sigma(t/\gamma)$ corresponding to the retarded strain history $\epsilon(\gamma t)$, $\gamma < 1$, is again given by (1.79). In this case, as γ becomes very small, (1.79) gives

$$\sigma(t/\gamma) = G_\alpha(\infty)\, \epsilon(t), \qquad \gamma \to 0. \tag{1.81}$$

Relation (1.81) shows that very slow viscoelastic processes are approximately elastic in nature, except where $G_\alpha(\infty) = 0$. As was observed in Section 1.3, $G_1(\infty) = 0$ corresponds to a viscoelastic fluid.

Although the results in this section are intuitively obvious, it is reassuring to have the rigorous analytical verification. More detailed considerations of this type involving the work done by a viscoelastic material in accelerated and retarded processes are given by Gurtin and Herrera [1.13].

1.8. ALTERNATIVE MECHANICAL PROPERTY FUNCTIONS

To be consistent with a common notation in elasticity, the symbols for the isotropic relaxation functions in simple shear and dilatation are frequently taken as $\mu(t)$ and $k(t)$, respectively, where

$$\begin{aligned} \mu(t) &= G_1(t)/2 \\ k(t) &= G_2(t)/3. \end{aligned} \tag{1.82}$$

It is also possible to define relaxation functions appropriate to other states of uniform stress in terms of $G_1(t)$ and $G_2(t)$. The analogy between the Laplace transformed isotropic viscoelastic stress strain relations (1.17) and the comparable isotropic elastic stress strain relations shows that relaxation functions appropriate to other states of stress are defined through the elasticity relationships simply by replacing elastic moduli by the s multiplied transform of the corresponding viscoelastic relaxation functions.

For example, using the comparable elasticity relations from Sokolnikoff [1.3], it follows that the relaxation function which characterizes a state of uniaxial extension is defined through its transform as

$$\bar{E}(s) = \frac{3\bar{G}_1(s)\,\bar{G}_2(s)}{\bar{G}_1(s) + 2\bar{G}_2(s)} = \frac{9\bar{\mu}(s)\,\bar{k}(s)}{\bar{\mu}(s) + 3\bar{k}(s)}. \tag{1.83}$$

Again, relative to uniaxial extension, viscoelastic Poisson's ratio $\nu(t)$ is defined as minus the ratio of the time dependent lateral strain to the constant axial strain, under stress relaxation conditions. Then the transform of this in terms of $\bar{G}_1(s)$ and $\bar{G}_2(s)$ is given by

$$\bar{\nu}(s) = \frac{\bar{G}_2(s) - \bar{G}_1(s)}{s[\bar{G}_1(s) + 2\bar{G}_2(s)]} = \frac{3\bar{k}(s) - 2\bar{\mu}(s)}{2s[\bar{\mu}(s) + 3\bar{k}(s)]} \tag{1.84}$$

which follows from the similar elasticity relation.

A relaxation function $\lambda(t)$ analogous to the first Lamé constant in elasticity can also be defined through

$$\bar{\lambda}(s) = \tfrac{1}{3}[\bar{G}_2(s) - \bar{G}_1(s)] = \bar{k}(s) - \tfrac{2}{3}\bar{\mu}(s). \tag{1.85}$$

The viscoelastic stress strain relation in terms of the $\lambda(t)$ and $\mu(t)$ relaxation functions is now given by

$$\sigma_{ij}(t) = \delta_{ij}\int_{-\infty}^{t} \lambda(t-\tau)\frac{\partial\epsilon_{kk}(\tau)}{d\tau}\,d\tau + 2\int_{-\infty}^{t} \mu(t-\tau)\frac{\partial\epsilon_{ij}(\tau)}{d\tau}\,d\tau. \tag{1.86}$$

Similar conversion procedures apply to the transforms of creep functions. The relations between the various viscoelastic complex moduli and the complex Poisson's ratio are obtained from the comparable elasticity forms by directly reinterpreting the elasticity constants as being comparable complex functions of frequency. For example, the uniaxial complex modulus which governs uniaxial deformation conditions is given through

$$E^*(i\omega) = \frac{3G_1{}^*(i\omega)\,G_2{}^*(i\omega)}{G_1{}^*(i\omega) + 2G_2{}^*(i\omega)} = \frac{9\mu^*(i\omega)\,k^*(i\omega)}{\mu^*(i\omega) + 3k^*(i\omega)}. \tag{1.87}$$

1.9. SPECTRA

Several different means of specifying viscoelastic mechanical properties have been given. Specifically, relaxation functions, creep functions, and complex moduli have been discussed as alternative means of describing properties, and mechanical models have been seen to provide a physical interpretation of mechanical behavior. In this section we introduce yet another means of denoting properties. However, in contrast to the previous property forms, the ones given here are not directly accessible by experimental means. These property forms, known as spectra, are an indirect characterization of properties; nevertheless, they will be shown to have certain advantages.

The various property forms were introduced in the preceding section. Here a simplified notation will be used for these properties. Specifically, the notations $E(t)$ and $J(t)$ will denote uniaxial relaxation and creep functions, respectively. The spectra properties to be introduced here will be related to $E(t)$ and $J(t)$, with the understanding that similar forms apply to properties relevant to any state of stress. Introduce the relaxation spectrum $H(\tau)$ through the following relationship with the relaxation function $E(t)$:

$$E(t) = \int_0^\infty H(\tau)\, e^{-t/\tau}\, d\tau + E(\infty) \tag{1.88}$$

where $E(\infty)$ is the long-time asymptote

$$E(\infty) = \lim_{t\to\infty} E(t).$$

The specification of the relaxation spectrum $H(\tau)$ thus determines the relaxation function $E(t)$, and in fact $H(\tau)$ is an alternative property form. Before considering the inversion of (1.88) to express $H(\tau)$ in terms of $E(t)$, we discuss the physical meaning of (1.88) and other parallel formalisms.

Consider the simplest possible form of (1.88), where $E(\infty) = 0$ and the relaxation spectrum is a simple delta function

$$H(\tau) = E_0 \delta(\tau - \tau_0). \tag{1.89}$$

Substituting (1.89) into (1.88) gives

$$E(t) = E_0\, e^{-t/\tau_0}$$

which is just the relaxation function corresponding to a Maxwell model (Section 1.5). Similarly, a series of delta functions in a representation for $H(\tau)$ corresponds directly to the generalized Maxwell model. Now we see the proper physical interpretation of $H(\tau)$. It represents a generalization of a discrete spectrum inherent in a mechanical model to a more general specification that involves a continuous spectrum. The use of a continuous spectrum form of $H(\tau)$ is a very useful means of specifying properties since the method shows directly the relative weighting of the different time ranges involved in the relaxation process.

An entirely similar situation exists relative to creep functions. Introduce the retardation (or creep) spectrum $L(\tau)$ through the relation

$$J(t) = \int_0^\infty L(\tau)(1 - e^{-t/\tau})\, d\tau + J(0) \tag{1.90}$$

where

$$J(0) = \lim_{t\to 0} J(t).$$

A single delta function representation for $L(\tau)$ along with $J(0) = 0$ corresponds to the Kelvin model. A series of delta functions for $L(\tau)$ corresponds to the generalized Kelvin model (Section 1.5). The continuous form of $L(\tau)$ provides a general description of the creep-type property of the material.

The relaxation and creep spectra are widely employed in studies of the interrelation of macroscopic properties and molecular motions. The spectra form of properties gives a direct relationship with the characteristic time range of molecular response; see, for example, Rouse [1.14] and Bueche [1.15]. In this connection, properties and spectra are often specified through the use of logarithmic rather than linear time scales. In such cases an alternative relaxation spectrum is specified through

$$E(t) = \int_0^\infty \tilde{H}(\tau)\, e^{-t/\tau}\, d(\log \tau) + E(\infty). \tag{1.91}$$

The interrelation of form (1.88) and (1.91) is simply

$$H(\tau) = \frac{\tilde{H}(\tau)}{\tau} \log e$$

where $\log = \log_{10}$ is commonly used. Molecular scale interpretations are not the only cases in which spectra provide a useful and convenient means of specifying viscoelastic properties. In Section 5.4 the use of spectra will be shown to facilitate theoretical derivations.

Now we turn to the problem of solving for $H(\tau)$ from $E(t)$. Let

$$\tau = 1/\lambda$$

in (1.88) to obtain

$$E(t) - E(\infty) = \int_0^\infty \frac{H(1/\lambda)}{\lambda^2} e^{-t\lambda}\, d\lambda. \tag{1.92}$$

Note that the integral in (1.92) is simply a Laplace transform with t the transform parameter. Using the inversion formula [see Appendix B, Eq. (B.4)] there results

$$\frac{H(1/\lambda)}{\lambda^2} = \frac{1}{2\pi i} \int_{\gamma-i\infty}^{\gamma+i\infty} e^{\lambda t}[E(t)-E(\infty)]\, dt. \tag{1.93}$$

A similar formula applies for the creep spectrum. Although Eq. (1.93) is the formal solution of the problem of determining $H(\tau)$, its implementation requires an analytical form for $E(t)$. Approximate means of determining $H(\tau)$ directly from experimental data on $E(t)$ will be considered in Section 4.6, along with the corresponding creep problem.

To complete the present formulations we seek a means to determine the relaxation spectrum from the complex modulus. From (1.54) we write

$$\sigma(t) = E(\infty)\, \epsilon_0 e^{i\omega t} + i\omega\epsilon_0 \int_{-\infty}^{t} \hat{E}(t-\tau)\, e^{i\omega\tau}\, d\tau. \tag{1.94}$$

If we let

$$t - \tau = \eta$$

(1.94) takes the form

$$\sigma(t) = \left[E(\infty) + i\omega \int_0^\infty \hat{E}(\eta)\, e^{-i\omega\eta}\, d\eta\right] \epsilon_0 e^{i\omega t}. \tag{1.95}$$

From (1.95), as in Section 1.6, the complex modulus is defined as

$$E^*(i\omega) = E(\infty) + i\omega \int_0^\infty \hat{E}(\eta)\, e^{-i\omega\eta}\, d\eta. \tag{1.96}$$

In the present notation we have

$$\hat{E}(t) = \int_0^\infty H(\tau)\, e^{-t/\tau}\, d\tau. \tag{1.97}$$

Substituting (1.97) into (1.96) and interchanging the order of integration we get

$$E^*(i\omega) = i\omega \int_0^\infty H(\tau) \left[\int_0^\infty e^{-(1/\tau+i\omega)\eta}\, d\eta\right] d\tau + E(\infty)$$

which reduces to

$$E^*(i\omega) = i\omega \int_0^\infty \frac{H(\tau)\,\tau}{1+i\omega\tau}\, d\tau + E(\infty). \tag{1.98}$$

Again change the variable, using

$$\tau = 1/\lambda$$

with which (1.98) becomes

$$E^*(i\omega) = i\omega \int_0^\infty \frac{H(1/\lambda)\, d\lambda}{\lambda^2(\lambda + i\omega)} + E(\infty). \tag{1.99}$$

Now replace ω in (1.99) by $i\omega + \epsilon$; then

$$E^*(-\omega + i\epsilon) = (-\omega + i\epsilon) \int_0^\infty \frac{H(1/\lambda)\, d\lambda}{\lambda^2[\lambda + (-\omega + i\epsilon)]} + E(\infty). \tag{1.100}$$

Next we take the limit of (1.100) as $\epsilon \to 0$. Equation (1.100) then has the form

$$\lim_{\epsilon\to 0} [E^*(-\omega + i\epsilon)] = \lim_{\epsilon\to 0} \left\{(-\omega + i\epsilon) \int_0^\infty \frac{H(1/\lambda)}{\lambda^2} \left[\frac{\lambda - \omega}{(\lambda - \omega)^2 + \epsilon^2}\right] d\lambda \right.$$
$$\left. + (-\omega + i\epsilon) \int_0^\infty \frac{H(1/\lambda)}{\lambda^2} \left[\frac{-i\epsilon}{(\lambda - \omega)^2 + \epsilon^2}\right] d\lambda \right\} + E(\infty). \tag{1.101}$$

The limit of the first integral in (1.101) is obviously given by

$$-\omega \int_0^\infty \frac{H(1/\lambda)}{\lambda^2(\lambda - \omega)}\, d\lambda.$$

The determination of the limit of the second integral in (1.101) is facilitated by the delta function identity

$$\lim_{\epsilon\to 0} \frac{\epsilon}{(\lambda - \omega)^2 + \epsilon^2} = \pi\delta(\lambda - \omega). \tag{1.102}$$

Thus the limit of the last integral term in (1.101) is found to be

$$(i\pi/\omega)\, H(1/\omega).$$

With these results (1.101) becomes

$$\lim_{\epsilon\to 0} E^*(-\omega + i\epsilon) = -\omega \int_0^\infty \frac{H(1/\lambda)}{\lambda^2(\lambda - \omega)}\, d\lambda + \frac{i\pi}{\omega} H\left(\frac{1}{\omega}\right) + E(\infty). \tag{1.103}$$

Equating imaginary parts in (1.103) gives

$$H(1/\omega) = (\omega/\pi)\, \mathrm{Im}[\lim_{\epsilon\to 0} E^*(-\omega + i\epsilon)].$$

Finally, letting

$$1/\omega = \tau$$

gives

$$H(\tau) = (1/\pi\tau)\, \mathrm{Im}[\lim_{\epsilon\to 0} E^*(-1/\tau + i\epsilon)]. \tag{1.104}$$

This is the formal solution of the relaxation spectrum in terms of the complex modulus. In an analytical specification for $E^*(i\omega)$, $i\omega$ is replaced by $-1/\tau + i\epsilon$, and the limit is taken. This and similar results are given by Gross [1.12]. Again, it is appropriate to note that approximate means of interrelating properties and spectra are given in Section 4.6. The direct use of relation (1.104) would be cumbersome, and approximate means of determining $H(\tau)$ are well motivated.

PROBLEMS

1.1. **The standard linear solid is a mechanical model relating components of stress and strain through the form**

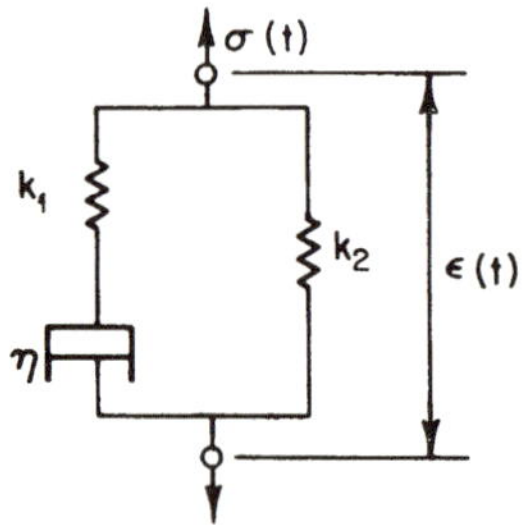

where k_1 and k_2 are the spring constants and η is the coefficient of viscosity. Write the differential operator relationship between stress and strain for this model. Show that the relaxation function, creep function, and real and imaginary parts of the complex modulus are given, respectively, by

$$E(t) = k_1 e^{-t/\tau} + k_2$$

where $\tau = \eta/k_1$,

$$J(t) = \frac{1}{k_2} - \frac{k_1}{k_2(k_1 + k_2)} e^{-t/t_1}$$

where $t_1 = \eta(k_1 + k_2)/k_1 k_2$, and

$$E' = k_2 + \frac{k_1\omega^2}{\omega^2 + (k_1/\eta)^2}, \qquad E'' = \frac{\omega k_1{}^2/\eta}{\omega^2 + (k_1/\eta)^2}.$$

Sketch the behavior of the real and imaginary parts of the complex modulus as a function of frequency.

1.2. Verify the relationship $\bar{G}(s) = [s^2 \bar{J}(s)]^{-1}$ between the Laplace transformed relaxation function and creep function for the standard linear solid of Problem 1.1.

1.3. The complex compliance $J^*(i\omega)$ is defined as the complex operator relating the steady state harmonic stress and strain through

$$\epsilon_0 e^{i\omega t} = J^*(i\omega)\,\sigma_0 e^{i\omega t}$$

where ϵ_0 and σ_0 are the amplitudes of the components of strain and stress, respectively. Using the relation $J^*(i\omega) = [G^*(i\omega)]^{-1}$, find the real and imaginary parts of the complex compliance from the complex modulus for the standard linear solid of Problem 1.1.

1.4. Another means of specifying mechanical properties is through the complex viscosity $\eta^*(i\omega)$ which relates steady state harmonic stress and strain rate through

$$\sigma_0 e^{i\omega t} = \eta^*(i\omega)(i\omega\epsilon_0 e^{i\omega t}).$$

Write $\eta^*(i\omega)$ in terms of the complex modulus $G^*(i\omega)$ and determine $\eta^*(i\omega)$ for the standard linear solid of Problem 1.1. Compare the resulting variation with frequency, of the real and imaginary parts of $\eta^*(i\omega)$ with those of the complex modulus $G^*(i\omega)$.

1.5. Obtain the representation for the relaxation function of a generalized Maxwell model and for the creep function of a generalized Kelvin model. Reason why the generalized Maxwell model is better suited to represent a relaxation function, than to represent a creep function, whereas the generalized Kelvin model is more suitable for creep functions.

1.6. Tell what form of the creep integral constitutive relation must be used in connection with the fading memory hypothesis of Section 1.3.

1.7. Tell what distinction in the analytical forms of creep functions must be made to distinguish between viscoelastic solids and fluids. For a viscoelastic fluid in a state of slow steady shearing flow, obtain the coefficient of viscosity in terms of the creep function, and relate this to the properties of the corresponding relaxation function.

1.8. Prove the delta function identity, Eq. (1.102), used in the derivation of the relaxation spectrum expressed in terms of the complex modulus.

1.9. Examine the behavior of the loss tangent,

$$\tan\varphi = G''(\omega)/G'(\omega),$$

as frequency $\omega \to 0$, in the two separate cases of viscoelastic solids and fluids.

1.10. Using the definition of the viscoelastic Poisson's ratio given in Section 1.8, derive the formula (1.84) from the viscoelastic stress strain relations involving $G_1(t)$ and $G_2(t)$.

REFERENCES

1.1. Boltzmann, L., "Zur Theorie der elastischen Nachwirkung," *Sitzungsber. Math. Naturwiss. Kl. Kaiserl. Akad. Wiss.* **70** (2), 275 (1874).
1.2. Volterra, V., "Sulle equazioni integrodifferenziali della teoria dell'elasticita," *Atti Reale Accad. Lincei* **18** (2), 295 (1909).
1.3. Sokolnikoff, I. S., *Mathematical Theory of Elasticity*, 2nd ed. McGraw-Hill, New York, 1956.
1.4. Riesz, F., and B. Sz.-Nagy, *Functional Analysis*. Ungar, New York, 1955.
1.5. Gurtin, M. E., and E. Sternberg, "On the Linear Theory of Viscoelasticity," *Arch. Ration. Mech. Anal.* **11**, 291 (1962).
1.6. Staverman, A. J., and F. Schwarzl, "Linear Deformation Behavior of High Polymers," *in Die Physik der Hochpolymeren* (H. A. Stuart, ed.), Vol. 4, Chapter 1. Springer, Berlin, 1956.
1.7. Volterra, V., *Theory of Functionals and of Integral and Integro-Differential Equations*. Dover, New York, 1959.
1.8. Coleman, B. D., and V. Mizel, "A General Theory of Dissipation in Materials with Memory," *Arch. Ration. Mech. Anal.* **27**, 255 (1967).
1.9. Coleman, B. D., "Thermodynamics of Materials with Memory," *Arch. Ration. Mech. Anal.* **17**, 1 (1964).
1.10. Truesdell, C., and W. Noll, *in Handbuch der Physik* (S. Flügge, ed.), Vol. 3, No. 3. Springer, Berlin, 1965.
1.11. Bland, D. R., *The Theory of Linear Viscoelasticity*. Pergamon Press, Oxford, 1960.
1.12. Gross, B., *Mathematical Structure of the Theories of Viscoelasticity*. Hermann, Paris, 1953.
1.13. Gurtin, M. E., and I. Herrera, "On Dissipation Inequalities and Linear Viscoelasticity," *Quart. Appl. Math.* **23**, 235 (1965).
1.14. Rouse, P. E., Jr., "A Theory of the Linear Viscoelastic Properties of Dilute Solutions of Coiling Polymers," *J. Chem. Phys.* **21** (7), 1272 (1953).
1.15. Bueche, F., "The Viscoelastic Properties of Plastics," *J. Chem. Phys.* **22** (4), 603 (1954).

Chapter II Isothermal Boundary Value Problems

In considering the means of solving isothermal, linear, viscoelastic boundary value problems it is of considerable utility to classify the types of problems according to certain characteristics which simplify the means of solution in these cases. Thus, after stating the basic equations, and establishing the uniqueness of solution, several of these special cases will be considered and illustrated with examples. Also, a class of problems will be considered which do not fit into any of the simplifying special cases, and examples of this type will be given. All problems and analyses in this chapter will be appropriate to isotropic materials under isothermal conditions.

Although this chapter is concerned with the general means of solving boundary value problems, very few detailed numerical results are displayed. It is not possible to give detailed numerical solutions and expect the results to have any degree of generality. Most viscoelastic solutions cannot be expressed in simple nondimensional forms involving unspecified mechanical properties, as can often be done in elasticity. Accordingly, numerical results appropriate to particular materials will be given only in the first few developments, and there only to illustrate that, as a practical matter, such specific results can be obtained. In this connection, the examples to be studied in this chapter assume a knowledge of the appropriate mechanical properties functions. The means of determining mechanical properties was considered briefly in Chapter 1 and is extensively treated in Chapter 4.

2.1. FORMULATION OF THE BOUNDARY VALUE PROBLEM

The mathematical set of conditions which govern the solution of isothermal, isotropic, linear viscoelastic boundary value problems will now be assembled. The linear theory stress strain relations, in several different forms, have already been displayed in Chapter 1. Of all the governing conditions, it is only the viscoelastic stress strain relations which differ from those of the corresponding infinitesimal elasticity theory. All other governing conditions follow directly from the theory of linear elasticity, but with proper cognizance taken of the time dependent nature of all variables.

Letting u_i designate the displacements, in rectangular Cartesian coordinates, the strain displacement relations are

$$2\epsilon_{ij}(t) = u_{i,j}(t) + u_{j,i}(t) \tag{2.1}$$

where a comma designates partial differentiation with respect to x_i. Two forms of the isotropic stress strain relations are repeated here for completeness, although all of the various forms in Chapter 1 may be used. The first form is

$$\sigma_{ij}(t) = \delta_{ij} \int_{-\infty}^{t} \lambda(t-\tau) \frac{\partial \epsilon_{kk}(\tau)}{\partial \tau} \, d\tau + 2 \int_{-\infty}^{t} \mu(t-\tau) \frac{\partial \epsilon_{ij}(\tau)}{\partial \tau} \, d\tau \tag{2.2}$$

where $\lambda(t)$ and $\mu(t)$ are the appropriate relaxation functions. The second form of the stress strain relations is given in terms of deviatoric and dilatational components through

$$s_{ij}(t) = 2 \int_{-\infty}^{t} \mu(t-\tau) \frac{\partial e_{ij}(\tau)}{\partial \tau} \, d\tau$$

and (2.3)

$$\sigma_{ii}(t) = 3 \int_{-\infty}^{t} k(t-\tau) \frac{\partial \epsilon_{jj}(\tau)}{\partial \tau} \, d\tau$$

where

$$\sigma_{ij}(t) = s_{ij}(t) + \tfrac{1}{3}\delta_{ij}\sigma_{kk}(t), \qquad s_{ii} = 0$$

and (2.4)

$$\epsilon_{ij}(t) = e_{ij}(t) + \tfrac{1}{3}\delta_{ij}\epsilon_{kk}(t), \qquad e_{ii} = 0.$$

The stress tensor has already been taken to be symmetric which satisfies the balance of angular momentum relationship. The balance of linear momentum is expressed by the equations

$$\sigma_{ij,j}(t) + F_i(t) = 0 \tag{2.5}$$

$$\sigma_{ij,j}(t) + F_i(t) = \rho \frac{\partial^2 u_i}{\partial t^2} \tag{2.6}$$

where F_i are the body forces per unit volume and ρ is the mass density. Relations (2.5) apply to quasi-static problems, and (2.6) apply to dynamic problems. One form of the six independent compatibility equations, from Sokolnikoff [2.1] is given by

$$\epsilon_{ij,kl}(t) + \epsilon_{kl,ij}(t) = \epsilon_{ik,jl}(t) + \epsilon_{jl,ik}(t). \tag{2.7}$$

Finally, the initial conditions, appropriate to an initial stress free state of rest, and the boundary conditions are given by

$$u_i(t) = \epsilon_{ij}(t) = \sigma_{ij}(t) = 0, \qquad -\infty < t < 0 \tag{2.8}$$

and

$$\begin{aligned} \sigma_{ij}(t)\, n_j &= S_i(t) \qquad \text{on } B_\sigma \\ u_i(t) &= \Delta_i(t) \qquad \text{on } B_u \end{aligned} \tag{2.9}$$

where n_j are the components of the unit outward normal to the boundary of the body. B_σ is that part of the boundary over which the components of the stress vector are prescribed, while B_u is that part of the boundary over which the components of the displacement vector are prescribed. B_u and B_σ are required to remain constant with time. That is, the type of the boundary condition at a point, whether it be prescribed stress or displacement, will not be allowed to change. The restriction to this particular type of boundary conditions will be relaxed in Section 2.12 wherein we will consider problems for which B_u and B_σ do not remain constant with time and which involve mixed type boundary conditions whereby the boundary condition at a point on the surface involves components of both the displacement and stress vectors. It is to be understood that all field variables in (2.1)–(2.9) in general also have a dependence upon coordinates x_i.

Sometimes it is convenient to have the equations of motion directly expressed in terms of displacements. Combining (2.1), (2.2), and (2.6) gives

$$\int_{-\infty}^{t} \mu(t-\tau)\frac{\partial u_{i,jj}(\tau)}{d\tau}\,d\tau + \int_{-\infty}^{t} [\lambda(t-\tau)+\mu(t-\tau)]\frac{\partial u_{k,ki}(\tau)}{\partial\tau}\,d\tau + F_i = \rho\frac{\partial^2 u_i}{\partial t^2}. \tag{2.10}$$

The complete solution of this set of equations, with inertia terms neglected, is referred to as a quasi-static solution. The means of obtaining quasi-static and dynamic solutions is the subject of interest in this chapter. Before proceeding to such considerations, however, we must establish the uniqueness of solution of the viscoelastic boundary value problem. Such an assurance of the uniqueness of solutions greatly increases their practical value.

2.2. UNIQUENESS OF SOLUTION

We now state and prove a uniqueness theorem associated with the viscoelastic boundary value problem posed in the preceding section. We shall only be concerned with the isotropic and quasi-static case here. In Section 7.1 we shall use a different method to prove a more general viscoelastic uniqueness theorem appropriate to dynamic, anisotropic, and nonisothermal conditions.

The uniqueness theorem is now stated:

The isotropic, quasi-static, viscoelastic boundary value problem governed by the strain displacement relations (2.1), *the stress constitutive relations* (2.3) *and* (2.4), *the equations of equilibrium* (2.5), *the initial conditions* (2.8), *and the boundary conditions* (2.9) *possesses a unique solution, provided the initial values of the relaxation functions satisfy*

$$\mu(0) > 0 \qquad \textit{and} \qquad k(0) > 0.$$

To prove this theorem we first assume that two separate solutions exist and designate these by

$$\begin{array}{ccc} u_i^{(1)} & & u_i^{(2)} \\ \epsilon_{ij}^{(1)} & \text{and} & \epsilon_{ij}^{(2)} \\ \sigma_{ij}^{(1)} & & \sigma_{ij}^{(2)} \end{array}$$

where these are assumed to be continuous functions of time and of the spatial coordinates. A difference solution is designated by

$$u_i^d(x_i, t) = u_i^{(1)}(x_i, t) - u_i^{(2)}(x_i, t),$$

$$\epsilon_{ij}^d(x_i, t) = \epsilon_{ij}^{(1)}(x_i, t) - \epsilon_{ij}^{(2)}(x_i, t),$$

and

$$\sigma_{ij}^d(x_i, t) = \sigma_{ij}^{(1)}(x_i, t) - \sigma_{ij}^{(2)}(x_i, t).$$

Since the difference solution satisfies homogeneous boundary conditions over the entire boundary B, we can write the following surface integral:

$$\int_B \sigma_{ij}^d(x_i, t)\, n_j(x_i)\, u_i^d(x_i, t)\, da = 0 \tag{2.11}$$

where n_j designate the components of the unit outward normal vector. The divergence theorem, the relations (2.4), and the symmetry of the stress tensor are used to write (2.11) as the volume integral

$$\int_V [s_{ij}^d(x_i, t)\, e_{ij}^d(x_i, t) + \tfrac{1}{3}\sigma_{ii}^d(x_i, t)\, \epsilon_{jj}^d(x_i, t)]\, dv = 0 \tag{2.12}$$

where $\sigma_{ij,j}^d = 0$ has also been used.

We integrate the stress constitutive relations (2.3) by parts, use the intiial conditions (2.8), and substitute the result into (2.12) to obtain

$$\begin{aligned} \int_V \Big[& k(0)\, \epsilon_{ii}^d(t)\, \epsilon_{jj}^d(t) + 2\mu(0)\, e_{ij}^d(t)\, e_{ij}^d(t) \\ & + \int_0^t \dot{k}(t-\tau)\, \epsilon_{ii}^d(t)\, \epsilon_{jj}^d(\tau)\, d\tau \\ & + 2\int_0^t \dot{\mu}(t-\tau)\, e_{ij}^d(t)\, e_{ij}^d(\tau)\, d\tau \Big]\, dv = 0 \end{aligned} \tag{2.13}$$

where

$$\dot{k}(t-\tau) = -\frac{d}{d\tau} k(t-\tau)$$

and

$$\dot{\mu}(t - \tau) = -\frac{d}{d\tau}\,\mu(t - \tau).$$

In (2.13), and henceforth, the dependence of the field variables on the spatial coordinates is implied.

The form of (2.13) is simplified by introducing the following notational changes. We let

$$\begin{aligned} [k(0)]^{1/2}\,\epsilon_{ii}^{d}(t) &= f_1(t), \\ [2\mu(0)]^{1/2}\,e_{ij}^{d}(t) &= f_n(t), \qquad i, j = 1, 2, 3, \quad n = 2, \ldots, 10 \\ \frac{\dot{k}(t - \tau)}{k(0)} &= \psi_1(t, \tau), \end{aligned} \tag{2.14}$$

and

$$\frac{\dot{\mu}(t - \tau)}{\mu(0)} = \psi_n(t, \tau), \qquad n = 2, \ldots, 10.$$

With these changes (2.13) becomes

$$\int_V \sum_{n=1}^{10} \left[f_n(t) f_n(t) + \int_0^t \psi_n(t, \tau) f_n(t) f_n(\tau)\, d\tau \right] dv = 0. \tag{2.15}$$

We now let the functions $\psi_n(t, \tau)$ be bounded whereby a constant B exists such that

$$|\,\psi_n(t, \tau)| \leqslant B/10, \qquad n = 1, \ldots, 10. \tag{2.16}$$

Using (2.16), we can relate the magnitudes of the two terms in relation (2.15) through the form

$$s(t) \leqslant \frac{B}{10} \int_V \sum_{n=1}^{10} \left[\int_0^t |\, f_n(t) f_n(\tau)|\; d\tau \right] dv \tag{2.17}$$

where

$$s(t) = \int_V \sum_{n=1}^{10} f_n^{\,2}(t)\; dv. \tag{2.18}$$

We designate by constant A, the bound on $s(t)$, such that

$$s(t) \leqslant A. \tag{2.19}$$

With these notational agreements we now seek to prove the following relationship:

$$s(t) \leqslant \frac{A(2Bt)^m}{(m+1)!}, \qquad m = 0, 1, 2, 3\ldots. \tag{2.20}$$

Obviously (2.20) is true for $m = 0$; we shall prove it is true for all higher integer values of m by an induction method. That is, we shall assume (2.20) is true for $m = r$ and prove that it is true for $m = r + 1$. To obtain this proof we start with the following inequality:

$$\int_V [\tau^{r/2} \,|f_n(t)| - t^{r/2} \,|f_n(\tau)|]^2 \, dv \geqslant 0.$$

This has the expanded form

$$\int_V |f_n(t) f_n(\tau)| \, dv \leqslant \frac{\tau^{r/2} t^{-r/2}}{2} \int_V f_n^2(t) \, dv + \frac{\tau^{-r/2} t^{r/2}}{2} \int_V f_n^2(\tau) \, dv. \tag{2.21}$$

Using (2.18) and (2.20) for $m = r$ in (2.21) gives us the form

$$\int_V |f_n(t) f_n(\tau)| \, dv \leqslant \frac{A(2B)^r}{(r+1)!} t^{r/2} \tau^{r/2}. \tag{2.22}$$

Substituting (2.22) into (2.17) and carrying out the integration gives

$$s(t) \leqslant \frac{A(2Bt)^{r+1}}{(r+2)!}. \tag{2.23}$$

Thus having assumed (2.20) is true for $m = r$, we have proved that it is true for $m = r + 1$. This along with the fact that (2.20) is true for $m = 0$ shows that (2.20) must be true for unboundedly large values of m, thus,

$$s(t) \leqslant 0. \tag{2.24}$$

But referring to the definition (2.18) of $s(t)$, the definition (2.14) of $f_n(t)$ and recalling the assumption $\mu(0) > 0$ and $k(0) > 0$ shows that the inequality in (2.24) is not possible and the equality part of (2.24) implies that

$$\epsilon_{ij}^d(x_i, t) = 0.$$

Thus, the difference solution for strains must vanish, and it follows that the solution of the viscoelastic boundary value problem which was posed is unique, to within a rigid body displacement field.

The uniqueness theorem just proved was originally due to Volterra, with the form given following that of Gurtin and Sternberg [2.2]. More general viscoelastic uniqueness theorems, still under isothermal conditions, have been given by Edelstein and Gurtin [2.3], Odeh and Tadjbakhsh [2.4] Barberan and Herrera [2.5], and Lubliner and Sackman [2.6]. A uniqueness theorem of comparable generality to that given here, but proved by a method using the Laplace transform, was given by Onat and Breuer [2.7].

2.3. SEPARATION OF VARIABLES CONDITIONS

In the quasi-static cases where separation of variables conditions prevail the solutions of boundary value problems may be obtained by special means. By separation of variables we mean the existence of the solution for all field variables in the form

$$\begin{aligned} u_i(x_i\,,t) &= \mathring{u}_i(x_i)\,u(t) \\ \epsilon_{ij}(x_i\,,t) &= \mathring{\epsilon}_{ij}(x_i)\,u(t) \\ \sigma_{ij}(x_i\,,t) &= \mathring{\sigma}_{ij}(x_i)\,F(t). \end{aligned} \tag{2.25}$$

First of all, it should be noted that, for a separation of variables solution to exist in the general form (2.25), it is necessary that viscoelastic Poisson's ratio be a real constant. This may be seen from either the equations of equilibrium or the compatibility equations. For example, substituting $u_i(x_i\,,t)$ from (2.25) into (2.10), neglecting inertia terms, and taking $F_i = 0$ shows that these equations of equilibrium can be satisfied by (2.25) only if

$$\lambda(t) + \mu(t) = K\mu(t)$$

where K is a constant. Using relations (1.84) and (1.85), the above relation implies that

$$\nu(t) = \nu, \quad \text{a constant.}^1$$

A further implication of this requirement of constant Poisson's ratio is that the ratio of the relaxation function in shear and the relaxation function in volume change must be a constant, thus

$$\frac{G_2(t)}{G_1(t)} = \frac{3k(t)}{2\mu(t)} = \frac{1+\nu}{1-2\nu}. \tag{2.26}$$

Finally it follows that the two isotropic creep functions $J_1(t)$ and $J_2(t)$ must be related by

$$\frac{J_2(t)}{J_1(t)} = \frac{1-2\nu}{1+\nu}. \tag{2.27}$$

[1] We are excluding from consideration the limited class of problems for which (2.25) applies with the special restriction that $\mathring{u}_{i,jj} = 0$ and $\mathring{u}_{k,k} =$ constant, in which case viscoelastic Poisson's ratio remains unrestricted. This was pointed out to the author by Professor M. M. Carroll.

There are other necessary conditions besides constant Poisson's ratio which must be met for a separation of variables solution to exist. But, before considering the most general case, it is instructive to consider the two following special cases.

First Boundary Value Problem

We let the entire boundary be subject to prescribed stresses through the separated form

$$\sigma_{ij}(x_i, t)\, n_j(x_i) = \mathring{S}_i(x_i)\, F(t) \qquad \text{on} \quad B \tag{2.28}$$

and we let the body forces be prescribed by

$$F_i(x_i, t) = \mathring{F}_i(x_i)\, F(t). \tag{2.29}$$

The compatibility equations (2.7) are written in terms of stresses using the creep integral relations (1.14) and (1.15), the ratio (2.27), and the equilibrium equations (2.5). The procedure for carrying this out is the same as in elasticity—see Sokolnikoff [2.1] for example—with the result that

$$\int_0^t J_1(t-\tau)\, \frac{d\theta_{ij}(\tau)}{d\tau}\, d\tau = 0 \tag{2.30}$$

where

$$\theta_{ij}(t) = \sigma_{ij,kk}(t) + \frac{1}{1+\nu}\, \sigma_{kk,ij}(t) + \frac{\nu}{1-\nu}\, \delta_{ij} F_{k,k}(t) + F_{i,j}(t) + F_{j,i}(t). \tag{2.31}$$

For (2.30) to be satisfied, it is necessary and sufficient that

$$\theta_{ij}(t) = 0. \tag{2.32}$$

Relations (2.31) and (2.32) are simply the Beltrami–Michell compatibility equations of elasticity theory. We assume a stress solution in the form

$$\sigma_{ij}(x_i, t) = \mathring{\sigma}_{ij}(x_i)\, F(t) \tag{2.33}$$

where $F(t)$ is known from the boundary condition and body force specification. Then the solution for $\mathring{\sigma}_{ij}(x_i)$ is obtained from the equilibrium equations (2.5), the compatibility equations in terms of stresses (2.31) and (2.32), and the boundary conditions (2.28). It is recognized that the problem of determining $\mathring{\sigma}_{ij}(x_i)$ is exactly the same as that in elasticity theory. With the stresses determined, the displacement solution will be assumed in the form

$$u_i(x_i, t) = \mathring{u}_i(x_i)\, u(t) \tag{2.34}$$

where both $\mathring{u}_i(x_i)$ and $u(t)$ are to be found. The strains which correspond to (2.34) are then

$$\epsilon_{ij}(x_i, t) = \mathring{\epsilon}_{ij}(x_i)\, u(t) \tag{2.35}$$

where

$$2\mathring{\epsilon}_{ij}(x_i) = \mathring{u}_{i,j}(x_i) + \mathring{u}_{j,i}(x_i). \tag{2.36}$$

By substituting (2.33) and (2.35) into the creep integral relations (1.14) and (1.15), we get

$$\mathring{e}_{ij}(x_i)\, u(t) = \mathring{s}_{ij}(x_i) \int_0^t J_1(t-\tau) \frac{dF(\tau)}{d\tau}\, d\tau \tag{2.37}$$

and

$$\mathring{\epsilon}_{kk}(x_i)\, u(t) = \mathring{\sigma}_{kk}(x_i) \left(\frac{1-2\nu}{1+\nu}\right) \int_0^t J_1(t-\tau) \frac{dF(\tau)}{d\tau}\, d\tau. \tag{2.38}$$

Letting

$$\tilde{J}(t) = J_1(t)/J_1(0), \tag{2.39}$$

then (2.37) and (2.38) give

$$u(t) = \int_0^t \tilde{J}(t-\tau) \frac{dF(\tau)}{d\tau}\, d\tau, \tag{2.40}$$

$$\mathring{e}_{ij}(x_i) = J_1(0)\, \mathring{s}_{ij}(x_i), \tag{2.41}$$

and

$$\mathring{\epsilon}_{kk}(x_i) = \frac{(1-2\nu)}{(1+\nu)} J_1(0)\, \mathring{\sigma}_{kk}(x_i). \tag{2.42}$$

Relation (2.40) determines the temporal part of the displacement solutions, and the spatial part is obtained by integrating (2.41) and (2.42) to find $\mathring{u}_i(x_i)$. This formally completes the solution of this type of boundary value problem.

Second Boundary Value Problem

In this problem there are no body forces and the displacements are prescribed over the entire boundary, in the form

$$u_i(x_i, t) = \mathring{\Delta}_i(x_i)\, u(t) \qquad \text{on} \quad B. \tag{2.43}$$

We assume a displacement solution in the form

$$u_i(x_i, t) = \mathring{u}_i(x_i)\, u(t). \tag{2.44}$$

The corresponding strains are

$$\epsilon_{ij}(x_i, t) = \mathring{\epsilon}_{ij}(x_i)\, u(t) \tag{2.45}$$

where $\mathring{\epsilon}_{ij}(x_i)$ is given by (2.36).

The relaxation integral relations (1.10) and (1.11) then give

$$s_{ij}(x_i, t) = 2\mathring{e}_{ij}(x_i) \int_0^t \mu(t-\tau) \frac{du(\tau)}{d\tau} d\tau \tag{2.46}$$

$$\sigma_{kk}(x_i, t) = \frac{2(1+\nu)}{1-2\nu} \mathring{\epsilon}_{jj}(x_i) \int_0^t \mu(t-\tau) \frac{du(\tau)}{d\tau} d\tau. \tag{2.47}$$

When σ_{ij} is obtained from these relations and substituted into the equations of equilibrium, $\sigma_{ij,j}(x_i, t) = 0$, it follows that

$$\mathring{u}_{i,kk}(x_i) + \frac{1}{1-2\nu} \mathring{u}_{k,ki}(x_i) = 0. \tag{2.48}$$

With the solution of these elasticity type equations, subject to boundary conditions (2.43), the displacement solution is known. Then the stress solution, which will have a separable form, follows from (2.36), (2.46), and (2.47) to complete the solution.

The results given here on these two separation of variables type problems and solutions follow the work of Tsien [2.8]. Comparable and earlier results for incompressible materials were given by Alfrey [2.9].

Mixed Boundary Value Problem

By combining the results of the two separate problems just considered, we can see how a separation of variables solution can exist in a problem with mixed boundary conditions. Specifically, the boundary displacement and tractions must have the separable form given by

$$\sigma_{ij}(x_i, t)\, n_j(x_i) = \mathring{S}_i(x_i)\, F(t) \qquad \text{on } B_\sigma$$
$$u_i(x_i, t) = \mathring{\Delta}_i(x_i)\, u(t) \qquad \text{on } B_u$$

with $F(t)$ and $u(t)$ related through

$$F(t) = \kappa \int_{-\infty}^t \mu(t-\tau) \frac{du(\tau)}{d\tau} d\tau$$

where κ is a constant. Of course the body forces must have the same time variation as the surface tractions, and viscoelastic Poisson's ratio must be a constant.

2.4. STEADY STATE HARMONIC CONDITIONS

If the boundary conditions and body forces of a viscoelastic boundary value problem are specified as steady state harmonic functions of time, the solution of all field variables will have this type of time dependence. The steady state stress strain relation (1.56) can be written as

$$\mathring{s}_{ij}(x_i)\, e^{i\omega t} = 2\mu^*(i\omega)\, \mathring{e}_{ij}(x_i)\, e^{i\omega t} \tag{2.49}$$

and

$$\mathring{\sigma}_{kk}(x_i)\, e^{i\omega t} = 3k^*(i\omega)\, \mathring{\epsilon}_{kk}(x_i)\, e^{i\omega t} \tag{2.50}$$

where ω is the specified frequency of oscillation. After dropping the $e^{i\omega t}$ terms in (2.49) and (2.50), we see that the stress strain relations have exactly the same form as in elasticity, except, now, complex quantities are involved. These two relations along with (2.1), (2.5), and (2.7)–(2.9), in suitable harmonic form, then pose an elasticity type problem for the determination of the spatial parts of the variables $\mathring{\sigma}_{ij}(x_i)$, $\mathring{\epsilon}_{ij}(x_i)$, and correspondingly $\mathring{u}_i(x_i)$.

Actually, this formulation is of a special type of separation of variables problem; however, it offers some generalization over the conditions assumed in the previous section. It is not required here that the two isotropic mechanical property functions be related by a real positive constant, as in (2.26), and they remain independent. Also, in this type of problem, the inclusion of inertia terms in (2.6), rather than (2.5) the quasi-static form, does not cause any undue extra complication. An example of this type of problem is given in Section 2.7.

2.5. INTEGRAL TRANSFORM METHODS

Neither a separation of variables type problem nor steady state conditions will be assumed here; rather, only the conditions implicit in the statement of the basic equations (2.1)–(2.9) are assumed, and it further is assumed that the Laplace transform of all time variables exists. The Laplace transform of (2.1), (2.3), (2.4), (2.5), and (2.9) are given by

$$2\bar{\epsilon}_{ij} = \bar{u}_{i,j} + \bar{u}_{j,i}\,, \tag{2.51}$$

$$\bar{s}_{ij} = 2s\bar{\mu}\bar{e}_{ij}\,, \tag{2.52}$$

$$\bar{\sigma}_{kk} = 3s\bar{k}\bar{\epsilon}_{kk}\,, \tag{2.53}$$

$$\bar{s}_{ij} = \bar{\sigma}_{ij} - \tfrac{1}{3}\delta_{ij}\bar{\sigma}_{kk}\,, \qquad \bar{s}_{ii} = 0, \tag{2.54}$$

$$\bar{e}_{ij} = \bar{\epsilon}_{ij} - \tfrac{1}{3}\delta_{ij}\bar{\epsilon}_{kk}\,, \qquad \bar{e}_{ii} = 0, \tag{2.55}$$

$$\bar{\sigma}_{ij,j} + \bar{F}_i = 0, \tag{2.56}$$

$$\bar{\sigma}_{ij}n_j = \bar{S}_i \qquad \text{on} \quad B_\sigma\,, \tag{2.57}$$

and

$$\bar{u}_i = \bar{\Delta}_i \qquad \text{on} \quad B_u \tag{2.58}$$

where a bar over a variable designates its Laplace transformed form, and s is the transform variable.

The set of conditions (2.51)–(2.58) has a form identical with that of linear elasticity if the transform of viscoelastic variables are associated with the corresponding elastic variables and if $s\bar{\mu}(s)$ and $s\bar{k}(s)$ are associated with the elastic moduli μ and k. It follows that for problems governed by (2.1)–(2.9) and the conditions stated there with, the Laplace transformed viscoelastic solution is obtained directly from the solution of the corresponding elastic problem by replacing μ and k with $s\bar{\mu}$ and $s\bar{k}$, respectively. The final solution will be realized upon inverting the transformed solution. In general, elastic solutions can be converted to Laplace transformed viscoelastic solutions through the replacement of the elastic moduli and elastic Poisson's ratio by the transform parameter multiplied transforms of the appropriate viscoelastic relaxation functions and viscoelastic Poisson's ratio. An entirely similar procedure also applies in the anisotropic case.

This association of the integral transform of the viscoelastic solution with the solution of the corresponding elastic problem is sometimes called the elastic-viscoelastic correspondence principle, or the elastic-viscoelastic analogy. Although the derivation given here uses the Laplace transform, an entirely similar procedure follows from the use of the Fourier transform. Read [2.10] was the first to recognize this association, through the Fourier transform, while Sips [2.11], Brull [2.12], and Lee [2.13] gave the corresponding Laplace transform results.

This correspondence procedure reveals that the vast catalog of static elastic solutions can be converted to quasi-static viscoelastic solutions. The entire procedure involves replacing elastic moduli by the appropriate forms of the transform of viscoelastic properties, and reinterpreting elastic field variables as transformed viscoelastic field variables, then inverting. The mechanics of the inversion process will be studied in detail in the examples of Sections 2.8 and 2.9. In these examples it will be shown that it is possible to specify mechanical properties in a general manner such that the inversion process can be completed to obtain analytical solutions. It should be noted that integral transform methods may be profitably applied to the class of problems admitting the separation of variables type solution discussed in Section 2.3. The integral transform methods supply a means of solving the convolution integral relations which result for these types of problems.

2.6. EFFECT OF INERTIA TERMS

The previous cases have been limited to quasi-static considerations, whereby the inertia terms in the equations of motion are neglected. The treatment of the separation of variables type problems given in Section 2.3 can only be applied when inertia terms are neglected, whereas for the steady state type problems discussed in Section 2.4 inertia terms can be included with no difficulty.

In the integral transform methods of the previous section, inertia terms can be included in the equations of motion, but, in this case, the analogy is not between elastic solutions and the transform of viscoelastic solutions but rather it is between the transform of elastic solutions and the transform of viscoelastic solutions. This latter analogy is a much less useful association than that of the quasi-static case, and, in general, for dynamic viscoelastic problems, it is necessary to construct the entire solution without relying upon static elasticity results. It should be mentioned that the inversion of the transformed dynamic viscoelastic solution is much more difficult than that in the quasi-static case. This difficulty is illustrated in the example of Section 2.10 and Chapter 6, on viscoelastic wave propagation.

2.7. STEADY STATE HARMONIC OSCILLATION EXAMPLE

Having assembled the governing equations for a theory of isothermal viscoelasticity, we can now face the problem of solving explicit boundary value problems. As a first example, we consider the torsional oscillation of a right circular cylinder, under steady state harmonic conditions. The relevant equation of motion in polar cylindrical coordinates r, θ, and z for an elastic cylinder is

$$\frac{\partial^2 u_\theta}{\partial r^2} + \frac{1}{r}\frac{\partial u_\theta}{\partial r} + \frac{\partial^2 u_\theta}{\partial z^2} - \frac{u_\theta}{r^2} = \frac{\rho}{\mu}\frac{\partial^2 u_\theta}{\partial t^2} \tag{2.59}$$

where $u_\theta = u_\theta(r, z, t)$, and ρ is the mass density. Let

$$u_\theta = r\Phi(z, t)$$

and

$$\Phi(z, t) = f(z)\, e^{i\omega t} \tag{2.60}$$

where ω is the frequency of harmonic oscillation. By combining (2.59) and (2.60), $f(z)$ can readily be determined as

$$f(z) = A \sin(\Omega z/h) + B \cos(\Omega z/h) \tag{2.61}$$

where

$$\Omega^2 = \rho\omega^2 h^2/\mu. \tag{2.62}$$

From the resulting displacement field, we find that the only nonzero component of stress is $\sigma_{\theta z}$ which may be integrated over the cross section to obtain the total twisting moment M as

$$M(z, t) = (\pi/2)(b^4 - a^4)(\rho\omega^2 h/\Omega)[A \cos(\Omega z/h) - B \sin(\Omega z/h)]\, e^{i\omega t} \tag{2.63}$$

where a and b are the inner and outer radii, respectively, h is the length of the cylinder, and A and B are arbitrary constants to be determined from end conditions.

Taking the end $z = h$ as being fixed with no displacement while the end $z = 0$ has the applied torque $\hat{M}e^{i\omega t}$, we find that the ratio of the twisting moment on the end $z = 0$ to the angle of twist is given by

$$\frac{M(0, t)}{\Phi(0, t)} = \frac{\hat{M}}{f(0)} = \frac{\pi(b^4 - a^4)}{2} \rho\omega^2 h \frac{\cot \Omega}{\Omega}. \tag{2.64}$$

Following the method of Section 2.4 we convert the elastic solution (2.64) to a steady state harmonic viscoelastic solution by replacing μ by $\mu^*(i\omega)$ in Ω (2.62). This procedure gives

$$\frac{\hat{M}}{f(0)} = \frac{\pi(b^4 - a^4)}{2} \rho\omega^2 h \frac{\cot \Omega^*}{\Omega^*} \tag{2.65}$$

where

$$\Omega^*(i\omega) = [\rho\omega^2 h^2/\mu^*(i\omega)]^{1/2}.$$

Knowing $\mu^*(i\omega)$, relation (2.65) allows one to determine $f(0)$, the angle of twist at $z = 0$, for a specified value of $\hat{M}$, the applied torque. We note from (2.65) that with $\hat{M}$ considered to be real, $f(0)$ is a complex number which is determined by the magnitude of $\Phi(0, t)$ and the phase angle by which it lags $M(0, t)$. Alternatively, if one considers an experimental apparatus of the type analyzed here, relation (2.65) permits the determination of the complex shear modulus through the observation of the applied torque and the angular response. This latter procedure was followed by Gottenberg and Christensen [2.14], and the results are given in detail in Section 7.3.

It is to be noted that inertia term effects are included in the present problem, and the result is a general steady state dynamic solution.

2.8. QUASI-STATIC RESPONSE EXAMPLE

We now give an example to illustrate the use of the integral transform method for obtaining solutions. In the problem considered in Section 2.7, the steady state torsional oscillation of a cylinder is analyzed and it is seen how to use that result in a program for determining the complex modulus $\mu^*(i\omega)$. In so doing it is implicitly assumed that any starting transients have "died out," such that only steady state conditions prevail. The same problem will again be analyzed but it will now include the transient starting effects due to the viscoelasticity of the material. That is, the right circular cylinder of inner radius a, outer radius b, and length h is taken with the end $z = h$ fixed, while the end $z = 0$ has a prescribed angle of rotation given by

$$\begin{aligned} \Phi(0, t) &= 0, & t &\leqslant 0 \\ \Phi(0, t) &= k \sin \omega t, & 0 &< t < \infty \end{aligned} \tag{2.66}$$

where k is the given amplitude of oscillation. The problem is to find the twisting moment $M(z, t)|_{z=0}$ required to produce the oscillation (2.66). In this problem, the frequency ω will be taken to be sufficiently small so that inertia terms may be neglected.

The appropriate static elasticity relationship between the twisting moment $M(0, t)$ on the cylinder and the angle of twist $\Phi(0, t)$ of the end $z = 0$ is given by

$$\frac{M(0, t)}{\Phi(0, t)} = \frac{\pi(b^4 - a^4)\mu}{2h}.$$

Following the integral transform method of Section 2.5, this elasticity relationship is reinterpreted as the Laplace transform of the corresponding viscoelastic relationship by replacing μ by $s\bar{\mu}(s)$, $M(0, t)$ by $\bar{M}(s)$, and $\Phi(0, t)$ by $\bar{\Phi}(0, s)$, where $\bar{\mu}(s)$ is the Laplace transform of the viscoelastic relaxation function $\mu(t)$. This gives

$$\bar{M}(s) = \frac{\pi(b^4 - a^4)\, s\bar{\mu}(s)\, k\omega}{2h(s^2 + \omega^2)} \tag{2.67}$$

where the Laplace transform of $\Phi(0, t)$, (2.66), has been used. In order to invert (2.67) it is necessary to specify $\bar{\mu}(s)$. To this end $\mu(t)$ is assumed in the form

$$\mu(t) = G_0 + \sum_{j=1}^{N} G_j e^{-t/t_j} \tag{2.68}$$

where G_0, G_j, and t_j, $j = 1 \ldots N$, are constants to be determined to represent any particular relaxation function of interest. The s multiplied Laplace transform of (2.68) is given by

$$s\bar{\mu}(s) = \frac{A(s)}{\prod_{j=1}^{N} (s + t_j^{-1})} \tag{2.69}$$

where a common denominator has been employed and $A(s)$ is an Nth degree polynominal in s, the coefficients of which are determined from the values of the G_j coefficients.

Substituting (2.69) in (2.67), the resulting expression is inverted to give

$$M(t) = \frac{\pi(b^4 - a^4)\, k\omega}{2h} \left\{ \frac{-e^{-i\omega t} A(-i\omega)}{2i\omega \prod_{j=1}^{N} (-i\omega + t_j^{-1})} + \frac{e^{i\omega t} A(i\omega)}{2i\omega \prod_{j=1}^{N} (i\omega + t_j^{-1})} \right.$$

$$\left. + \sum_{n=1}^{N} \lim_{s \to -t_n^{-1}} \left[\frac{A(s)(s + t_n^{-1})}{\prod_{j=1}^{N} (s + t_j^{-1})} \right] \frac{e^{-t/t_n}}{(t_n^{-2} + \omega^2)} \right\}. \qquad (2.70)$$

See Appendix B for the method of inversion.

The first two terms in (2.70) represent the steady state response, while the last term delineates the transient part of the response.

The complete solution (2.70) has been used by Gottenberg and Christensen [2.15] to predict the starting transient of the specimen which was used to determine the complex shear modulus, as described in Section 2.7. The relaxation function in shear for the material of the specimen is as shown in Fig. 2.1 and

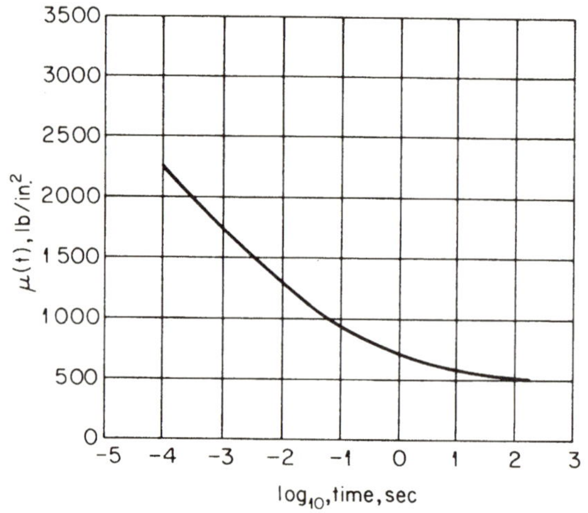

Fig. 2.1. Relaxation function in shear, polyurethane matrix containing salt crystals and aluminum powder at 79.0°F (after Gottenberg and Christensen [2.14]).

was determined from the corresponding complex modulus. The interconversion procedure is discussed further in Section 4.3. The relaxation function of Fig. 2.1 is represented by (2.68), to within plotting accuracy, with $N = 8$ and G_j and t_j given in Table 2.1.

The comparison of the results predicted by (2.70), for the data from Table 2.1, with the experimentally observed response is given in Fig. 2.2, which is from Ref. [2.15]. The agreement is seen to be complete.

Table 2.1 RELAXATION FUNCTION DATA

j	G_j	t_j
0	500	
1	997	1.5×10^{-5}
2	538	1.5×10^{-4}
3	494	1.5×10^{-3}
4	392	1.5×10^{-2}
5	306	1.5×10^{-1}
6	154	1.5
7	119	1.5×10
8	20	1.5×10^{2}

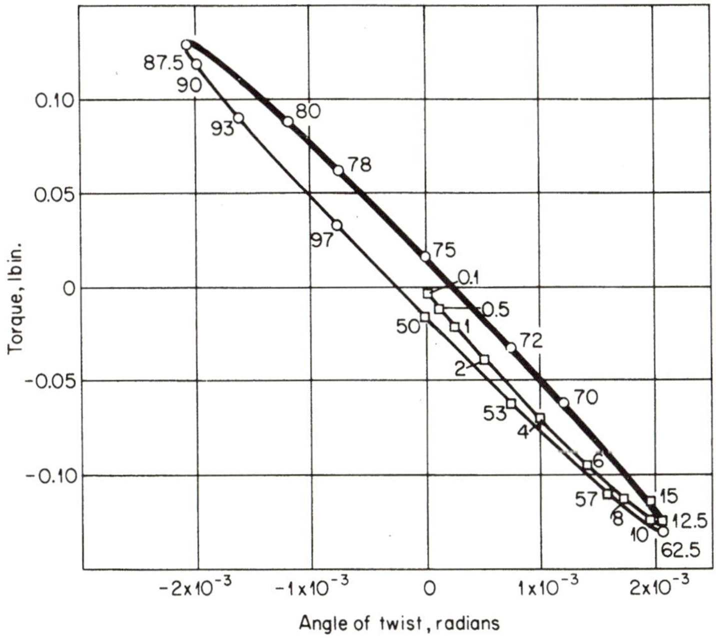

Fig. 2.2. Transient response example at a frequency of 0.02Hz and a temperature of 79.0°F (after Gottenberg and Christensen [2.15]). □ denotes times in seconds for first cycle in analysis. ○ denotes times in seconds for second cycle in analysis. Solid line denotes experimental results.

With regard to the procedure of fitting relaxation function data by a series of decaying exponentials, as has been done here, the following observations are pertinent. In general at least one term for each decade of graphical data is required. The relaxation times involved in these separate terms should be spaced at one decade intervals, with the largest and smallest relaxation times roughly corresponding to the largest and smallest times at which the relaxation data are to be modeled. Of course, any variation of this procedure is acceptable, so long as the analytical representation actually fits the data over the desired time range, there is always this final check. Problem 2.10 at the end of this chapter involves a modeling procedure of the type discussed here. In recognition of the ad hoc nature of this and other similar procedures, Gradowczyk and Moavenzadeh [2.16] have proposed a systematic means of representing relaxation and creep functions. The determination of mechanical properties is considered more fully in Chapter 4.

2.9. PRESSURIZATION OF A CYLINDER

Another, more complicated, example of the usage of the Laplace transform in obtaining viscoelastic solutions is now given. The problem of the pressurization of a viscoelastic cylinder constrained by an elastic case is of technical importance. The considerations here are for sufficiently long cylinders such that plane strain conditions can be assumed. First, the analysis is given for an incompressible viscoelastic cylinder, including inertia term effects. After this, the corresponding analysis for a compressible cylinder with inertia terms neglected is given.

Incompressible Cylinder, Dynamic Response

The pressurization of an incompressible viscoelastic cylinder, under axisymmetric, plane strain conditions, is the problem of interest. Under these conditions the dependence of the radial component of displacement upon the radial coordinate is determined directly from the condition of incompressibility. With this distribution of displacements considered to be known, the problem reduces to that of only one independent variable, time, and it is effectively a single degree of freedom problem. Under these very special conditions imposed by the incompressibility constraint, inertia terms can be included with no undue complications.

Proceeding along the line of reasoning mentioned above, the condition of incompressibility is given by

$$\frac{\partial u}{\partial r} + \frac{u}{r} = 0 \tag{2.71}$$

where $u = u(r, t)$ is the radial component of displacement in polar cylindrical coordinates r, θ, and z. The solution of (2.71) is simply

$$u = C/r \tag{2.72}$$

where $C = C(t)$ is a function of time which is yet to be determined.

The incompressible form of the elastic stress strain relations which are appropriate to the conditions of this problem is given by

$$\begin{aligned} \sigma_{rr} &= 2\mu \frac{\partial u}{\partial r} + S \\ \sigma_{\theta\theta} &= 2\mu \frac{u}{r} + S \\ \sigma_{zz} &= S \end{aligned} \tag{2.73}$$

where the hydrostatic stress S is a basic unknown of the problem. These relations are converted to the Laplace transforms of the viscoelastic stress strain relations, giving

$$\begin{aligned} \bar{\sigma}_{rr}(r, s) &= 2s\bar{\mu}(s) \frac{\partial \bar{u}(r, s)}{\partial r} + \bar{S}(r, s) \\ \bar{\sigma}_{\theta\theta}(r, s) &= 2s\bar{\mu}(s) \frac{\bar{u}(r, s)}{r} + \bar{S}(r, s) \\ \bar{\sigma}_{zz}(r, s) &= \bar{S}(r, s). \end{aligned} \tag{2.74}$$

The single nontrivial equation of motion has the form

$$\frac{\partial \sigma_{rr}}{\partial r} + \frac{\sigma_{rr} - \sigma_{\theta\theta}}{r} = \rho \frac{\partial^2 u}{\partial t^2}. \tag{2.75}$$

Taking the Laplace transform of this equation with conditions of initial rest and using (2.72) and (2.74), we find that the solution is given by

$$\bar{S} = \rho s^2 \bar{C} \ln r + \bar{D} \tag{2.76}$$

where $\bar{D} = \bar{D}(s)$.

Using (2.72) and (2.76), the transformed stresses are then given by

$$\begin{aligned} \bar{\sigma}_{rr}(r, s) &= -2s\bar{\mu}(\bar{C}/r^2) + \rho s^2 \bar{C} \ln r + \bar{D} \\ \bar{\sigma}_{\theta\theta}(r, s) &= 2s\bar{\mu}(\bar{C}/r^2) + \rho s^2 \bar{C} \ln r + \bar{D} \\ \bar{\sigma}_{zz} &= \rho s^2 \bar{C} \ln r + \bar{D}. \end{aligned} \tag{2.77}$$

The boundary conditions of this problem are specified by

$$\begin{aligned} \sigma_{rr}(r, t) &= -p(t) \quad && \text{at} \quad r = a \\ \sigma_{rr}(r, t) &= -q(t) \quad && \text{at} \quad r = b \\ u(r, t) &= u_c(t) \quad && \text{at} \quad r = b \end{aligned} \tag{2.78}$$

where $q(t)$ and $u_c(t)$ are the pressure and displacement of the constraining thin elastic case, and are related by

$$u_c(t) = q(t)[b^2(1 - \nu_c^2)/E_c h] \tag{2.79}$$

where ν_c and E_c are the elastic properties of the case, h is its thickness, and b is its radius. The last two equations of (2.78) are the conditions of continuity between the viscoelastic cylinder and the elastic case.

The satisfaction of the boundary conditions (2.78), subject to (2.79), determines $\bar{C}(s)$ and $\bar{D}(s)$. Using a representation for the relaxation function in shear the same as that in (2.68), then $s\bar{\mu}$ is written as

$$s\bar{\mu} = A(s)/B(s) \tag{2.80}$$

where $A(s)$ and $B(s)$ are polynomials in s. With (2.80), $\bar{C}(s)$ and $\bar{D}(s)$ are

$$\bar{C}(s) = -s\bar{p}B(s)/F(s) \tag{2.81}$$

$$\bar{D}(s) = -s\bar{p}E(s)/F(s) \tag{2.82}$$

where

$$F(s) = -2sA(s)\left(\frac{1}{a^2} - \frac{1}{b^2}\right) + \rho s^3 B(s) \ln\frac{a}{b} - \frac{sE_c h B(s)}{b^3(1-\nu_c^2)} \tag{2.83}$$

$$E(s) = \frac{2A(s)}{b^2} - \rho s^2 B(s) \ln b - \frac{E_c h B(s)}{b^3(1-\nu_c^2)}. \tag{2.84}$$

Taking, for example, the applied pressure

$$p(t) = p_0 h(t), \tag{2.85}$$

the Laplace transform of this when substituted into (2.81) and (2.82) gives the complete solution for $\bar{C}(s)$ and $\bar{D}(s)$. The denominator $F(s)$ in $\bar{C}(s)$ and $\bar{D}(s)$ may be factored to give

$$F(s) = \kappa(s - a_1)(s - a_2) \cdots (s - a_m) \tag{2.86}$$

where the a_j's are in general complex roots, κ is a constant, and the integer m is determined by the number of terms included in the representation of the type (2.68) for $\mu(t)$. The stresses (2.77) when inverted by the method of Appendix B give

$$\sigma_{rr}(r, t) = -p_0 \sum_{j=1}^{m} \frac{e^{a_j t}}{\lim\limits_{s\to a_j} [F(s)/(s-a_j)]} \left[-\frac{2}{r^2} A(a_j) + \rho \ln r (a_j)^2 B(a_j) + E(a_j)\right]$$

$$\sigma_{\theta\theta}(r, t) = -p_0 \sum_{j=1}^{m} \frac{e^{a_j t}}{\lim\limits_{s\to a_j} [F(s)/(s-a_j)]} \left[\frac{2}{r^2} A(a_j) + \rho \ln r (a_j)^2 B(a_j) + E(a_j)\right] \tag{2.87}$$

$$\sigma_{zz}(r, t) = -p_0 \sum_{j=1}^{m} \frac{e^{a_j t}}{\lim\limits_{s\to a_j} [F(s)/(s-a_j)]} [\rho \ln r (a_j)^2 B(a_j) + E(a_j)]$$

where the complex roots a_j of $F(s) = 0$ can be determined by standard digital computer programs, when the characteristics of a particular material are specified in the representation of $\mu(t)$.

Compressible Cylinder, Quasi-Static Response

For the compressible viscoelastic cylinder in polar cylindrical coordinates r, θ, and z with axial symmetry, the single nontrivial equation of (quasi-static) equilibrium is given by

$$\frac{\partial^2 \bar{u}}{\partial r^2} + \frac{1}{r}\frac{\partial \bar{u}}{\partial r} - \frac{\bar{u}}{r^2} = 0 \tag{2.88}$$

where $\bar{u} = \bar{u}(r, s)$ is the Laplace transform of the radial displacement. The solution of (2.88) is given by

$$\bar{u} = \bar{C}r + (\bar{D}/r) \tag{2.89}$$

where $\bar{C}(s)$ and $\bar{D}(s)$ are functions of the transform variable, and are to be determined.

Using the strain–displacement relations and the transform of the relaxation integral stress strain relations, it follows that

$$\bar{\sigma}_{rr} = 2s\bar{\mu}\left[\frac{\partial \bar{u}}{\partial r} + \frac{s\bar{\nu}}{1 - 2s\bar{\nu}}\bar{e}\right] \tag{2.90}$$

$$\bar{\sigma}_{\theta\theta} = 2s\bar{\mu}\left[\frac{\bar{u}}{r} + \frac{s\bar{\nu}}{1 - 2s\bar{\nu}}\bar{e}\right] \tag{2.91}$$

$$\bar{\sigma}_{zz} = \frac{2s^2\bar{\nu}\bar{\mu}\bar{e}}{1 - 2s\bar{\nu}} \tag{2.92}$$

where

$$\bar{e} = (\partial \bar{u}/\partial r) + (\bar{u}/r). \tag{2.93}$$

The boundary conditions of this problem are the same as in the previous one, and are given by (2.78) subject to (2.79). The constants $\bar{C}$ and $\bar{D}$ are evaluated from the Laplace transform of (2.78) and (2.79). It is found that

$$\bar{\sigma}_{rr} = C_1 - (b^2/r^2)\, C_2 \tag{2.94}$$

$$\bar{\sigma}_{\theta\theta} = C_1 + (b^2/r^2)\, C_2 \tag{2.95}$$

$$\bar{\sigma}_{zz} = 2s\bar{\nu}C_1 \tag{2.96}$$

where

$$C_1 = \frac{-\bar{p}(K - s\bar{\mu})}{(K - s\bar{\mu}) + (b^2/a^2)[K(1 - 2s\bar{\nu}) + s\bar{\mu}]} \tag{2.97}$$

$$C_2 = \frac{\bar{p}[K(1 - 2s\bar{\nu}) + s\bar{\mu}]}{(K - s\bar{\mu}) + (b^2/a^2)[K(1 - 2s\bar{\nu}) + s\bar{\mu}]} \tag{2.98}$$

and

$$K = \frac{E_c h}{2b(1 - \nu_c^2)}. \tag{2.99}$$

When $\mu(t)$ and $k(t)$ are taken in forms of the type of (2.68), it follows that $\bar{\mu}$ and $\bar{\nu}$ are ratios of polynomials in the transform parameter s. We take $p(t)$, for example, as a step function in time (2.85), and substitute its transform and the transforms $\bar{\mu}(s)$ and $\bar{\nu}(s)$ into (2.94)–(2.99). This results in transformed stresses which involve the ratios of polynomials in s and can be inverted by the method given in Appendix B. We carry out this procedure with viscoelastic Poisson's ratio taken as a real constant and $s\bar{\mu}(s)$ taken in the form (2.80), which gives an expression for $\bar{\sigma}_{rr}$ of the type

$$\bar{\sigma}_{rr} = p_0[C(s) - (b^2/r^2)\, D(s)]/F(s) \tag{2.100}$$

where

$$C(s) = -KB(s) + A(s), \qquad D(s) = (1 - 2\nu)\, KB(s) + A(s),$$
$$F(s) = s\{KB(s) - A(s) + (b^2/a^2)[(1 - 2\nu)\, KB(s) + A(s)]\}.$$

$F(s)$ can be factored as in (2.86). The inversion of (2.100) gives

$$\sigma_{rr}(r, t) = p_0 \sum_{j=1}^{m} \frac{[C(a_j) - (b^2/r^2)\, D(a_j)]\, e^{a_j t}}{\lim_{s \to a_j} [F(s)/(s - a_j)]}. \tag{2.101a}$$

The other two stress components from (2.95) and (2.96) are given by

$$\sigma_{\theta\theta}(r, t) = p_0 \sum_{j=1}^{m} \frac{[C(a_j) + (b^2/r^2)\, D(a_j)]\, e^{a_j t}}{\lim_{s \to a_j} [F(s)/(s - a_j)]}$$

and

$$\sigma_{zz}(r, t) = p_0 \sum_{j=1}^{m} \frac{2\nu C(a_j)\, e^{a_j t}}{\lim_{s \to a_j} [F(s)/(s - a_j)]}. \tag{2.101b}$$

A specific evaluation of the stresses by (2.101) has been given by Christensen and Schreiner [2.17]. In that study the relaxation function $\mu(t)$ is represented in the form (2.68) of decaying exponentials, and the corresponding roots of $F(s)$ are found. Also included in Ref. [2.17] is a particular means of including the effect of a boundary which is changing due to the burning or erosion of the material. There have been many studies on the pressurization of viscoelastic cylinders. One approach, typified by the work of Huang *et al.* [2.18] involves reducing the problem to that of solving an integral equation, or equations, which can easily be accomplished numerically using direct measurements of mechanical properties. Another related study is that given by Ting [2.19]. This reference also includes the possible effect of an eroding boundary.

2.10. PRESSURIZATION OF A SPHERICAL CAVITY

As a final example of the use of integral transform methods, the response to pressure change on a spherical cavity in an infinite medium is studied. In contrast to the previous examples, however, this problem is used to illustrate the use of the Fourier transform method. Inertia terms are included; therefore, this represents a dynamic response problem. As such, this problem involves wave propagation effects; however, these will not be examined in detail here, rather merely the method of solution is studied. Specific consideration of wave propagation effects is deferred until Chapter 6. The solution given here is taken from the work of Lockett [2.20].

In spherical coordinates r, θ, φ with spherical symmetry the single nontrivial elastic equation of motion is given by

$$\frac{\partial^2 u}{\partial r^2} + \frac{2}{r}\frac{\partial u}{\partial r} - \frac{2u}{r^2} = \left(\frac{\rho}{\lambda + 2\mu}\right)\frac{\partial^2 u}{\partial t^2} \tag{2.102}$$

where $u = u(r, t)$ is the radial component of displacement and λ and μ are the Lamé constants. Following the method of Section 2.5, this equation of motion is converted to the Fourier transformed equation of motion of the viscoelastic medium by replacing $u(r, t)$ by $\bar{u}(r, \omega)$ and λ and μ by $\lambda^*(i\omega)$ and $\mu^*(i\omega)$. This gives

$$\frac{\partial^2 \bar{u}}{\partial r^2} + \frac{2}{r}\frac{\partial \bar{u}}{\partial r} - \frac{2\bar{u}}{r^2} = -\Omega^2\omega^2\bar{u} \tag{2.103}$$

where

$$\Omega^2 = \rho/[\lambda^*(i\omega) + 2\mu^*(i\omega)]. \tag{2.104}$$

The solution of (2.103) is

$$\bar{u} = \frac{\partial}{\partial r}\left[\frac{C(\omega)\, e^{i\Omega\omega r} + D(\omega)\, e^{-i\Omega\omega r}}{r}\right] \tag{2.105}$$

where $C(\omega)$ and $D(\omega)$ are to be determined. Since there is no outer boundary, there can be no reflections causing incoming waves; so, for only outgoing waves $C(\omega)$ must vanish. This leaves (2.105) as

$$\bar{u} = -(D/r^2)(1 + i\Omega\omega r)\, e^{-i\Omega\omega r}. \tag{2.106}$$

The only nonzero stresses in this problem are σ_{rr}, and $\sigma_{\theta\theta} = \sigma_{\varphi\varphi}$. These may be derived from (2.106), using the transformed stress strain relations and the strain displacement relations, to get

$$\bar{\sigma}_{rr} = (D/r^3)[4\mu^*(1 + i\Omega\omega r) - \rho\omega^2 r^2]\, e^{-i\Omega\omega r} \tag{2.107}$$

and

$$\bar{\sigma}_{\theta\theta} = -(D/r^3)[2\mu^*(1 + i\Omega\omega r) + \Omega^2\omega^2 r^2\lambda^*]\, e^{-i\Omega\omega r}. \tag{2.108}$$

At the cavity radius $r = a$ the boundary condition is

$$\bar{\sigma}_{rr}((r, \omega) = \bar{p}(\omega), \qquad r = a \tag{2.109}$$

where $\bar{p}$ is the Fourier transform of the given pressure on the cavity surface. Using (2.107) and (2.109) we find that

$$D = \frac{a^3\bar{p}e^{i\Omega\omega a}}{4\mu^*(1 + i\Omega\omega a) - \rho\omega^2 a^2}. \tag{2.110}$$

By substituting (2.110) into (2.106)–(2.108), the resulting forms are suitable for inversion. The Fourier inversion formula (1.73) can be written in the alternate form

$$f(t) = \frac{1}{\pi}\,\mathrm{Re}\int_0^\infty \bar{f}(\omega)\, e^{i\omega t}\, d\omega \tag{2.111}$$

where Re designates the real part. For example, applying (2.111) to the inversion of the displacement of the cavity $u(a, t)$, it becomes

$$u(a, t) = -\frac{a}{\pi}\,\mathrm{Re}\int_0^\infty \frac{(1 + i\Omega\omega a)\,\bar{p}e^{i\omega t}\, d\omega}{4\mu^*(1 + i\Omega\omega a) - \rho\omega^2 a^2}. \tag{2.112}$$

The integration in (2.112) can be performed when the pressure $p(t)$ is specified and when $\mu^*(i\omega) = \mu'(\omega) + i\mu''(\omega)$ and $\lambda^*(i\omega) = \lambda'(\omega) + i\lambda''(\omega)$ are specified as functions of frequency ω for a particular viscoelastic material. In general, this integration cannot be evaluated analytically; however, it is a practical matter to evaluate it numerically. A particular pressure will now be specified for which the numerical evaluation of (2.112) is further simplified.

In the further example the pressure on the spherical cavity is taken as

$$p(t) = \frac{p_0}{\pi}\int_{\omega_1}^{\omega_2} \frac{\sin \omega t}{\omega}\, d\omega. \tag{2.113}$$

Using the definition of the sine integral

$$\mathrm{Si}(u) = \int_0^u \frac{\sin x}{x}\, dx \tag{2.114}$$

then (2.113) can be written as

$$p(t) = \frac{p_0}{\pi}\,[\mathrm{Si}(\omega_2 t) - \mathrm{Si}(\omega_1 t)]. \tag{2.115}$$

The Fourier transform of this, defined through (1.72), is given by

$$\begin{aligned} \bar{p}(\omega) &= 0, && 0 \leqslant \omega < \omega_1 \\ &= p_0/i\omega, && \omega_1 \leqslant \omega \leqslant \omega_2 \\ &= 0, && \omega_2 < \omega < \infty. \end{aligned} \tag{2.116}$$

In obtaining $p(t)$ from (2.115) the sine integrals may be evaluated from any standard reference, such as Janhke and Emde [2.21]. A schematic plot of $p(t)$ vs t is given by Fig. 2.3 where $\omega_1 = 5$, $\omega_2 = 1000$. From this we see that the

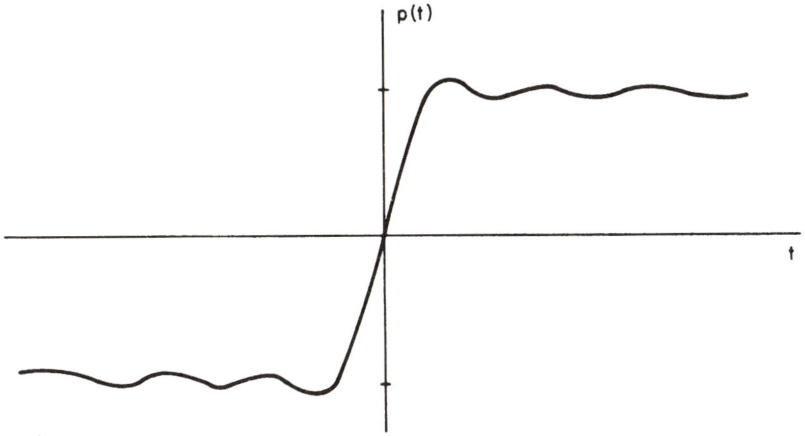

Fig. 2.3. Pressure on spherical cavity.

pressure $p(t)$ from (2.115) is an approximation to a discontinuous change in pressure. The advantage of using (2.115) rather than a step function pressure history is that the characteristics revealed by (2.116) enable us to use an inversion integral over a finite frequency range, rather than the usual infinite frequency range. Now using $\bar{p}$ from (2.116), (2.112) becomes

$$u(a, t) = \frac{p_0 a}{\pi} \operatorname{Re} \int_{\omega_1}^{\omega_2} \frac{(1 + i\Omega\omega a)\, ie^{i\omega t}\, d\omega}{4\omega\mu^*(1 + i\Omega\omega a) - \rho\omega^3 a^2}\cdot \tag{2.117}$$

The integration in (2.117) involves finite limits, and it may be evaluated directly when $\mu^*(i\omega)$ and $\lambda^*(i\omega)$ are specified as functions of frequency.

In this dynamic problem analyzed through the use of the Fourier transform, the inversion process is very difficult to accomplish in analytical form. However, in the case of quasi-static problems, the Fourier transform technique can be used to obtain analytical solutions. The procedure in such quasi-static problems is entirely similar to that already outlined using the Laplace transform, where it was shown to be practical to obtain analytical solutions for a general characterization of the mechanical properties. Also, the use of Fourier transform method in both quasi-static and dynamic problems often leads to a final form of the solution involving a simple numerical integration.

In general, when we seek an analytical solution in transient dynamic viscoelastic problems, using integral transform methods, the inversion process involves branch points. In the present example, Ω in (2.117) introduces such branch points. Because of this complication, analytical solutions are seldom obtained. This situation is explored more fully in Chapter 6.

2.11. FREE VIBRATION

In the problem of the preceding section, boundary conditions involving time histories are specified and the dynamic response of the viscoelastic body is determined. Another type of problem in which inertia term effects are included and in fact provide the entire nature of the response is the class of free vibration problems. This class of problems is defined such that the work of all boundary forces, where they exist, is identically zero; thus, the boundary conditions are

$$\begin{aligned} u_i &= 0 \qquad \text{on} \quad B_u \\ \sigma_{ij} n_j &= 0 \qquad \text{on} \quad B_\sigma . \end{aligned} \tag{2.118}$$

Also, the externally applied body forces must be taken as vanishing identically, except for possibly their temporary application to initially start the motion. Thus, these problems are of initial value type, and the analysis of viscoelastic bodies subjected to such conditions, is the item of study here.

The Laplace transformed equations of motion are given by

$$\bar{u}_{i,jj} + (1 - 2s\bar{\nu})^{-1}\,\bar{u}_{j,ji} = \frac{\rho}{s\bar{\mu}(s)}\,\mathscr{L}\left(\frac{\partial^2 u_i}{\partial t^2}\right) \tag{2.119}$$

It is the solution to (2.119), satisfying the boundary conditions (2.118) and subject to certain initial conditions, which we are seeking. First, the special case whereby the viscoelastic Poisson's ratio is a constant will be treated, since this case affords great simplification.

Poisson's Ratio ν = Constant, Special Case

It is convenient to first review the free vibration analysis used for elastic bodies. The equations of motion are given by

$$u_{i,jj} + (1 - 2\nu)^{-1}\,u_{j,ji} = \frac{\rho}{\mu}\,\frac{\partial^2 u_i}{\partial t^2} \tag{2.120}$$

and the boundary conditions (2.118) still apply. A solution in the form

$$u_i(x_i, t) = U_i^E(x_i)\, e^{i\omega t} \tag{2.121}$$

is assumed. The spatial functions $U_i^E(x_i)$ are evaluated so that the equations of motion (2.120) are satisfied, and the satisfaction of the boundary conditions (2.118) leads to the specification of an eigenvalue problem. The corresponding eigenvalues and eigenvectors determine the natural frequencies and the corresponding mode shapes, which will be designated by

$$\omega_n \quad \text{and} \quad U_i^{E,n}(x_i),$$

respectively. The spectrum of frequencies ω_n can either be discrete or continuous, depending upon whether or not the elastic body is of finite extent. Here, and henceforth in this section, only the case of discrete eigenvalues are considered; however, the results may easily be generalized to the nondiscrete case. The complete solution then involves superposition of the eigenvectors; thus, from (2.121)

$$u_i(x_i, t) = \sum_n U_i^{E,n}(x_i)\, e^{i\omega_n t}. \tag{2.122}$$

The arbitrary scaling constants, implicit in each eigenvector, are evaluated such that initial conditions are satisfied. Formally, the procedure for evaluating these complex constants uses the orthogonality of eigenvectors in the same manner as is followed in evaluating the constants in a complex Fourier series representation.

Now the corresponding viscoelastic problem is considered. We assume the viscoelastic solution in the form

$$u_i(x_i, t) = U_i^{E,n}(x_i) f^n(t) \tag{2.123}$$

where $f^n(t)$ are unknown functions of time and $U_i^{E,n}(x_i)$ are the elastic eigenvectors for the corresponding elastic problem, which has the elastic Poisson's ratio equal to that of the viscoelastic problem. Substituting (2.123) into the equations of motion (2.119) gives

$$[U_{i,jj}^{E,n} + (1 - 2\nu)^{-1} U_{j,ji}^{E,n}]\bar{f}^n = (\rho/s\bar{\mu})[s^2\bar{f}^n - \dot{f}^n(0) - sf^n(0)]\, U_i^{E,n} \tag{2.124}$$

where

$$f^n(0) = f^n(t)\,|_{t=0}$$

and

$$\dot{f}^n(0) = \left.\frac{df^n(t)}{dt}\right|_{t=0} \tag{2.125}$$

with $f^n(0)$ and $\dot{f}^n(0)$ to be determined from the initial conditions. But, from the corresponding elastic solution, (2.122) and (2.120) give

$$U^{E,n}_{i,jj} + (1 - 2\nu)^{-1} U^{E,n}_{j,ji} = (-\omega_n{}^2\rho/\mu)\, U^{E,n}_i. \tag{2.126}$$

Substitute (2.126) into (2.124) and evaluate the corresponding values of $\bar{f}^n(s)$. This gives

$$\bar{f}^n = \frac{\dot{f}^n(0) + sf^n(0)}{s^2 + (\omega_n{}^2 s\bar{\mu}/\mu)}. \tag{2.127}$$

Since $\omega_n{}^2$ from the corresponding elastic problem is proportional to μ, the solution given by (2.127) is actually independent of the assumed elastic shear modulus μ. Taking $\mu(t)$, the viscoelastic relaxation function in shear in the form of a series of decaying exponentials as in (2.68), we see that (2.127) involves only simple poles, the roots of which are easily found using standard digital computer programs, and the inversion of $\bar{f}(s)$ may be carried out directly. This is exactly the same situation as was encountered for quasi-static response problems, and where it was found to be practical to take as many terms as were needed in the representation (2.68) for $\mu(t)$ to accurately characterize it. A simple example involving the inversion of (2.127) is given as a problem at the end of the chapter.

The complete solution to the free vibration problem is then given by

$$u_i(x_i\,, t) = \sum_n U^{E,n}_i(x_i) f^n(t) \tag{2.128}$$

subject to the initial conditions

$$u_i(x_i\,, t)\,|_{t=0} = W_i(x_i)$$

and

$$\left.\frac{\partial u_i(x_i\,, t)}{\partial t}\right|_{t=0} = V_i(x_i). \tag{2.129}$$

We substitute the initial conditions (2.129) into the left-hand side of (2.128) and its first derivative, and multiply the resulting forms by $U^{E,n}_i(x_i)$. We then integrate this over the region V to give

$$f^n(0) = \int_V W_i(x_i)\, U^{E,n}_i(x_i)\, dv$$

and

$$\dot{f}^n(0) = \int_V V_i(x_i)\, U^{E,n}_i(x_i)\, dv \tag{2.130}$$

where the orthogonality of the normalized elastic eigenvectors $U^{E,n}_i$ has been used.

The analysis just described is from the work of Hunter [2.22]. This completes the free vibration analysis in the case where Poisson's ratio is a constant; now, this restriction will be relaxed.

General Viscoelastic Poisson's Ratio Case

In this more general case it is convenient to proceed directly by assuming a harmonic oscillation, rather than by using the Laplace transform, as was done for $\nu =$ constant. Thus, let

$$u_i(x_i\,, t) = U_i(x_i)\, e^{i\omega t}. \tag{2.131}$$

Using this assumed displacement form in the equations of motion we get

$$U_{i,jj} + \frac{U_{j,ji}}{1 - 2\nu^*(i\omega)} = \frac{-\rho\omega^2 U_i}{\mu^*(i\omega)}. \tag{2.132}$$

The means of solving (2.132), subject to the boundary conditions (2.118) is formally the same as is involved in the corresponding elasticity problem. In particular, the frequency equation, symbolically represented by

$$g\big(\rho\omega^2/\mu^*(i\omega), \nu^*(i\omega)\big) = 0 \tag{2.133}$$

follows from that of the corresponding free vibration elasticity problem wherein μ and ν are replaced by μ^* and ν^*, respectively.

The frequency equation (2.133), in contrast to the corresponding elasticity frequency equation, is a complex function of frequency, and the roots will, in general, be complex and will be denoted by

$$\omega_n^* \text{ complex roots of (2.133).}$$

In writing $\mu^*(i\omega) = \mu'(\omega) + i\mu''(\omega)$, $\mu'(\omega)$ and $\mu''(\omega)$ will now each be complex when ω is complex, with a similar situation being true for $\nu^*(i\omega)$. Thus, the analytical forms of $\mu'(\omega)$, $\mu''(\omega)$, $\nu'(\omega)$, and $\nu''(\omega)$ are needed for use in (2.133).

The complete free vibration solution is now written as

$$u_i(x_i\,, t) = \sum_n U_i^{(n)}\big(x_i\,, \rho(\omega_n^*)^2/\mu^*, \nu^*\big)\, e^{i\omega_n^* t} \tag{2.134}$$

where $U_i^{(n)}$ are the spatial parts, or mode shapes, of the solution corresponding to the roots ω_n^*. The solution (2.134) can be written in the alternative form

$$u_i(x_i\,, t) = \sum_n R_i^{(n)}(x_i) \exp(-\omega_n'' t) \exp i[\omega_n' t + \varphi_i^{(n)}(x_i)] \tag{2.135}$$

where

$$R_i^{(n)}(x_i) = [(U'_{i(n)})^2 + (U''_{i(n)})^2]^{1/2}$$

and

$$\varphi_i^{(n)}(x_i) = \tan^{-1} \frac{U''_{i(n)}(x_i)}{U'_{i(n)}(x_i)} \tag{2.136}$$

with $U_i^{(n)}$ and $\omega_n{}^*$ being decomposed into real and imaginary parts through

$$U_i^{(n)}(x_i\,,\rho(\omega_n{}^*)^2/\mu^*, \nu^*) = U'_{i(n)}(x_i) + iU''_{i(n)}(x_i)$$

and

$$\omega_n{}^* = \omega_n{}' + i\omega_n'' \,. \tag{2.137}$$

The character of the free vibration solution (2.135) can now be assessed. Quite obviously, there is a damped exponential dependence. However, the most interesting feature of the solution is that there are no unique, observable mode shapes. This possibility is precluded by the presence of the space dependent phase angles $\phi_i^{(n)}(x_i)$. This situation is in contrast to that for Poisson's ratio equal to a constant, where there is a unique, observable shape which can be associated with each separate frequency of oscillation.

Although the procedure just outlined for determining the free vibration characteristics of a general viscoelastic body is, in principle, straightforward, the actual details of carrying out the procedure are rather complicated. These complications are directly due to the frequency dependency of the viscoelastic mechanical properties.

It is instructive to investigate the simplifications which occur in the present procedure when viscoelastic Poisson's ratio $\nu^*(i\omega)$ is taken to be a real constant ν. In this case the frequency equation (2.133) is written as

$$g(\mathscr{L}^2, \nu) = 0 \tag{2.138}$$

where

$$\mathscr{L}^2 = \rho\omega^2/\mu^*(i\omega). \tag{2.139}$$

The roots $\mathscr{L}_n{}^2$ of the frequency equation (2.138) are exactly the same as in the comparable elasticity problem, and correspondingly are real. The values of ω are found from (2.139) for each real root $\mathscr{L}_n{}^2$. These roots, designated by $\omega_n{}^*$, are complex, and in finding them from (2.139) it is necessary that $\mu^*(i\omega)$ be specified analytically, as in the preceding case for a general viscoelastic Poisson's ratio. The free vibration solution in the present case is given by

$$u_i(x_i\,, t) = \sum_n U_i^{(n)}(x_i\,, \mathscr{L}_n\,, \nu)\, e^{i\omega_n{}^*t}. \tag{2.140}$$

The mode shapes $U_i^{(n)}$ are real, in contrast to those of (2.134), and these mode shapes, or eigenvectors, are simply those of the corresponding elasticity problem. The present procedure, for $\nu =$ real constant, gives a means of determining viscoelastic free vibration characteristics alternative to the method outlined using the Laplace transform.

In free vibration problems where $\nu(t)$ is not a constant, it can often be approximated by a constant value, and the procedures outlined for $\nu =$ constant can be followed. Also, it should be noted that in viscoelastic free vibration problems which involve only one mechanical property, as in the vibration of beams and plates governed by the classical theory, the methods outlined for $\nu =$ real constant can be employed.

Approximate Evaluation Method

There is a widely used approximation in studying the free vibration of viscoelastic bodies, which is illustrated here for a single degree of freedom viscoelastic system. For a single degree of freedom system, the frequency equation, corresponding to (2.133), has the form

$$g(\omega, \mu^*) = \omega^2 - \xi^2\mu^*(i\omega) = 0 \tag{2.141}$$

with ξ^2 designating a shape factor. For example, a vibrating viscoelastic cantilever beam having a large mass at the tip, may be idealized as a one degree of freedom system, and ξ^2 involves the length and moment of inertia of the beam and the mass of the end object. Also, in this example, the appropriate mechanical property for use in (2.141) is not $\mu^*(i\omega)$, but rather is the complex modulus $E^*(i\omega)$.

Although the exact solution for the free vibration could easily be obtained by the method outlined using the Laplace transform, we proceed in a different manner since it will lead us to a convenient approximate solution. The term $(\mu^*(i\omega))^{1/2} = (\mu' + i\mu'')^{1/2}$ is expanded so that (2.141) has the form

$$\left\{\omega + \xi\left[(\mu')^{1/2}\left(1 + \frac{i}{2}\frac{\mu''}{\mu'} + \cdots\right)\right]\right\}\left\{\omega - \xi\left[(\mu')^{1/2}\left(1 + \frac{i}{2}\frac{\mu''}{\mu'} + \cdots\right)\right]\right\} = 0 \tag{2.142}$$

where the remaining terms in the expansion are of order $O((\mu''/\mu')^2)$. We assume $\mu''/\mu' \ll 1$ such that the additional terms in the above expansion can be neglected. This gives the roots of (2.142) as

$$\omega \cong \pm\xi\left[(\mu')^{1/2}\left(1 + \frac{i}{2}\frac{\mu''}{\mu'}\right)\right]. \tag{2.143}$$

We let the complex roots ω be designated by

$$\omega = \omega' + i\omega''. \tag{2.144}$$

Substituting the complex frequency, (2.144) into μ' and μ'' in (2.143), gives the real and imaginary parts of (2.143) as

$$\omega' = \pm\xi[\mathrm{Re}(\mu')^{1/2} - \tfrac{1}{2}\,\mathrm{Im}((\mu')^{1/2}\,(\mu''/\mu'))] \tag{2.145}$$

and

$$\omega'' = \pm\xi[\mathrm{Im}(\mu')^{1/2} + \tfrac{1}{2}\,\mathrm{Re}((\mu')^{1/2}\,(\mu''/\mu'))] \tag{2.146}$$

where $\mu' = \mu'(\omega' + i\omega'')$ and $\mu'' = \mu''(\omega' + i\omega'')$. But, consistent with the approximation $\mu''/\mu' \ll 1$, we can now take $\omega'' \ll \omega'$. By using this latter restriction in (2.145) and (2.146) we get

$$\omega' \cong \pm\xi[\mu'(\omega')]^{1/2} = \pm\Omega' \tag{2.147}$$

$$\omega'' \cong \pm\frac{\xi}{2}[\mu'(\omega')]^{1/2}\frac{\mu''(\omega')}{\mu'(\omega')} = \frac{\xi}{2}[\mu'(\Omega')]^{1/2}\frac{\mu''(\Omega')}{\mu'(\Omega')} \tag{2.148}$$

where from (1.58) and (1.59) $\mu'(\omega)$ and $\mu''(\omega)$ are even and odd functions of ω, respectively. Using (2.131) for the single degree of freedom system, along with (2.147) and (2.148), we get the approximate solution

$$u(t) \cong [A\exp(i\xi(\mu')^{1/2}\,t) + B\exp(-i\xi(\mu')^{1/2}\,t)]\exp[(-\xi/2)(\mu')^{1/2}\,(\mu''/\mu')t] \tag{2.149}$$

where $\mu' = \mu'(\Omega')$ and $\mu'' = \mu''(\Omega')$ with Ω' being determined from (2.147). The constants A and B in (2.149) are arbitrary complex constants. The solution (2.149) based upon $\mu''/\mu' \ll 1$ offers a simple interpretation of the frequency of oscillation and the rate of decay in terms of the real and imaginary parts of $\mu^*(i\omega)$. A small modification allows us to proceed in cases where use is to be made of this free vibration method of determining mechanical properties for materials for which $\mu''/\mu' \ll 1$ is not valid. In this case, an additional elastic element must be used in conjunction with the viscoelastic stiffening element, so that the effective complex stiffness of the combination has a very small ratio of imaginary to real parts. The general problem of the free vibration of single degree of freedom viscoelastic systems has been extensively considered by Struik [2.23].

2.12. LIMITATIONS OF INTEGRAL TRANSFORM METHODS

The basic equations governing the viscoelastic boundary value problems, given in Section 2.1, are stated subject to the condition that B_u and B_σ are independent of time, where these are the parts of the boundary upon which displacement components and stress vector components respectively are prescribed. If this were not true, at some points on the boundary there would be prescribed displacements for some values of time and prescribed stress vector components for the other values of time. In such cases, the use of integral transform methods fails because their usage requires that the type of boundary condition (displacement or stress vector components) at a point be time invariant. There are also other types of problems for which the direct application of integral transform methods are inapplicable. For example, ablation problems, through a phase change, cause the boundaries of a viscoelastic body to change size and shape. Nonisothermal problems in which the mechanical properties are taken to be temperature dependent are another example. Problems of this latter type will be considered in Section 3.6.

There are practical problems which involve boundary conditions of a changing type. As an example, the time dependent indentation of a viscoelastic half space by a curved rigid indentor is in this class. As the indentor is depressed into the half space, there are some points on the boundary of the region which at first have traction free boundary conditions but later have a displacement boundary condition such that the half space boundary conforms to the geometry of the indentor in the contact region.

For problems of this class, the set of conditions given in Section 2.1 govern the solution of boundary value problems. Now, however, it cannot be assumed that B_u and B_σ are independent of time; therefore, the integral transform methods of Section 2.5 do not apply. For problems of this type there are no completely general methods of solution available, and such problems are to be expected to be more difficult than comparable problems for which integral transform methods apply. However, some problems in this class have been found to be tractable, and one such example is now given. This general class of problems for which integral transform methods do not apply is treated fully in Chapter 5.

Indentor on Beam

Perhaps the simplest problem of this type is that of the deformation of a uniform viscoelastic beam by a curved rigid indentor. Figure 2.4 shows the nature of the problem with the coordinate convention, where $2a(t)$ is the length

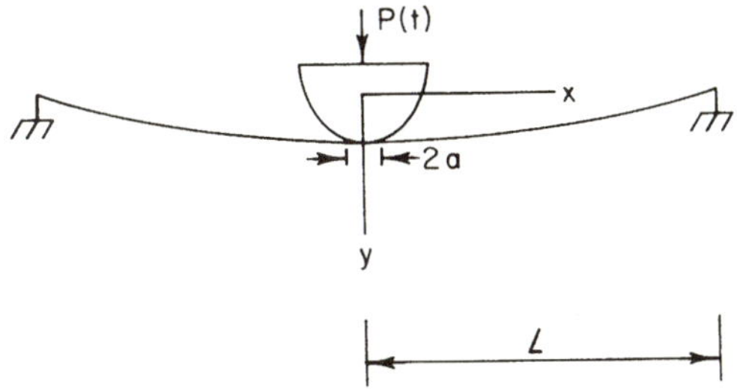

Fig. 2.4. Indentor on beam.

of the contact region, and $P(t)$ is the force applied to the rigid indentor. The indentor is taken to have the cubic profile given by

$$y = d(t) - c\,|\,x\,|^3 \tag{2.150}$$

where $d(t)$ is the displacement of the indentor, $|\,x\,|$ is the magnitude of x, and c is a given constant.

Classical beam theory conditions with simply supported end conditions are assumed. With inertia terms neglected, the elastic theory gives

$$EI\,\frac{d^4w}{dx^4} = q(x) \tag{2.151}$$

while the corresponding assumptions applied to a viscoelastic beam theory give

$$I\int_0^t E(t-\tau)\,\frac{\partial}{\partial\tau}\left[\frac{\partial^4 w(x,\tau)}{\partial x^4}\right]d\tau = q(x,t) \tag{2.152}$$

where I is the moment of inertia of the cross section, $E(t)$ is the uniaxial relaxation function, w is the transverse displacement, and q is the lateral load. Contact is assumed to begin at $t = 0$. From symmetry, only $x \geqslant 0$ needs to be considered. In the contact region $x < a(t)$, where $a(t)$ is a basic unknown of the problem, the deflection of the beam is assumed to conform to the geometry of the indentor; thus,

$$w(x,t) = d(t) - cx^3, \qquad x < a(t), \qquad t \geqslant 0. \tag{2.153}$$

Outside the contact region, the lateral load vanishes and (2.152) is satisfied by

$$w = C_1 + C_2x + C_3x^2 + C_4x^3, \qquad x > a(t) \tag{2.154}$$

where C_j are functions of time to be determined.

The condition that the shear resultant on the ends of the beam balance the applied load $P(t)$ gives, using (2.154),

$$12I\int_0^t E(t-\tau)\,\frac{dC_4(\tau)}{d\tau}\,d\tau = P(t). \tag{2.155}$$

The end conditions, that at $x = L$, $w = \partial^2 w/\partial x^2 = 0$, are specified by

$$C_1 + C_2L + C_3L^2 + C_4L^3 = 0 \tag{2.156}$$

and

$$2C_3 + 6C_4L = 0. \tag{2.157}$$

The continuity conditions at the edge of the contact region $x = a$ are that w, $\partial w/\partial x$, and $\partial^2 w/\partial x^2$ are continuous. Using (2.153) and (2.154), these conditions give

$$C_1 + C_2a + C_3a^2 + C_4a^3 = d - ca^3, \tag{2.158}$$

$$C_2 + 2C_3a + 3C_4a^2 = -3ca^2, \tag{2.159}$$

and

$$2C_3 + 6C_4 a = -6ca. \tag{2.160}$$

Relations (2.155)–(2.160) give six nonlinear equations. If the load $P(t)$ is considered as specified then the six unknowns are $C_1(t)$, $C_2(t)$, $C_3(t)$, $C_4(t)$, $a(t)$, and $d(t)$. Alternatively, if the displacement of the indentor $d(t)$ is considered as specified, then the six unknowns are $C_1(t)$, $C_2(t)$, $C_3(t)$, $C_4(t)$, $a(t)$, and $P(t)$. These two cases are considered separately.

$P(t)$ Specified

In this situation, with $P(t)$ specified, $C_4(t)$ is most conveniently solved from (2.155) through the use of the Laplace transform. It is found from this that

$$C_4(t) = f(t)/12I \tag{2.161}$$

where

$$f(t) = \mathscr{L}^{-1}[\bar{P}(s)/s\bar{E}(s)] \tag{2.162}$$

and $\mathscr{L}^{-1}$ stands for the inversion operation. This use of Laplace transform must not be confused with the general method of solution through the integral transform of all field equations. For typical histories of $P(t)$, the inversion indicated in (2.162) can be carried out directly. Now considering $f(t)$ and thus $C_4(t)$ as known, the other unknowns of the problem are solved algebraically from (2.156)–(2.160). We find that

$$C_1(t) = \frac{L^3 f(t)[24Ic - f(t)]}{12I[12Ic + f(t)]}, \tag{2.163}$$

$$C_2(t) = \frac{L^2 f^2(t)}{4I[12Ic + f(t)]}, \tag{2.164}$$

$$C_3(t) = -\frac{Lf(t)}{4I}, \tag{2.165}$$

$$a(t) = \frac{Lf(t)}{12Ic + f(t)}, \tag{2.166}$$

and

$$d(t) = \frac{cL^3 f(t)[24Ic + f(t)]}{[12Ic + f(t)]^2} \tag{2.167}$$

where $f(t)$ is given by (2.162). This completes the solution for the case having $P(t)$ specified. Now, we consider the alternative situation.

$d(t)$ Specified

When $d(t)$ rather then $P(t)$ is considered as being specified, the solution of Eqs. (2.156)–(2.160) gives

$$C_1(t) = cL^3 \left[-3\gamma(t) - \frac{1}{\gamma(t)} + 4\right], \tag{2.168}$$

$$C_2(t) = 3cL^2 \left[\gamma(t) + \frac{1}{\gamma(t)} - 2\right], \tag{2.169}$$

$$C_3(t) = 3cL \left[-\frac{1}{\gamma(t)} + 1\right], \tag{2.170}$$

$$C_4(t) = c \left[\frac{1}{\gamma(t)} - 1\right], \tag{2.171}$$

and

$$a(t)/L = 1 - \gamma(t) \tag{2.172}$$

where

$$\gamma(t) = [1 - (d(t)/cL^3)]^{1/2}. \tag{2.173}$$

The force $P(t)$ applied to the indentor is found from (2.155) as

$$\frac{P(t)}{12cI} = E(0)\left[\frac{1}{\gamma(t)} - 1\right] - \int_0^t \frac{dE(t-\tau)}{d\tau}\left[\frac{1}{\gamma(t)} - 1\right] d\tau \tag{2.174}$$

where an integration by parts has been performed, and $\gamma(t)$ is given by (2.173) with $d(t)$ specified. This completes the solution for this particular case.

A key assumption in the preceding derivation is that the beam corresponds to the shape of the indentor. An examination of the resulting solution shows that there is a pair of compressive forces at the edges of the contact region, whereas there is a tensile force at the center of the contact region. Thus the problem corresponds to that of an indentor bonded at the center to the beam. A more general analysis, which does not require the bonding assumption, has been given by Granstam [2.24]. The present example suffices to show the practicality of approaching problems for which integral transform methods do not apply. Problems of this general class are treated in Chapter 5.

A similar analysis, using a quadratic indentor rather than the cubic indentor, can be performed. In so doing, we find that the load-deflection curve for the indentor has a discontinuity at the origin. This unrealistic result is due to the use of the classical beam theory which neglects shear deformation effects. A somewhat more complicated type of analysis also can be carried out for the problem of a rigid indentor in contact with a viscoelastic plate, governed by the classical plate theory assumptions. As with the beam problem, the solution here includes some physically unrealistic effects, which are due to the neglect of shear deformation. A viscoelastic analysis of beams and plates, including shear deformation effects, can be formulated. This would comprise a generalization of the elastic contact problem analysis method given by Essenburg [2.25].

2.13. SUMMARY AND CONCLUSIONS

Some practical methods of analysis of viscoelastic boundary value problems have been given. In cases where the nature of the material and the nature of the boundary value problem is such that a separation of variables solution is permissible, the spatial part of the solution is obtained from an elasticity type analysis, while the temporal part follows from simple convolution integral relationships.

Steady state harmonic problems, either with or without inertia terms in the equations of motion, can be solved in a manner formally the same as that for comparable elasticity type problems. In fact, steady state harmonic elastic solutions can be converted to corresponding viscoelastic solutions through the replacement of the elastic moduli by the corresponding complex viscoelastic moduli. The actual computation of solution variables involves complex number arithmetic which accounts for phase angle shift effects.

A more general class of problems are those for which the integral transform methods apply. The main restriction in these problems is that the parts of the boundary (B_u and B_σ), on which displacement type and stress type boundary conditions are prescribed, do not change with time. For these problems, the utilization of integral transform techniques reduces the problem to a two step process, whereby the spatial part of the solution is solved in the manner of the comparable elasticity problem, followed by the inversion process. The procedure is quite general. The quasi-static examples, which have been worked out, demonstrate that analytical solutions can be obtained with the viscoelastic mechanical properties being specified in a general manner, with no need to use particular and restrictive mechanical model representations, despite frequent technical assertions to the contrary. In order to carry out the transform inversion process to obtain the quasi-static solution it is only necessary that input variables—boundary conditions and body forces—be of such a type the transformed solution involves only simple poles.

In the further application of analytical solutions to obtain numerical values for the field variables, we find it necessary to assign numerical values to the mechanical properties representations. In the example of Section 2.8, it is shown to be practical to represent the relaxation function by a sequence of decaying exponential terms with one term per time decade in the time range in which the experimentally determined relaxation function is represented. The details of carrying out the numerical evaluation of the field variables then involves finding the complex roots of a polynomial in the transform variable. Standard digital computer programs are available for carrying out this step.

The question arises as to the limits on the time or frequency range for which mechanical properties must be represented accurately, for use in a given problem. First, with regard to relaxation and creep functions, these need only be represented up to the maximum time for which the solution is desired. With respect to the short time limit, these forms must be accurately represented for times of at least a decade or two shorter than the earliest time in which the solution is desired. This, of course, also depends on the nature of the input variables, the boundary conditions and body forces. The reason for this short time requirement is due to the fading memory nature of the material, whereby recent events still influence the current state of the field variables. Relative to the specification of the complex moduli as a function of frequency, the practical range in which properties must be represented is determined by the predominate frequency content of the input variables—again, the boundary conditions and body forces—and the time range in which the solution is desired. Although no precise rules in this regard can be specified, the short time range of observation requires good high frequency properties representations, while the long time range of observations obviously requires good low frequency data.

In the examples considered, it is possible to obtain analytical inversions. But there are situations in which this may not possible. As an example, we consider an elastic problem for which only a numerical solution can be obtained with the elastic properties having specified values. It follows that the Laplace (or Fourier) transform of the corresponding viscoelastic problem is known from the elastic solution, but only for particular values of the transform variable. This circumstance will necessitate performing a separate numerical elasticity solution for each different value of the viscoelastic transform variable. To invert the resulting viscoelastic solution will require the use of approximate methods of the inversion of the Laplace transform, if it is used, rather than the Fourier transform. Problems of this type are considered in Chapter 4. It should also be mentioned that in some cases viscoelastic boundary value problems can be reduced to the standard types of integral equations. Although integral transform techniques may be employed to obtain analytical solutions, in the manner already described, a solution may alternatively be obtained by direct numerical integration of the integral equation. In another approach, Robotnov [2.26] has developed an operator algebra that is intended to avoid the use of integral transform methods.

Although the general means of treating dynamical problems is considered in Chapter 4, in the special case of free vibration problems, a general procedure is outlined for solving the resulting eigenvalue problem. Even though the procedure is formally similar to that in elasticity, the details of accomplishing it are, in general, quite involved. In the special case where viscoelastic Poisson's ratio ν is a real constant, a simple and direct free vibration analysis procedure is formulated utilizing the mode shapes of the corresponding elasticity problem. By using the Laplace transformation technique, the transient behavior of the solution is shown to be determined in exactly the same manner as in the general quasi-static problem analysis discussed previously. In the case where the imaginary part of the governing complex modulus compared with the real part is small, it is shown how these two quantities can be directly related to the natural frequency and the attenuation characteristics of a vibrating single degree of freedom system.

Finally, for those problems in which the portions of the boundary B_u having prescribed displacements and those portions B_σ having prescribed stresses are time dependent, no completely general method of analysis can be given, and the appropriate method of analysis must be determined for each individual problem. One such problem was posed and the solution constructed, showing that it is possible and practical to proceed in such cases. More general problems of this type are treated in Chapter 5.

PROBLEMS

2.1. Formulate the procedure for solving the first and second boundary value problems for incompressible viscoelastic materials, with the separation of variables technique being applicable. Assume quasi-static conditions.

2.2. A long viscoelastic cylinder of internal radius a and external radius b is pressurized by an internal pressure $p(t)$, while the outer boundary is stress free. Using the Fourier transform, obtain the quasi-static solution for the displacement and stress components in terms of semi-infinite integrals. Assume plane strain conditions.

2.3. A sphere of internal radius a and external radius b is pressurized by an internal pressure $p(t) = p_0h(t)$, while the outer boundary is stress free. Obtain the Laplace transformed solution for the displacement and stress components for a general viscoelastic material, assuming quasi-static conditions. Perform the inversion under the further assumption that volumetric relaxation function is elastic, while the shear relaxation function is that of a single decaying exponential term.

2.4. Obtain the quasi-static stress response of a viscoelastic half space acted upon by a suddenly applied concentrated normal load $p(t) = p_0h(t)$ at a point on its otherwise stress free boundary. This corresponds to the Boussinesq problem of elasticity theory. Assume that the volumetric relaxation function is elastic while the shear relaxation function is that for a generalized Maxwell model. Obtain specific results when the shear relaxation function is specialized to that of a Maxwell model.

2.5. An elastic beam of section modulus $E_e I_e$ has a viscoelastic coating on it which contributes additionally to the stiffness of the beam. The Laplace transform of the viscoelastic contribution to the section modulus is given by $\bar{E}_v(s) I_v$. Modify the indentation analysis of Section 2.12 to include beams of this type. Obtain the displacement of the beam at its center when the indentor's load is given in the form $p(t) = p_0 h(t)$ and when $\bar{E}_v$ corresponds to a single term decaying exponential relaxation function.

2.6. Carry out an indentation analysis similar to that of Section 2.12 but consider, instead of a viscoelastic beam, a uniform viscoelastic circular plate with the classical elastic plate load-deflection equation modified to represent the viscoelastic plate. Assume that the edge of the plate is simply supported.

2.7. It is proposed to measure the complex modulus through the use of a free vibration experiment. Specifically, a viscoelastic cantilever beam has a mass attached to the end and is set into a condition of free vibration. By neglecting the mass of the beam compared with the end mass, the system is idealized as that of a single degree of freedom system. Determine the real and imaginary parts of the complex modulus $E^*(i\omega)$ in terms of the observed frequency of oscillation ω and the logarithmic decrement of the decay. Let I represent the moment of inertia of the beam, L its length, m the end mass, and use classical beam theory. It may be assumed that $E''/E' \ll 1$. Obtain the modifications which must be made in this procedure in the case where the stiffness element is that of an elastic beam with a viscoelastic coating. In this situation do not assume $E''/E' \ll 1$ for the viscoelastic material.

2.8. Invert relation (2.127) for $\mu(t)$ being represented by a single decaying exponential term, $\mu e^{-t/\tau}$, with relaxation time τ. Interpret the resulting free vibration solution in the two cases where $\omega_n^2 \gtrless 1/4\tau^2$. Verify that the resulting solution satisfies the initial conditions and examine the long time nature of the solution for the two separate cases of prescribed initial displacements and prescribed initial velocities.

2.9. Using the results of Problem 2.8, but with $\mu(t)$ replaced by $E(t)$, obtain the free vibration response of a simply supported viscoelastic beam subjected to an initial state of uniform velocity. Use classical elastic beam theory, with Young's modulus E replaced by the transformed relaxation function $s\bar{E}(s)$.

2.10. Assume that a relaxation function for a particular material can be adequately described by the form

$$G(t) = \hat{G}/(3 + \log_{10} t), \qquad 0.01 \leqslant t \leqslant 10.$$

In order to use this property in analytical studies, it is desired to fit $G(t)$ in the range $-1 \leqslant \log_{10} t \leqslant 1$ by a series of decaying exponentials. To this end, take

$$\frac{G(t)}{\hat{G}} = 0.20 + \sum_{i=1}^{4} G_i \exp\{-t/[15 \times 10^{(-4+i)}]\}.$$

The first term 0.20 is an estimate of the long time asymptotic value of $G(t)/\hat{G}$. The relaxation times are taken in the range of the description of $G(t)$. Evaluate G_i, $i = 1, 2, 3, 4$, by matching $G(t)/\hat{G}$ at $\log_{10} t = -2, -1, 0, 1$, and compare the given form with that evaluated.

REFERENCES

2.1. Sokolnikoff, I. S., *Mathematical Theory of Elasticity*, 2nd ed. McGraw-Hill, New York, 1956.

2.2. Gurtin, M. E., and E. Sternberg, "On the Linear Theory of Viscoelasticity," *Arch. Ration. Mech. Anal.* **11**, 291 (1962).

2.3. Edelstein, W. S., and M. E. Gurtin, "Uniqueness Theorems in the Linear Dynamic Theory of Anisotropic Viscoelastic Solids," *Arch. Ration. Mech. Anal.* **17**, 47 (1964).

2.4. Odeh, F., and I. Tadjbakhsh, "Uniqueness in the Linear Theory of Viscoelasticity," *Arch. Ration. Mech. Anal.* **18**, 244 (1965).

2.5. Barberan, J., and I. Herrera, "Uniqueness Theorems and the Speed of Propagation of Signals in Viscoelastic Materials," *Arch. Ration. Mech. Anal.* **23**, 173 (1966).

2.6. Lubliner, J., and J. L. Sackman, "On Uniqueness in General Linear Viscoelasticity," *Quart. Appl. Math.* **25**, 129 (1967).

2.7. Onat, E. T., and S. Breuer, "On Uniqueness in Linear Viscoelasticity," *in Progress in Applied Mechanics* (D. C. Drucker, ed.) (The Prager Anniversary Volume), p. 349. Macmillan, New York, 1963.

2.8. Tsien, H. S., "A Generalization of Alfrey's Theorem for Viscoelastic Media," *Quart. Appl. Math.* **8**, 104 (1950).

2.9. Alfrey, T., "Nonhomogeneous Stresses in Viscoelastic Media," *Quart. Appl. Math.* **2**, 113 (1944).

2.10. Read, W. T., "Stress Analysis for Compressible Viscoelastic Materials," *J. Appl. Phys.* **21**, 671 (1950).

2.11. Sips, R., "General Theory of Deformation of Viscoelastic Substances," *J. Polym. Sci.* **9**, 191 (1951).

2.12. Brull, M. A., "A Structural Theory Incorporating the Effect of Time-Dependent Elasticity," *Proc. 1st Midwestern Conf. Solid Mech.* 141 (1953).

2.13. Lee, E. H., "Stress Analysis in Viscoelastic Bodies," *Quart. Appl. Math.* **13**, 183 (1955).

2.14. Gottenberg, W. G., and R. M. Christensen, "An Experiment for Determination of the Mechanical Property in Shear for a Linear Isotropic Viscoelastic Solid," *Int. J. Eng. Sci.* **2**, 45 (1964).

2.15. Gottenberg, W. G., and R. M. Christensen, "Prediction of the Transient Response of a Linear Viscoelastic Solid," *J. Appl. Mech.* **33**, 449 (1966).

2.16. Gradowczyk, M. H., and F. Moavenzadeh, "Characterization of Linear Viscoelastic Materials," *Trans. Soc. Rheol.* **13**, 173 (1969).

2.17. Christensen, R. M., and R. N. Schreiner, "Response to Pressurization of a Viscoelastic Cylinder with an Eroding Internal Boundary," *AIAA J.* **3**, 1451 (1965).
2.18. Huang, N. C., E. H. Lee, and T. G. Rogers, "On the Influence of Viscoelastic Compressibility in Stress Analysis," *in Proc. 4th Int. Cong. Rheol.* (E. H. Lee, ed.). Wiley, New York, 1965.
2.19. Ting, T. C. T., "Remarks on Linear Viscoelastic Stress Analysis in Cylinder Problems," *in Developments in Theoretical and Applied Mechanics* (T. C. Huang and M. W. Johnson, Jr., eds.). Wiley, New York, 1965.
2.20. Lockett, F. J., "Interpretation of Mathematical Solutions in Viscoelasticity Theory Illustrated by a Dynamic Spherical Cavity Problem," *J. Mech. Phys. Solids* **9**, 215 (1961).
2.21. Jahnke, E., and F. Emde, *Tables of Functions.* Dover, New York, 1945.
2.22. Hunter, S. C., "The Solution of Boundary Value Problems in Linear Viscoelasticity," *Proc. 4th Symp. Nav. Structural Mech.* 257. Pergamon Press, Oxford, 1965.
2.23. Struik, L. C. E., "Free Damped Vibrations of Linear Viscoelastic Materials," *Rheol. Acta* **6**, 119 (1967).
2.24. Granstam, I., "Contact between a Curved Viscoelastic Beam and a Rigid Plane," *J. Strain Anal.* **8**, 58 (1973).
2.25. Essenburg, F., "On Surface Constraints in Plate Problems," *J. Appl. Mech.* **29**, 340 (1962).
2.26. Rabotnov, Y. N., *Elements of Hereditary Solid Mechanics.* Mir, Moscow, 1980.

Chapter III Thermoviscoelasticity

The material presented in the previous two chapters has assumed isothermal conditions. Since viscoelastic materials do not conserve energy (the significance of this will be made precise in this chapter), we will now concern ourselves with such matters as heat flow, temperature states, and the restrictions implied in deducing an isothermal theory. Accordingly, a consistent thermodynamical derivation of a general linear theory of thermoviscoelasticity will be given. Some of the implications and restrictions implied by this linear theory will be examined. Also some nonisothermal effects which are outside the scope of the linear theory of thermoviscoelasticity will be examined.

3.1. THERMODYNAMICAL DERIVATION OF CONSTITUTIVE RELATIONS

In this section, we derive a linear theory of thermoviscoelasticity. This derivation is based upon two fundamental thermodynamical postulates, the balance of energy and the entropy production inequality. These two postulates, along with two further constitutive assumptions, lead to a complete theory. The derivation we give here is taken from that of Christensen and Naghdi [3.1]. Thermodynamical derivations which proceed from a point of view different from that to be followed here have been given by Biot [3.2], Eringen [3.3], Hunter [3.4], and Schapery [3.5].

We will soon state the local relations expressing the balance of energy and the entropy production inequality. For a derivation of these local relations from the general global forms see Truesdell and Toupin [3.6]. In writing these relations we introduce several new field variables, including temperature and entropy. Although it is not possible, within the present framework, to give precise definitions of these new quantities, they are familiar terms in that they are extensions to irreversible thermodynamical conditions of the corresponding well-defined terms in thermal equilibrium states.

The local balance of energy equation for infinitesimal theory is given by

$$\rho r - \rho[\dot{A} + \dot{T}S + T\dot{S}] + \sigma_{ij}\dot{\epsilon}_{ij} - Q_{i,i} = 0 \tag{3.1}$$

where ρ is the mass density, r is the heat supply function per unit mass, A is the Helmholtz free energy per unit mass, T is the absolute temperature, S is the entropy per unit mass, and Q_i are the Cartesian components of the heat flux vector measured per unit area per unit time. The function r in (3.1) provides the means of either adding or removing heat by an external agency. In (3.1), as well as in the following, when the time variable is not stated we understand it to be current time t and a superimposed dot designates a time derivative. The form (3.1) of the conservation of energy relation is appropriate to infinitesimal theory since the involved stress and strain are those defined in the infinitesimal theory context. Also, since in the infinitesimal theory, the mass density ρ remains unchanged to the first order, the free energy per unit volume ρA can be considered with no individual reference to mass density and free energy per unit mass. The reason for the inclusion of the forms involving mass density in (3.1) and in the following work is for notational consistency with the nonlinear theory developments in Chapter 8. The balance of energy relation is often written in the form involving the internal energy rather than the free energy. The relationship of these two possible forms is obtained through the Legendre transformation $\rho U = \rho A + T\rho S$, where U is the internal energy per unit mass.

The related local entropy production inequality is given by

$$\rho T\dot{S} - \rho r + Q_{i,i} - Q_i(T_{,i}/T) \geqslant 0 \tag{3.2}$$

and is often referred to as the Clausius–Duhem inequality.

We will show that if constitutive relations for ρA and Q_i are postulated, in an appropriate form, then all other constitutive relations are derived in the process of satisfying (3.1) and (3.2). The starting point here, therefore, is to obtain a form for the free energy ρA.

The procedure to be followed here for establishing a form for the free energy parallels the means of deriving the stress constitutive equation given in Section 1.2. There the current stress is postulated to depend not only upon current strain, but also upon the complete past history of strain, which allows stress to be expressed as a functional of strain history. Under the assumption that the functional is a linear transformation, it is possible to invoke a representation theorem to write the functional as a Stieltjes integral. From this, the usual form of the stress strain relations is obtained. The situation is more complicated here in dealing with the free energy. The free energy not only depends upon the strain history, but also depends upon the temperature history, since isothermal conditions are not assumed. The free energy is thus a functional of both strain and temperature history and, as is immediately seen from comparable elasticity derivations, cannot be taken to be a linear functional. The free energy can, however, be derived through the use of an approximation theorem in the following manner.

Assume that $\epsilon_{ij}(t)$ and $T(t)$ are continuous on the interval $-\infty < t < \infty$, and assume further that $\epsilon_{ij}(t)$ tend to zero and $T(t) \to T_0$ as $t \to -\infty$. With these continuity assumptions, it follows from the Stone–Weierstrass theorem, that a real continuous scalar valued (or tensor valued) functional of $\epsilon_{ij}(\tau)$ and

$T(\tau)$, $-\infty < \tau \leqslant t$, may be uniformly approximated by a polynomial in a set of real, continuous, linear functionals of $\epsilon_{ij}(\tau)$ and $T(\tau)$. Now, using the Riesz representation theorem, these linear functionals are expressed in terms of Stieltjes integrals in which the integrating functions are of bounded variation. Following the procedure of Section 1.2, these Stieltjes integrals are written in the form of (1.5) and a corresponding form involving the history of the infinitesimal temperature deviation from the base temperature T_0. Henceforth, we let $\theta(t)$ denote this temperature difference from T_0, and assume $\epsilon_{ij}(\tau)$ and $\theta(\tau)/T_0$ are infinitesimals of $O(\epsilon)$, as in Section 1.2. The polynomial expansion of ρA in terms of these linear functionals then gives us the form

$$\begin{aligned}\rho A = \rho A_0 &+ \int_{-\infty}^{t} D_{ij}(t-\tau)\frac{\partial \epsilon_{ij}(\tau)}{\partial \tau}\,d\tau - \int_{-\infty}^{t} \beta(t-\tau)\frac{\partial \theta(\tau)}{\partial \tau}\,d\tau \\ &+ \frac{1}{2}\int_{-\infty}^{t}\int_{-\infty}^{t} G_{ijkl}(t-\tau, t-\eta)\frac{\partial \epsilon_{ij}(\tau)}{\partial \tau}\frac{\partial \epsilon_{kl}(\eta)}{\partial \eta}\,d\tau\,d\eta \\ &- \int_{-\infty}^{t}\int_{-\infty}^{t} \varphi_{ij}(t-\tau, t-\eta)\frac{\partial \epsilon_{ij}(\tau)}{\partial \tau}\frac{\partial \theta(\eta)}{\partial \eta}\,d\tau\,d\eta \\ &- \frac{1}{2}\int_{-\infty}^{t}\int_{-\infty}^{t} m(t-\tau, t-\eta)\frac{\partial \theta(\tau)}{\partial \tau}\frac{\partial \theta(\eta)}{\partial \eta}\,d\tau\,d\eta + O(\epsilon^3)\end{aligned} \tag{3.3}$$

where A_0 is the mean free energy, and the integrating functions (mechanical properties) are assumed to be continuous for arguments $\tau_i \geqslant 0$ and are assumed to vanish identically for $\tau_i < 0$; i.e.,

$$\begin{aligned}&\beta(\tau_1) = 0, \qquad D_{ij}(\tau_1) = 0, \qquad G_{ijkl}(\tau_1, \tau_2) = 0, \\ &\varphi_{ij}(\tau_1, \tau_2) = 0, \qquad m(\tau_1, \tau_2) = 0 \qquad \text{for} \quad \tau_1 < 0, \tau_2 < 0.\end{aligned} \tag{3.4}$$

For the theory under consideration here, the terms of $O(\epsilon^3)$ in (3.3) are neglected and the integrating functions in (3.3) are necessarily independent of strain and temperature.

The heat supply function r can be eliminated between (3.1) and (3.2) to get

$$-\rho S\dot{\theta} - \rho\dot{A} + \sigma_{ij}\dot{\epsilon}_{ij} - Q_i(\theta_{,i}/T_0) \geqslant 0 \tag{3.5}$$

where $T = T_0 + \theta$ has been used and only terms of $O(\epsilon)$ have been retained in $T_{,i}/T$.

We substitute for ρA from (3.3) into (3.5) and carry out the indicated differentiation with respect to t, using Leibnitz's rule.

This operation gives

$$\Big\{-D_{ij}(0) - \int_{-\infty}^{t} G_{ijkl}(t-\tau, 0)\,\frac{\partial\epsilon_{kl}(\tau)}{\partial\tau}\,d\tau$$
$$+ \int_{-\infty}^{t} \varphi_{ij}(0, t-\tau)\,\frac{\partial\theta(\tau)}{\partial\tau}\,d\tau + \sigma_{ij}\Big\}\,\dot{\epsilon}_{ij}(t)$$
$$+ \Big\{\beta(0) + \int_{-\infty}^{t} m(t-\tau, 0)\,\frac{\partial\theta(\tau)}{\partial\tau}\,d\tau$$
$$+ \int_{-\infty}^{t} \varphi_{ij}(t-\tau, 0)\,\frac{\partial\epsilon_{ij}(\tau)}{\partial\tau}\,d\tau - \rho S\Big\}\,\dot{\theta}(t)$$
$$+ \Big\{-\int_{-\infty}^{t} \frac{\partial}{\partial t}\,D_{ij}(t-\tau)\,\frac{\partial\epsilon_{ij}(\tau)}{\partial\tau}\,d\tau + \int_{-\infty}^{t} \frac{\partial}{\partial t}\,\beta(t-\tau)\,\frac{\partial\theta(\tau)}{\partial\tau}\,d\tau$$
$$+ \Lambda - Q_i\,\frac{\theta,i}{T_0}\Big\} \geqslant 0 \tag{3.6}$$

where

$$\Lambda = -\frac{1}{2}\int_{-\infty}^{t}\int_{-\infty}^{t} \frac{\partial}{\partial t}\,G_{ijkl}(t-\tau, t-\eta)\,\frac{\partial\epsilon_{ij}(\tau)}{\partial\tau}\,\frac{\partial\epsilon_{kl}(\eta)}{\partial\eta}\,d\tau\,d\eta$$
$$+ \int_{-\infty}^{t}\int_{-\infty}^{t} \frac{\partial}{\partial t}\,\varphi_{ij}(t-\tau, t-\eta)\,\frac{\partial\epsilon_{ij}(\tau)}{\partial\tau}\,\frac{\partial\theta(\eta)}{\partial\eta}\,d\tau\,d\eta$$
$$+ \frac{1}{2}\int_{-\infty}^{t}\int_{-\infty}^{t} \frac{\partial}{\partial t}\,m(t-\tau, t-\eta)\,\frac{\partial\theta(\tau)}{\partial\tau}\,\frac{\partial\theta(\eta)}{\partial\eta}\,d\tau\,d\eta \tag{3.7}$$

and where we have used the following symmetry properties:

$$G_{ijkl}(t-\tau, t-\eta) = G_{klij}(t-\eta, t-\tau)$$

and

$$m(t-\tau, t-\eta) = m(t-\eta, t-\tau). \tag{3.8}$$

The inequality (3.6) must hold for all arbitrary values of $\dot{\epsilon}_{ij}(t)$ and $\dot{\theta}(t)$; therefore, it is necessary that the coefficients of $\dot{\epsilon}_{ij}(t)$ and $\dot{\theta}(t)$ in (3.6) vanish. Hence,

$$\sigma_{ij} = D_{ij}(0) + \int_{-\infty}^{t} G_{ijkl}(t-\tau, 0)\,\frac{\partial\epsilon_{kl}(\tau)}{\partial\tau}\,d\tau - \int_{-\infty}^{t} \varphi_{ij}(0, t-\tau)\,\frac{\partial\theta(\tau)}{\partial\tau}\,d\tau \tag{3.9}$$

and

$$\rho S = \beta(0) + \int_{-\infty}^{t} \varphi_{ij}(t-\tau, 0)\,\frac{\partial\epsilon_{ij}(\tau)}{\partial\tau}\,d\tau + \int_{-\infty}^{t} m(t-\tau, 0)\,\frac{\partial\theta(\tau)}{\partial\tau}\,d\tau \tag{3.10}$$

which leaves the inequality (3.6) as

$$-\int_{-\infty}^{t} \frac{\partial}{\partial t} D_{ij}(t-\tau)\frac{\partial \epsilon_{ij}(\tau)}{\partial \tau}\,d\tau + \int_{-\infty}^{t} \frac{\partial}{\partial t} \beta(t-\tau)\frac{\partial \theta(\tau)}{\partial \tau}\,d\tau$$
$$+ \Lambda - Q_i \frac{\theta,i}{T_0} \geqslant 0. \tag{3.11}$$

Relations (3.9) and (3.10) are the constitutive relations for stress and entropy, respectively. From these it is clear that $D_{ij}(0)$ is the initial stress and that $\beta(0)$ is the initial entropy, ρS_0. The integrating functions $G_{ijkl}(t-\tau, 0)$, $\varphi_{ij}(0, t-\tau)$, $\varphi_{ij}(t-\tau, 0)$, and $m(t-\tau, 0)$ are the appropriate relaxation function forms of the mechanical properties. It should be noted that if $G_{ijkl}(\tau, \eta)$ is considered as a surface in τ, η space, then the relaxation functions involved in (3.9) and (3.10) are curves on these surfaces. It is the relaxation function $G_{ijkl}(t, 0)$ in this theory which corresponds to the relaxation function $G_{ijkl}(t)$ in the isothermal theory (1.5).

The first two terms in (3.11) are of first order, whereas the last two are of second order, assuming Q_i suitably nondimensionalized is of first order, so to satisfy the inequality for all processes, it is necessary that

$$\frac{\partial D_{ij}(t)}{\partial t} = 0, \quad \frac{\partial \beta(t)}{\partial t} = 0 \tag{3.12}$$

as well as

$$\Lambda - Q_i(\theta_{,i}/T_0) \geqslant 0. \tag{3.13}$$

By considering a particular process such that $\theta_{,i} = 0$, which is a uniform temperature field, for which (3.13) must apply, there results

$$\Lambda \geqslant 0. \tag{3.14}$$

This is called the dissipation inequality, where Λ given by (3.7) represents the rate of dissipation of energy. Using (3.14), it is sufficient that (3.13) be satisfied by

$$Q_i(\theta_{,i}/T_0) \leqslant 0. \tag{3.15}$$

To complete the development of the constitutive relations we find it necessary to postulate a constitutive relation for the heat flux vector Q_i. Consistent with the earlier developments, this will be taken as

$$Q_i = -\int_{-\infty}^{t} k_{ij}(t-\tau)\frac{\partial \theta_{,j}(\tau)}{\partial \tau}\,d\tau \tag{3.16}$$

such that Q_i depends linearly upon the history of the temperature gradient $\theta_{,j}$. We combine (3.15) and (3.16) to obtain

$$\theta_{,i}\int_{-\infty}^{t} k_{ij}(t-\tau)\frac{\partial \theta_{,j}(\tau)}{\partial \tau}\,d\tau \geqslant 0. \tag{3.17}$$

However, for a fixed value of t, the instantaneous value of $\theta_{,i}$ and the value of the functional may, in general, be of opposite sign, since the functional depends on the entire past history of the temperature gradient. In fact, for a given value of t, $\theta_{,i}$ and the functional will have the same sign only if the tensor k_{ij} is positive definite and constant with respect to time, in which case (3.16) must reduce to

$$Q_i = -k_{ij}\theta_{,j} \tag{3.18}$$

where k_{ij} are now constants and k_{ij} is symmetric in i and j.

The energy equation (3.1) is now rewritten with the use of (3.3), (3.9), (3.10), (3.12), and (3.18) as

$$\rho r + \Lambda - T_0 \frac{\partial}{\partial t}\left[\int_{-\infty}^{t} \varphi_{ij}(t-\tau, 0)\frac{\partial \epsilon_{ij}(\tau)}{\partial \tau}\,d\tau\right.$$
$$\left. + \int_{-\infty}^{t} m(t-\tau, 0)\frac{\partial \theta(\tau)}{\partial \tau}\,d\tau\right] + (k_{ij}\theta_{,j})_{,i} = 0 \tag{3.19}$$

where Λ is given by (3.7). Λ, the rate of dissipation of energy, is a second order term and must be neglected from (3.19) in the consistent development of a first order theory. Equation (3.19) becomes

$$\rho r - T_0 \frac{\partial}{\partial t}\left[\int_{-\infty}^{t} \varphi_{ij}(t-\tau, 0)\frac{\partial \epsilon_{ij}(\tau)}{\partial \tau}\,d\tau\right.$$
$$\left. + \int_{-\infty}^{t} m(t-\tau, 0)\frac{\partial \theta(\tau)}{\partial \tau}\,d\tau\right] + (k_{ij}\theta_{,j})_{,i} = 0. \tag{3.20}$$

The integral involving strain history in (3.20) gives rise to a coupling between thermal and mechanical effects. Without this term (3.20) would be the uncoupled equation governing heat conduction.

It will be useful to have the corresponding equations for isotropic theory. These are now given. For isotropic theory, φ_{ij} must be taken as

$$\varphi_{ij}(\tau, \eta) = \delta_{ij}\varphi(\tau, \eta) \tag{3.21}$$

as well as representing G_{ijkl} in isotropic form through (1.9). Using the definitions of deviatoric stress and strain, (1.12) and (1.13) the free energy for isotropic materials is written as

$$\rho A = \frac{1}{2}\int_{-\infty}^{t}\int_{-\infty}^{t} G_1(t-\tau, t-\eta)\frac{\partial e_{ij}(\tau)}{\partial \tau}\frac{\partial e_{ij}(\eta)}{\partial \eta}\,d\tau\,d\eta$$
$$+ \frac{1}{6}\int_{-\infty}^{t}\int_{-\infty}^{t} G_2(t-\tau, t-\eta)\frac{\partial \epsilon_{kk}(\tau)}{\partial \tau}\frac{\partial \epsilon_{jj}(\eta)}{\partial \eta}\,d\tau\,d\eta$$
$$- \int_{-\infty}^{t}\int_{-\infty}^{t} \varphi(t-\tau, t-\eta)\frac{\partial \epsilon_{kk}(\tau)}{\partial \tau}\frac{\partial \theta(\eta)}{\partial \eta}\,d\tau\,d\eta$$
$$- \frac{1}{2}\int_{-\infty}^{t}\int_{-\infty}^{t} m(t-\tau, t-\eta)\frac{\partial \theta(\tau)}{\partial \tau}\frac{\partial \theta(\eta)}{\partial \eta}\,d\tau\,d\eta \tag{3.22}$$

where the initial stress and initial entropy effect in (3.3) has been dropped.

The stress constitutive relations for isotropic materials are now given by

$$s_{ij} = \int_{-\infty}^{t} G_1(t-\tau, 0) \frac{\partial e_{ij}(\tau)}{\partial \tau} d\tau \tag{3.23}$$

and

$$\sigma_{kk} = \int_{-\infty}^{t} G_2(t-\tau, 0) \frac{\partial \epsilon_{kk}(\tau)}{\partial \tau} d\tau - 3 \int_{-\infty}^{t} \varphi(0, t-\tau) \frac{\partial \theta(\tau)}{\partial \tau} d\tau. \tag{3.24}$$

The entropy constitutive relation, for isotropic materials, is the same as (3.10) except with φ_{ij} and ϵ_{ij} replaced by φ and ϵ_{kk}, respectively.

The rate of dissipation of energy Λ for isotropic materials becomes

$$\begin{aligned} \Lambda = &-\frac{1}{2} \int_{-\infty}^{t} \int_{-\infty}^{t} \frac{\partial}{\partial t} G_1(t-\tau, t-\eta) \frac{\partial e_{ij}(\tau)}{\partial \tau} \frac{\partial e_{ij}(\eta)}{\partial \eta} d\tau\, d\eta \\ &-\frac{1}{6} \int_{-\infty}^{t} \int_{-\infty}^{t} \frac{\partial}{\partial t} G_2(t-\tau, t-\eta) \frac{\partial \epsilon_{kk}(\tau)}{\partial \tau} \frac{\partial \epsilon_{jj}(\eta)}{\partial \eta} d\tau\, d\eta \\ &+ \int_{-\infty}^{t} \int_{-\infty}^{t} \frac{\partial}{\partial t} \varphi(t-\tau, t-\eta) \frac{\partial \epsilon_{kk}(\tau)}{\partial \tau} \frac{\partial \theta(\eta)}{\partial \eta} d\tau\, d\eta \\ &+\frac{1}{2} \int_{-\infty}^{t} \int_{-\infty}^{t} \frac{\partial}{\partial t} m(t-\tau, t-\eta) \frac{\partial \theta(\tau)}{\partial \tau} \frac{\partial \theta(\eta)}{\partial \eta} d\tau\, d\eta. \end{aligned} \tag{3.25}$$

Finally, for isotropic materials, a representation similar to (3.21) must be taken for k_{ij}; then the coupled heat conduction equation (3.20) becomes

$$\begin{aligned} \rho r + k\theta_{,ii} - T_0 \frac{\partial}{\partial t} \Big[&\int_{-\infty}^{t} \varphi(t-\tau, 0) \frac{\partial \epsilon_{kk}(\tau)}{\partial \tau} d\tau \\ &+ \int_{-\infty}^{t} m(t-\tau, 0) \frac{\partial \theta(\tau)}{\partial \tau} d\tau \Big] = 0. \end{aligned} \tag{3.26}$$

This completes the development of the linear thermoviscoelastic theory. These relations are collected and examined in Section 3.4, with regard to the formulation of boundary value problems. The corresponding thermodynamical derivation using creep functions is given in Section 3.7.

3.2. RESTRICTIONS AND SPECIAL CASES

We now consider some consequences of the dissipation inequality $\Lambda \geqslant 0$ for the isotropic theory. In so doing, it is necessary to investigate the effects of discontinuous strain and temperature histories. Strictly speaking, this use of discontinuous histories violates the continuity assumptions used in the derivation, and the rigorous justification for doing this must appeal to the line of reasoning utilized in Section 1.2, whereby the discontinuous history is represented by a uniformly convergent sequence of continuous functions.

We take a history for strain and temperature of the type

$$\begin{aligned} \epsilon_{ij}(t) &= \mathring{\epsilon}_{ij} h(t) \\ \theta(t) &= \mathring{\theta} h(t) \end{aligned} \tag{3.27}$$

where $\mathring{\epsilon}_{ij}$ and $\mathring{\theta}$ are the magnitudes of the discontinuities in ϵ_{ij} and θ at $t = 0$.

With (3.27) for $\mathring{\epsilon}_{ij} = \mathring{\epsilon}_{kk}\delta_{ij}$ substituted in (3.25), the dissipation inequality becomes

$$-\frac{1}{6}\frac{\partial}{\partial t} G_2(t, t)\, \mathring{\epsilon}_{ii}\mathring{\epsilon}_{jj} + \frac{\partial}{\partial t} \varphi(t, t)\, \mathring{\epsilon}_{ii}\mathring{\theta} + \frac{1}{2}\frac{\partial}{\partial t} m(t, t)(\mathring{\theta})^2 \geqslant 0. \tag{3.28}$$

From this general inequality, the following restrictions are obtained:

$$\frac{\partial}{\partial t} G_2(t, t) \leqslant 0, \tag{3.29}$$

$$\frac{\partial}{\partial t} m(t, t) \geqslant 0, \tag{3.30}$$

and

$$\left[\frac{\partial}{\partial t} \varphi(t, t)\right]^2 \leqslant \left[\frac{1}{3}\frac{\partial}{\partial t} G_2(t, t)\right]\left[-\frac{\partial}{\partial t} m(t, t)\right]. \tag{3.31}$$

Another special case of (3.27) is for $\mathring{\theta} = 0$ and $\mathring{\epsilon}_{ij} = e_{ij}$; in this case the dissipation inequality gives

$$\frac{\partial}{\partial t} G_1(t, t) \leqslant 0. \tag{3.32}$$

We note that the restriction (3.32), for example, does not necessarily imply a similar restriction upon the corresponding stress relaxation function $G_1(t, 0)$. But, if the function $G_1(\tau, \eta)$ has the special form $G_1(\tau + \eta)$, such that the relaxation function $G(\tau, 0) = G(\tau)$ determines the entire function, it follows from (3.32) that $\partial G_1(t, 0)/\partial t = dG_1(t)/dt \leqslant 0$. In this case, (3.29) and (3.32) reveal the thermodynamical requirement that the slope of the relaxation functions $G_1(t)$ and $G_2(t)$ must be negative. This slope requirement should be compared with the restrictions upon relaxation functions implied by the fading memory hypothesis of Section 1.3. We next show that the special case of $G_\alpha(\tau, \eta) = G_\alpha(\tau + \eta)$, $\alpha = 1, 2$ is consistent with the corresponding theory implied by mechanical models.

Restrictions (3.29)–(3.32) are necessary conditions for the satisfaction of the dissipation inequality. Now a sufficient condition for its satisfaction will be discussed. In this case only isothermal conditions are considered, and we further assume

$$G_\alpha(\tau, \eta) = \sum_{n=1}^{N} G_{\alpha n} e^{-(\tau+\eta)/t_{\alpha n}}, \qquad \alpha = 1, 2 \tag{3.33}$$

where

$$G_{\alpha n} \geqslant 0, \qquad t_{\alpha n} > 0.$$

We observe that the forms (3.33) imply $G_\alpha(\tau, \eta) = G_\alpha(\tau + \eta)$ and correspondingly imply $G_\alpha(\tau, 0) = G_\alpha(\tau)$. Thus, $G_\alpha(\tau)$ can be represented in the form of (3.33), that is, in the form of a series of decaying exponentials, and this can be shown to be the form of relaxation functions implied by mechanical model representations (see Section 1.5). After substituting (3.33) into the isothermal form of (3.25) we can verify that the dissipation inequality $\Lambda \geqslant 0$ is satisfied for all strain histories. In this case, the free energy and the rate of dissipation of energy forms will involve G_α in the forms $G_1(2t - \tau - \eta)$ and $G_2(2t - \tau - \eta)$, which is consistent with mechanical model formulations.

The question of attempting to establish realistic requirements from which $G_\alpha(\tau, \eta)$ must have the reduced form $G_\alpha(\tau + \eta)$ has received some attention. See, for example, Breuer and Onat [3.7]. In the further work here and in the applications of these results, the reduced forms of the relaxation functions with arguments added, $G_\alpha(\tau, \eta) = G_\alpha(\tau + \eta)$, are employed. This implies that when the relaxation functions, involved in the stress and entropy constitutive relations, are known or determined, the analytical means is available for determining the corresponding free energy in a particular process. Otherwise, the free energy could not be determined since it involves $G_\alpha(\tau, \eta)$, whereas only $G_\alpha(\tau, 0)$ would be known from considerations of stress.

Another special case of interest here concerns the relationship of the coefficient of thermal expansion to the mechanical properties involved in the present theory. We let a stress free sample of the material be subjected to the change in uniform temperature specified by

$$\theta = \theta_0 h(t). \tag{3.34}$$

From relation (3.24), the resulting volume change is determined by

$$\int_0^t G_2(t - \tau) \frac{\partial \epsilon_{kk}(\tau)}{\partial \tau} \, d\tau = 3\varphi(t)\, \theta_0 \tag{3.35}$$

where the arguments in the integrating functions have been taken in the additive form described previously. Now if $\varphi(t)$ can be written in the form

$$\varphi(t) = \alpha G_2(t) \tag{3.36}$$

where α is a constant, the solution of (3.35) is given by

$$\epsilon_{kk} = 3\alpha\theta_0 h(t). \tag{3.37}$$

In this special case, α is simply the time independent coefficient of thermal expansion. But the assumption (3.36) is required to obtain the volume expansion relation (3.37). In general, $\varphi(t)$ is not expected to be proportional to $G_2(t)$, in which situation the volume time response to (3.34) must be determined by (3.35). In this general case, there is no simple interpretation of volumetric effects in terms of a time independent coefficient of thermal expansion. The coefficient of thermal expansion is treated further in Section 3.7.

3.3. RELATIONSHIP TO NONNEGATIVE WORK REQUIREMENTS

Nonnegative work requirements are common place in continuum mechanics. In the incremental theory of plasticity the nonnegative work requirement leads to Drucker's postulate which is of fundamental importance in the formulation of constitutive relations. In linear isothermal viscoelasticity, Gurtin and Herrera [3.8] have employed the requirement that the work done to deform the material from the virgin state must satisfy the nonnegative work relation

$$\int_0^t \sigma_{ij}(\tau) \frac{\partial \epsilon_{ij}(\tau)}{\partial \tau} \, d\tau \geqslant 0. \tag{3.38}$$

They deduce from this the requirements that

$$G_{ijkl}(0)\,\gamma_{ij}\gamma_{kl} \geqslant 0 \qquad G_{ijkl}(0) = G_{klij}(0) \tag{3.39}$$

and

$$G_{ijkl}(\infty)\,\gamma_{ij}\gamma_{kl} \geqslant 0 \qquad G_{ijkl}(\infty) = G_{klij}(\infty) \tag{3.40}$$

where γ_{ij} is any symmetric tensor. Relations (3.39) are restrictions upon the initial values of the relaxation functions, while (3.40) are restrictions upon the asymptotic long time values. The corresponding results for isotropic materials are $G_\alpha(0) \geqslant 0$, $G_\alpha(\infty) \geqslant 0$, $\alpha = 1, 2$.

It is reasonable to inquire into the connection of the nonnegative work requirement (3.38) with the thermodynamical developments considered in the previous two sections. Under isothermal conditions, the inequality (3.5) becomes

$$-\rho \dot{A} + \sigma_{ij}\dot{\epsilon}_{ij} \geqslant 0. \tag{3.41}$$

The integral of this gives

$$-\rho A + \int_0^t \sigma_{ij}(\tau) \frac{\partial \epsilon_{ij}(\tau)}{\partial \tau} \, d\tau \geqslant 0. \tag{3.42}$$

But the left-hand side of (3.41) is the rate of dissipation of energy Λ and (3.42) can be written as

$$\int_0^t \Lambda(\tau)\, d\tau \geqslant 0 \tag{3.43}$$

which simply states that the total dissipated energy must be nonnegative.

We consider now the consequences of (3.42) and (3.43). If the free energy is required to be nonnegative, $\rho A \geqslant 0$, then, for the satisfaction of (3.42), it is necessary but not sufficient that the work done be nonnegative. Sufficiency requires that (3.43) also be satisfied. Alternatively, if the work done and the dissipated energy both are required to be nonnegative, nothing can be said about the sign of ρA. Accepting the thermodynamical requirement that $\Lambda \geqslant 0$, and correspondingly, (3.43), we see that $\rho A \geqslant 0$ implies nonnegative work; but nonnegative work does not imply $\rho A \geqslant 0$. It therefore seems that requiring $\rho A \geqslant 0$ is herein a stronger and more physically meaningful condition than requiring nonnegative work.

Some immediate consequences are now established from a nonnegative free energy requirement. We shall use step function histories in the isotropic form of free energy (3.22) specialized to isothermal conditions and using special forms $G_\alpha(\tau, \eta) = G_\alpha(\tau + \eta)$. Now the requirement $\rho A \geqslant 0$ becomes

$$G_\alpha(t) \geqslant 0, \qquad \alpha = 1, 2. \tag{3.44}$$

This requirement on the relaxation functions, along with the special forms of (3.29) and (3.32) which give $dG_\alpha(t)/dt \leqslant 0$, shows that the relaxation functions must be nonnegative, continuously decreasing functions of time. This is in complete agreement with experimental results. The fading memory hypothesis of Section 1.3 provides additional restrictions upon the derivatives of the relaxation functions, $dG_\alpha(t)/dt$, namely the requirement that $d^2G_\alpha(t)/dt^2 \geqslant 0$. Note also that the tensor symmetries in (3.39) and (3.40) are true for all values of time in representation (3.3).

3.4. FORMULATION OF THE THERMOVISCOELASTIC BOUNDARY VALUE PROBLEM

The relevant linear equations which govern the coupled thermoviscoelasticity theory are now collected. The equations of (quasi-static) equilibrium or the equations of motion, the strain displacement relations, and the stress strain relations are respectively given by

$$\sigma_{ij,j} + F_i = 0, \rho \frac{d^2u_i}{\partial t^2}, \tag{3.45}$$

$$\epsilon_{ij} = \tfrac{1}{2}(u_{i,j} + u_{j,i}), \tag{3.46}$$

$$\sigma_{ij} = \int_0^t G_{ijkl}(t-\tau)\frac{\partial \epsilon_{kl}(\tau)}{\partial \tau}\,d\tau - \int_0^t \varphi_{ij}(t-\tau)\frac{\partial \theta(\tau)}{\partial \tau}\,d\tau \quad \text{(anisotropic)}, \tag{3.47}$$

$$\left.\begin{aligned} s_{ij} &= \int_0^t G_1(t-\tau)\frac{\partial e_{ij}(\tau)}{\partial \tau}\,d\tau, \\ \sigma_{kk} &= \int_0^t G_2(t-\tau)\frac{\partial \epsilon_{kk}(\tau)}{\partial \tau}\,d\tau - 3\int_0^t \varphi(t-\tau)\frac{\partial \theta(\tau)}{\partial \tau}\,d\tau \end{aligned}\right\} \quad \text{(isotropic)} \tag{3.48}$$

where the deviatoric components s_{ij} and e_{ij} are defined by (1.12) and (1.13) and it is recalled that $\theta(\tau)$ denotes the infinitesimal temperature deviation from the base temperature T_0. The heat conduction equation is given by

$$\begin{aligned} \frac{k_{ij}}{T_0}\theta_{,ij} &= \frac{\partial}{\partial t}\int_0^t m(t-\tau)\frac{\partial \theta(\tau)}{\partial \tau}\,d\tau \\ &\quad + \frac{\partial}{\partial t}\int_0^t \varphi_{ij}(t-\tau)\frac{\partial \epsilon_{ij}(\tau)}{\partial \tau}\,d\tau \qquad \text{(anisotropic)} \end{aligned} \tag{3.49}$$

or

$$\begin{aligned} \frac{k}{T_0}\theta_{,ii} &= \frac{\partial}{\partial t}\int_0^t m(t-\tau)\frac{\partial \theta(\tau)}{\partial \tau}\,d\tau \\ &\quad + \frac{\partial}{\partial t}\int_0^t \varphi(t-\tau)\frac{\partial \epsilon_{kk}(\tau)}{\partial \tau}\,d\tau \qquad \text{(isotropic)} \end{aligned} \tag{3.50}$$

where k_{ij} or k, $m(t)$, and $\varphi_{ij}(t)$ or $\varphi(t)$ are mechanical properties of the material, with the arguments taken in the additive form described in Section 3.2. The initial conditions are taken as

$$\theta(t) = u_i(t) = \sigma_{ij}(t) = 0 \qquad \text{for} \quad t < 0. \tag{3.51}$$

The boundary conditions are

$$\begin{aligned} \sigma_{ij}n_j &= S_i(x_i, t) & &\text{on } B_\sigma, & t &\geqslant 0, \\ u_i &= \Delta_i(x_i, t) & &\text{on } B_u, & t &\geqslant 0, \\ \theta &= \hat{\theta}(x_i t) & &\text{on } B_1, & t &\geqslant 0, \end{aligned} \tag{3.52}$$

and

$$k_{ij}\theta_{,i}n_j = 0 \qquad \text{on } B_2, \quad t \geqslant 0$$

where B_1 is that part of the boundary upon which the temperature is prescribed, and B_2 is the complementary part of the boundary over which the surface is considered to be perfectly insulated against heat flow. More general heat flow boundary conditions can easily be formulated. The uniqueness of solution of the boundary value problem posed by relations (3.45)–(3.52) is established in Section 7.1.

It is instructive to examine the foregoing equations after they are Laplace transformed. In the general anisotropic case, these are given by

$$\bar{\sigma}_{ij,j} + \bar{F}_i = 0, \rho s^2 \bar{u}, \tag{3.53}$$

$$\bar{\epsilon}_{ij} = \tfrac{1}{2}(\bar{u}_{i,j} + \bar{u}_{j,i}), \tag{3.54}$$

$$\bar{\sigma}_{ij} = s\bar{G}_{ijkl}\bar{\epsilon}_{kl} - s\bar{\varphi}_{ij}\bar{\theta}, \tag{3.55}$$

$$(k_{ij}/T_0)\,\bar{\theta}_{,ij} = s^2\bar{m}\bar{\theta} + s^2\bar{\varphi}_{ij}\bar{\epsilon}_{ij}\,, \tag{3.56}$$

$$\begin{aligned} \bar{\sigma}_{ij}n_j &= \bar{S}_i \quad \text{on} \quad B_\sigma\,, & \bar{T} &= \hat{\bar{\theta}} \quad \text{on} \quad B_1\,, \\ \bar{u}_i &= \bar{\Delta}_i \quad \text{on} \quad B_u\,, & k_{ij}\bar{\theta}_{,i}n_j &= 0 \quad \text{on} \quad B_2 \end{aligned} \tag{3.57}$$

where s is the transform variable. The boundary value problem posed by relations (3.53)–(3.57) can be solved to obtain the spatial coordinate dependence in the same manner as in treating coupled thermoelastic problems. After completing this step, the complete solution is obtained upon inverting the transformed solution. This procedure is exactly the same as is formulated for isothermal problems in Chapter 2. In fact, it follows that thermoelastic solutions can be converted to the transform of thermoviscoelastic solutions by replacing elastic mechanical properties with the s multiplied transform of viscoelastic relaxation functions. Furthermore, the procedure for inverting solutions obtained by this means is identical with that formulated for isothermal problems and for the particulars, reference should be made to the boundary value problem solutions obtained in Chapter 2. Similar results can be argued in the case when the Fourier transform has been employed.

In problems where the coupling term involving ϵ_{ij} in (3.49) or (3.50), the heat conduction equation, can be neglected the mechanical response and thermal response problems are separated. After obtaining the temperature distribution, either by solving the heat conduction equation or from experimental results, the mechanical response problem is posed by relations (3.45)–(3.48) and (3.51) and (3.52). Integral transform methods are again a useful tool in solving such problems.

When steady state harmonic variation with time is specified for all field variables the thermoviscoelastic constitutive relations can be written in terms of complex functions of frequency, in exactly the same manner as is done in the isothermal case, Chapter 2. Problems of this type already have the specified time dependence, and the nature of the problems is to obtain the spatial coordinate variation solution, as in comparable thermoelastic problems.

The preceding discussion has been limited to quasi-static type problems. We now consider the type problems in which we include inertia term effects. These general dynamic response problems are much more difficult than are the quasi-static types. If integral transform methods are employed the resulting inversions involve branch points, the same as in the isothermal case. It is only in problems having a steady state harmonic variation with time that the inclusion of inertia terms causes no undue complication over the quasi-static case. An example of this type is given in Section 6.3 where the propagation of coupled harmonic thermoviscoelastic waves are analyzed.

3.5. TEMPERATURE DEPENDENCE OF MECHANICAL PROPERTIES

The thermoviscoelastic theory derivation of Section 3.1 allows the mechanical properties to depend upon the fixed base temperature T_0, but any dependence upon the infinitesimal temperature deviation from T_0 is necessarily neglected. In this section we examine some characteristic features of the dependence of mechanical properties upon temperature. This information applies directly to give the temperature dependence of mechanical properties upon base temperature T_0 for use in the theory of Section 3.1. In the next section we consider the necessary modifications which must be made in the thermoviscoelastic theory in order to include a temperature dependence effect based upon a continuously changing total temperature.

Some of the mechanical properties for typical polymeric viscoelastic materials have a very strong dependence upon temperature.

Such a temperature dependence is schematically shown for the real part of a complex modulus in Fig. 3.1. The symbol G in this section is used to designate any viscoelastic mechanical property. Similar temperature dependence curves

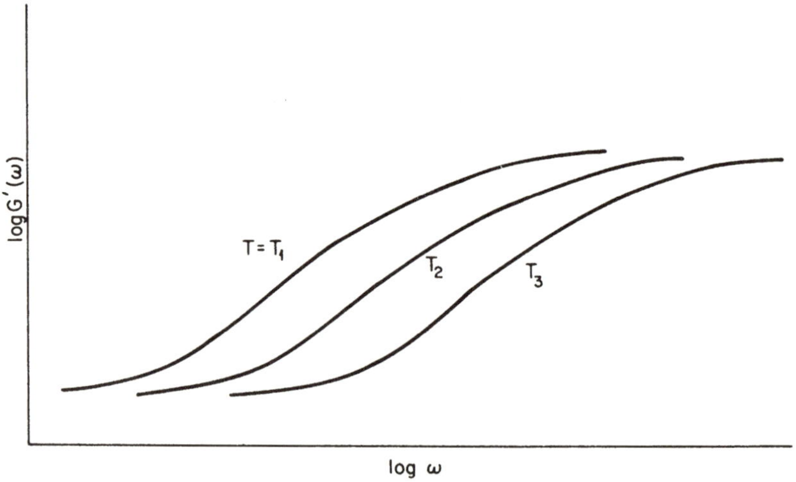

Fig. 3.1. Temperature dependence of complex moduli.

are shown schematically for relaxation functions in Fig. 3.2. The symbol T is remembered to designate absolute temperature.

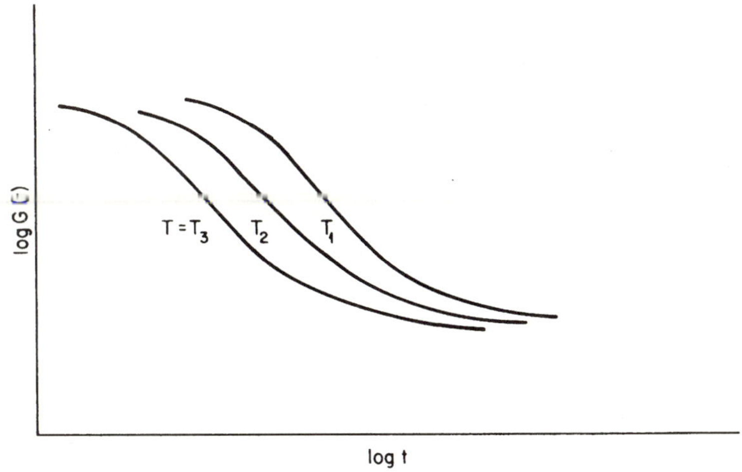

Fig. 3.2. Temperature dependence of relaxation functions.

There is some conventional terminology associated with the characteristic forms shown in Figs. 3.1 and 3.2. The high frequency or short-time range is referred to as the glassy region. Likewise, the low frequency or long-time range is referred to as the rubbery region. Data, such as that in Figs. 3.1 and 3.2, is frequently cross plotted to give modulus data as a function of temperature at a particular time or frequency. Such cross plots are as shown in Figs. 3.3 and 3.4. With respect to these plots, the high and low temperature ranges are again referred to as the rubbery and glassy regions, respectively.

For a particular material, it is convenient to have a means of distinguishing between temperatures at which there is a low-modulus rubbery-type behavior and temperatures at which there is a high-modulus glassy-type behavior. The region where the modulus, such as those in Figs. 3.3 and 3.4, has the greatest dependence upon temperature (maximum magnitude of slope), is called the transition region. The corresponding temperature, or narrow temperature range, is designated as the glass transition temperature T_g. It must be noted that the glass transition temperature necessarily depends upon the time of measurement $\hat{t}$ of the relaxation function, or upon the frequency of measurement $\hat{\omega}$ of the complex modulus.

The concept of a glass transition temperature is of broad significance and usefulness in a variety of applications of viscoelastic materials under changing temperature conditions. Accordingly, two further means of defining T_g will be considered.

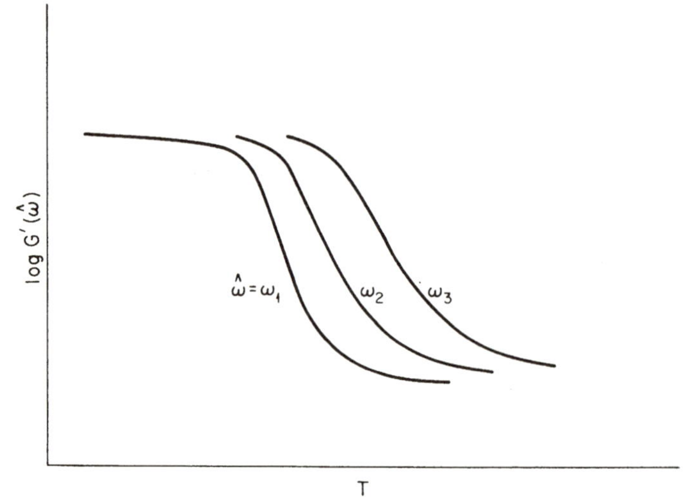

Fig. 3.3. Complex modulus versus temperature.

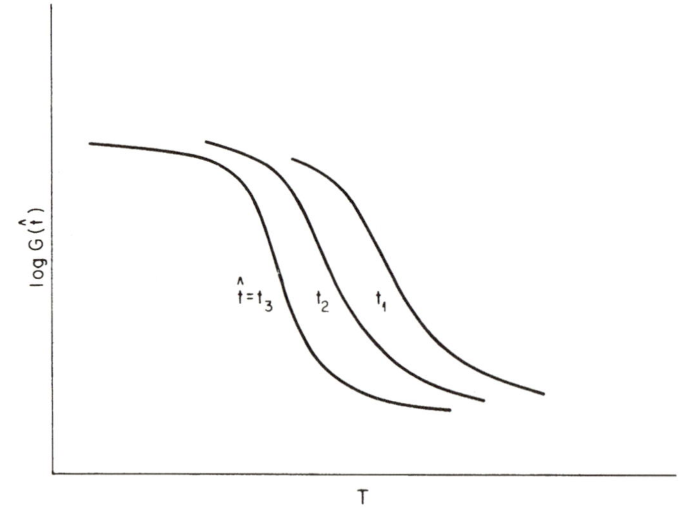

Fig. 3.4. Relaxation function versus temperature.

A somewhat more physically meaningful way of defining T_g than that above is as follows. If a stress free sample of the material is subjected to a uniform and changing temperature field, a measurement of the volume change ΔV from some initial volume has the characteristic shown in Fig. 3.5. The temperature at which the slope has a discontinuity, referred to as a second order transition in chemical terminology, is defined as the glass transition temperature T_g.

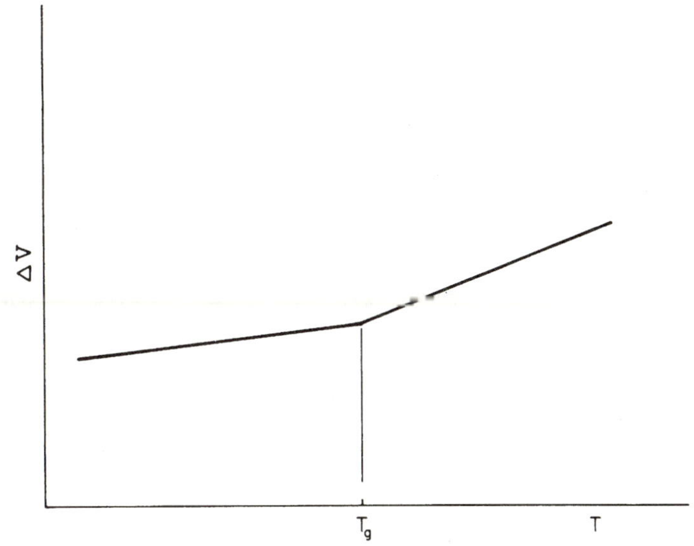

Fig. 3.5. Volume change versus temperature.

The behavior shown in Fig. 3.5 affords a simple molecular interpretation for the existence of the glassy and rubbery regions of viscoelastic behavior. At high temperatures, the volume expansion has given the individual molecules great mobility with comparatively little constraint, which is manifested macroscopically by a low value of the modulus. On the other hand, at low temperatures, due to the volumetric shrinking, great constraints are imposed upon the motion of the molecules, and, correspondingly, the modulus has a high value or it is relatively stiff. The slope discontinuity in ΔV vs T at T_g accounts for the usually rather sharp transition between the rubbery and glassy regions. The difficulty we recognize with this definition of T_g, based upon volume changes, is that it is also a time dependent measurêment. After changing the temperature of the specimen, and thereafter holding it constant, we find that the volume change is continuously changing in the manner of a creep function. This time dependent behavior is confirmed by the experimental evidence of Kovacs [3.9]. Furthermore, this time dependent behavior for ΔV is to be expected, as follows from the discussion in Section 3.2 of the significance of the coefficient of thermal expansion for viscoelastic materials.

The final definition of T_g to be described here involves the loss tangent of a complex modulus, that is, the ratio of the imaginary to the real parts $G''(\omega)/G'(\omega)$ which is nondimensional. Although there may be more than one maximum in $G''(\omega)/G'(\omega)$ versus ω or $\log \omega$, there typically is a local maximum at nearly the same frequency as that at which the real part of the complex modulus has its greatest variation with frequency. Now we consider mechanical property

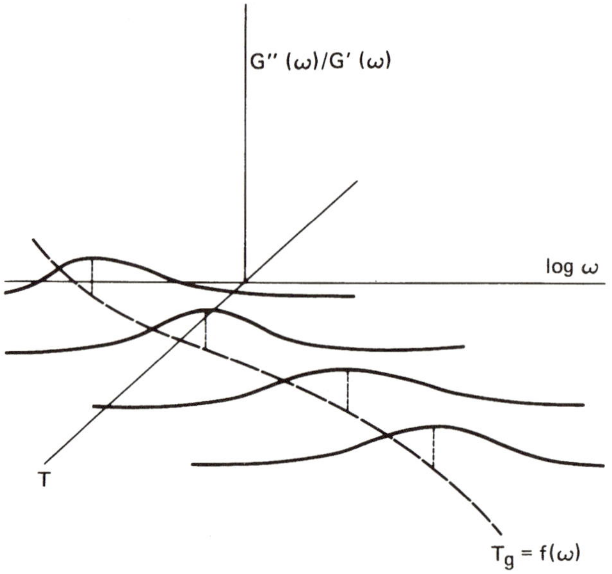

Fig. 3.6. Temperature dependence of $G''(\omega)/G'(\omega)$.

data plotted three dimensionally, as shown in Fig. 3.6. There is a single space curve which connects the relative maxima of $G''(\omega)/G'(\omega)$ versus log ω. The projection of this curve upon the temperature–frequency plane defines a curve which can be taken as that of the glass transition temperature as a function of frequency

$$T_g = f(\omega).$$

A relationship of this type provides a simple means of assessing whether a viscoelastic material will respond in a rubbery or glassy manner at a particular frequency, or predominate frequency of excitation. Alternatively at a given frequency the temperature can be varied to find the maximum loss tangent and accordingly define T_g vs ω.

All of these practical methods of defining T_g inherently contain a degree of arbitrariness involving the time or frequency of measurement. This difficulty is circumvented through the use of a thermodynamic method to define the glass transition temperature. The derivation is given in Section 3.7.

3.6. THERMORHEOLOGICALLY SIMPLE MATERIALS

Constant Temperature States

There is a special type of temperature dependence of mechanical properties which is amenable to analytical description and which applies to a wide variety of materials. This special class of materials is referred to as being thermorheologically simple, and the corresponding description of the temperature dependent properties was first proposed by Leadermann [3.10] and Ferry [3.11]. Early application of such descriptions was given by Schwarzl and Staverman [3.12].

The mathematical description of the temperature dependence of this class of materials will now be formulated for constant temperature states. We designate the isotropic relaxation and creep functions at the base temperature $T = T_0$ by

$$G_\alpha(t) \quad \text{and} \quad J_\alpha(t), \qquad T = T_0$$

where $\alpha = 1, 2$ throughout. We let the temperature field be changed uniformly (independent of space coordinates) and designate the corresponding relaxation functions by $\mathscr{G}_\alpha(t, T)$, where T denotes absolute temperature. Then

$$\mathscr{G}_\alpha(t, T_0) = G_\alpha(t). \tag{3.58}$$

We change the independent variable in $G_\alpha(t)$ such that

$$G_\alpha(t) = L_\alpha(\log t). \tag{3.59}$$

The basic postulate of thermorheologically simple materials is that of the relationship

$$\mathscr{G}_\alpha(t, T) = L_\alpha[\log t + \psi(T)] \tag{3.60}$$

where the shift function $\psi(T)$ obeys the relations

$$\psi(T_0) = 0, \qquad \frac{d\psi(T)}{dT} > 0. \tag{3.61}$$

The meaning of (3.60) is that changes in temperature cause the relaxation function to be shifted to the right or left when plotted against $\log t$, as abscissa. We introduce a change of variable in the shift function by setting

$$\psi(T) = \log \chi(T). \tag{3.62}$$

Relations (3.61) now imply that

$$\chi(T_0) = 1, \qquad \frac{d\chi(T)}{dT} > 0. \tag{3.63}$$

With the change (3.62), (3.60) becomes

$$\mathscr{G}_\alpha(t, T) = L_\alpha[\log(t\chi(T))]. \tag{3.64}$$

Now, recalling (3.59), (3.64) is written as

$$\mathscr{G}_\alpha(t, T) = G_\alpha(\xi) \tag{3.65}$$

where

$$\xi = t\chi(T). \tag{3.66}$$

Thus, the relaxation function $\mathscr{G}_\alpha(t, T)$ at any temperature can directly be obtained from the relaxation function $G_\alpha(t)$ at base temperature T_0 by replacing t with ξ from (3.66).[1] The shift function $\chi(T)$, (3.66), is a basic property of the material, and must in general, be determined experimentally.[2] Also, the same shift function $\chi(T)$ must apply to both G_1 and G_2, and also to φ in (3.48), if the other mechanical properties, which could be derived from these, are to satisfy the thermorheologically simple postulate. Similar kinds of derivations can be formulated appropriate to the description of the temperature dependence of creep functions and complex moduli. In fact, for temperature dependent relaxation functions specified by $G_\alpha(\xi)$ with $\xi = t\chi(T)$, using (1.58) and (1.59), the corresponding temperature dependent complex modulus is seen to be given by $G_\alpha{}^*[i(\omega/\chi(T))]$. That is, the complex modulus at various temperatures is obtained from the complex modulus $G_\alpha{}^*(i\omega)$ at the base temperature through replacing ω by $\omega/\chi(T)$. A similar shift procedure can be established for creep functions.

This shifting of the mechanical properties data parallel to the time or frequency axis has some interesting physical implications. It is recalled from Section 1.6 that the asymptotic values of the zero and infinite frequency complex modulus are actually just elastic moduli. These asymptotic values do not change under the shifting hypothesis for thermorheologically simple materials. Consequently, it is not the limiting case elastic behavior of viscoelastic materials which is affected by the shifting hypothesis here; rather, it is the intermediate behavior or the relaxation times, which is affected.

The thermorheologically simple postulate is also sometimes referred to as the time–temperature superposition principle, or the method of reduced variables.

Nonconstant Temperature States

It is of interest to attempt to extend the results just described for dependence of mechanical properties upon constant temperature states, to model material behavior under nonconstant, nonuniform temperature states. The reason for

[1] The relationship (3.65) is sometimes taken in the expanded form $\mathscr{G}_\alpha(t, T) = (\rho T/\rho_0 T_0)\, G_\alpha(\xi)$, where ρ_0 and T_0 are the initial density and temperature. Whereas the variable ξ accounts for a shifting of the data along the time axis, this expanded form for $\mathscr{G}_\alpha$ also gives a shifting of the data along the G_α axis, due to temperature and density changes. Normally these latter effects are small compared with the time scale shifting, and they are not considered here. For further information on these effects, see Ferry [3.13].

[2] An empirically derived form of the shift function $\chi(T)$ is that of the WLF equation. See Ferry [3.13] for the derivation and limitations of this expression.

considering this extension would be as an application in solving thermoviscoelastic boundary value problems. But this introduces a seeming inconsistency with the results of Section 3.1. There it is found to be a consequence of the first order theory that the mechanical properties depend only on the uniform base temperature T_0 with the dependence upon the infinitesimal temperature deviation from T_0 neglected. Here it is proposed to include a dependence upon absolute temperature. This is not actually an inconsistency; rather, it simply means that the effects to be studied here are outside the scope of the first order linear theory, and consequently, a coupled thermoviscoelastic theory which includes the temperature dependence of mechanical properties is necessarily nonlinear.

We are thus motivated to develop a general nonlinear theory of thermoviscoelasticity which upon the usual linearization of stress and strain still retains a general nonlinear dependence upon temperature. This must be accomplished with a consistent set of smallness assumptions. A general theory of this type has been given by Crochet and Naghdi [3.14, 3.15]. Guided by their work, we here derive the special results appropriate to the stress strain constitutive relation without consideration of the other field variables such as energy, entropy, and the heat flux vector, which necessarily are involved in the general theory. In so doing we shall be extending the uncoupled theory of linear thermoviscoelasticity to account for the temperature dependence of the relevant mechanical properties. The nonconstant, nonuniform temperature history is considered to be known.

The starting point of our derivation is the statement of a general nonlinear functional which expresses the dependence of the current value of stress upon the current values and histories of strain and temperature. This is given by

$$\mathbf{\Sigma}(t) = \underset{s=0}{\overset{\infty}{\Upsilon}} (\mathbf{E}(t-s), T(t-s), \mathbf{E}(t), T(t)) \tag{3.67}$$

where $T(t)$ is the absolute temperature, $\mathbf{E}(t)$ is the nonlinear strain measure defined by

$$\mathbf{E} = E_{KL} = \tfrac{1}{2}(x_{k,K}x_{k,L} - \delta_{KL}) \tag{3.68}$$

with $x_i = x_i(t, X_K)$ where x_i are the coordinates of a point in the deformed position and X_K are the coordinates of the same point in the reference configuration, and $\mathbf{\Sigma}(t)$ represents an appropriate nonlinear theory definition of stress, here taken to be the Piola–Kirchhoff stress tensor, as defined in Ref. [3.14]. We denote the isothermal constitutive relation corresponding to (3.67) through

$$\mathbf{\Sigma}(t)\,|_{T(t)=T_0} = \underset{s=0}{\overset{\infty}{\Upsilon^*}} (\mathbf{E}(t-s), \mathbf{E}(t)) \tag{3.69}$$

where T_0 is some fixed base temperature.

Assume the functional relationship (3.67) can be inverted such that the following functional exists:

$$\mathbf{E}(t) = \underset{s=0}{\overset{\infty}{\boldsymbol{\epsilon}'}} (\boldsymbol{\Sigma}(t-s), T(t-s), \boldsymbol{\Sigma}(t), T(t)). \tag{3.70}$$

For the material being in a stress free state, but with nonconstant temperature history we write

$$\mathbf{E}(t) \,|_{\boldsymbol{\Sigma}=\mathbf{0}} = \underset{s=0}{\overset{\infty}{\boldsymbol{\epsilon}'}} (\mathbf{0}, T(t-s), \mathbf{0}, T(t)) \tag{3.71}$$

which we can express as a separate functional of temperature only, through

$$\mathbf{E}(t) \,|_{\boldsymbol{\Sigma}=\mathbf{0}} = \underset{s=0}{\overset{\infty}{\boldsymbol{\epsilon}''}} (T(t-s), T(t)). \tag{3.72}$$

Under the assumptions of the general theory, Crochet and Naghdi [3.14] have shown that the functional in (3.72) must be restricted to express strain at zero stress as a function of current temperature, thus (3.72) reduces to

$$\underset{s=0}{\overset{\infty}{\boldsymbol{\epsilon}''}} (T(t-s), T(t)) = \boldsymbol{\alpha}(T(t)). \tag{3.73}$$

This behavior, which expresses strain in the stress free state as a function of current temperature, is in contrast to the most general type of behavior allowed in the infinitesimal theory derivation in Section 3.1. The results of that derivation permit the strain at zero stress to be a functional of temperature history. As such, the present type of behavior is a special case of the type of behavior allowed in Section 3.1.

We now decompose $\mathbf{E}(t)$ as expressed by (3.70) into two parts such that

$$\underset{s=0}{\overset{\infty}{\boldsymbol{\epsilon}'}} (\boldsymbol{\Sigma}(t-s), T(t-s), \boldsymbol{\Sigma}(t), T(t)) = \underset{s=0}{\overset{\infty}{\boldsymbol{\epsilon}''}} (T(t-s), T(t)) + \underset{s=0}{\overset{\infty}{\boldsymbol{\epsilon}}} (\boldsymbol{\Sigma}(t-s), T(t-s), \boldsymbol{\Sigma}(t), T(t)) \tag{3.74}$$

where from (3.72) we must take

$$\underset{s=0}{\overset{\infty}{\boldsymbol{\epsilon}}} (\mathbf{0}, T(t-s), \mathbf{0}, T(t)) = 0. \tag{3.75}$$

Using (3.70) and (3.73) in (3.74) gives the form

$$\mathbf{E}(t) - \boldsymbol{\alpha}\,(t) = \underset{s=0}{\overset{\infty}{\boldsymbol{\epsilon}}} (\boldsymbol{\Sigma}(t-s), T(t-s), \boldsymbol{\Sigma}(t), T(t)). \tag{3.76}$$

One more item is needed to complete our preliminary matters. Introduce a modified time scale which depends upon the history of temperature through the functional form

$$\zeta_s = \underset{\lambda=0}{\overset{\infty}{\gamma}} \left(T(t-\lambda), s\right) \tag{3.77}$$

where $\underset{\lambda=0}{\overset{\infty}{\gamma}}(\ )$ has properties such that

$$\zeta_s \mid_{s=0} = 0,$$

$$\frac{\partial \zeta_s}{\partial s} > 0, \tag{3.78}$$

and

$$\zeta_s \mid_{T=T_0} = s$$

where T_0 designates some constant base temperature.

We can now introduce the main hypothesis. It is assumed that the nonisothermal stress constitutive relation is determined by the corresponding isothermal functional with $\mathbf{E}$ replaced by $\mathbf{E} - \boldsymbol{\alpha}$ and with a modified time scale to account for the history of temperature. The assumption concerning the replacement of $\mathbf{E}$ by $\mathbf{E} - \boldsymbol{\alpha}$ is motivated by and consistent with the forms (3.75) and (3.76). The assumption concerning the modified time scale is suggested by the shifting properties we have already observed in connection with constant temperature states. With this hypothesis we use the isothermal functional (3.69) to express the nonisothermal functional (3.67) as

$$\boldsymbol{\Sigma}(t) = \underset{s=0}{\overset{\infty}{\boldsymbol{\Upsilon}}} \left([\mathbf{E}(t-\zeta_s) - \boldsymbol{\alpha}(t-\zeta_s)], \mathbf{E}(t) - \boldsymbol{\alpha}(t)\right) \tag{3.79}$$

where ζ_s is given by (3.77). The conditions under which the functional in (3.79) can be taken to be linear can now be assessed. In fact, the situation is identical with that which occurs in obtaining isothermal infinitesimal theory constitutive relations. Specifically, taking $\mathbf{E}(t) = O(\epsilon)$ and $\boldsymbol{\alpha}(t) = O(\epsilon)$, where ϵ is the smallness parameter defined by

$$\epsilon = \sup_{\tau} \mid x_{i,K}(\tau) - \delta_{iK} \mid,$$

we can write (3.79) in the infinitesimal theory form simply by modifying the isothermal infinitesimal theory form (1.4) in accordance with our basic hypothesis. This procedure gives the appropriate infinitesimal theory form of (3.79) as

$$\begin{aligned} \sigma_{ij}(t) = G_{ijkl}(0)[\epsilon_{kl}(t) - \alpha_{kl}(t)] \\ + \int_0^\infty [\epsilon_{kl}(t-\zeta_s) - \alpha_{kl}(t-\zeta_s)] \frac{\partial G_{ijkl}(s)}{\partial s}\, ds \end{aligned} \tag{3.80}$$

where the usual infinitesimal theory definition of stress is taken, and the upper limit in (1.4) has been converted from t to ∞ to account for deformation events prior to $t = 0$.

A key observation should be made at this point. Note that although we have required the strain $\mathbf{E}(t)$ and the strain due to temperature changes in a stress free state, designated by $\boldsymbol{\alpha}(t)$, to be small of order ϵ, nowhere has it been necessary for us to require that temperature change from a base temperature be small. The form (3.80) is thus the desired infinitesimal theory constitutive relation obtained from a nonlinear theory, wherein a general nonlinear dependence upon temperature is retained. We next convert the relation (3.80) into a form which permits an obvious identification with the type of characteristics we have associated with thermorheologically simple materials under constant temperature states.

Integrating (3.80) by parts and using the change of variable $t - \zeta_s = \tau$ gives

$$\sigma_{ij}(t) = \int_{-\infty}^{t} G_{ijkl}(s) \frac{\partial}{\partial \tau} [\epsilon_{kl}(\tau) - \alpha_{kl}(\tau)]\, d\tau \tag{3.81}$$

where the characteristics expressed in (3.78) have been used and it still remains to relate the s variable in (3.81) to the τ variable. To get this relation between s and τ assume that the functional form in (3.77) can be inverted to give

$$s = \underset{\lambda=0}{\overset{\infty}{\beta}} \left(T(t - \lambda), \zeta_s\right). \tag{3.82}$$

In the interests of obtaining some specific results, specialize the form in (3.82) to have the following representation:

$$s = \int_0^{\zeta_s} \chi(T(t - \lambda))\, d\lambda \tag{3.83}$$

where χ is a shift function such that

$$\frac{d\chi(T)}{dT} > 0$$

and

$$\chi(T_0) = 1$$

at some base temperature T_0, and it is assumed that $T(t) \to T_0$ for prior times sufficiently distant from the current time. Now, again using the change of variable $\zeta_s = t - \tau$, this time in (3.83), gives with a further change of variable the desired relation between s and τ as

$$s = \int_0^{t} \chi(T(\eta))\, d\eta - \int_0^{\tau} \chi(T(\eta))\, d\eta. \tag{3.84}$$

Using the form (3.84), (3.81) may be written as

$$\sigma_{ij}(x_i, t) = \int_{-\infty}^{t} G_{ijkl}(\xi - \xi') \frac{\partial}{\partial \tau} [\epsilon_{kl}(x_i, \tau) - \alpha_{kl}(x_i, \tau)] \, d\tau \tag{3.85}$$

where

$$\begin{aligned} \xi &= \int_0^t \chi(T(x_i, \eta)) \, d\eta \\ \xi' &= \int_0^\tau \chi(T(x_i, \eta)) \, d\eta. \end{aligned} \tag{3.86}$$

The isotropic form of (3.85) follows immediately as

$$s_{ij}(x_i, t) = \int_{-\infty}^{t} G_1(\xi - \xi') \frac{\partial e_{ij}(x_i, \tau)}{\partial \tau} \, d\tau \tag{3.87}$$

$$\sigma_{kk}(x_i, t) = \int_{-\infty}^{t} G_2(\xi - \xi') \frac{\partial}{\partial \tau} [\epsilon_{kk}(x_i, \tau) - \alpha(x_i, \tau)] \, d\tau \tag{3.88}$$

where the deviatoric components are defined by (1.12) and (1.13), and $\alpha(t) = \alpha(T(t))$ represents the generally nonlinear volume change due to temperature change under stress free conditions.

We can now identify the shift function in (3.86) with that involved in (3.66) for constant temperature states. In fact, under constant temperature states, relations (3.86) reduce to (3.66). Accordingly, relations (3.85)–(3.88) provide the infinitesimal theory constitutive relations appropriate to thermorheologically simple materials. These forms were first given by Morland and Lee [3.16]. The present derivation from the work of Crochet and Naghdi [3.14, 3.15] reveals, through a nonlinear theory derivation, the basic nature of this "mixed" theory involving infinitesimal strain but noninfinitesimal temperature changes. In the general theory of Refs. [3.14, 3.15] a full set of constitutive relations are derived and the appropriate nonlinear form of the heat conduction equation is found. The preceding derivation is for viscoelastic solids. A unified approach, appropriate to both solids and fluids, has been given by Crochet and Naghdi [3.17]. In further work, Crochet and Naghdi [3.18] have modified their theory to accommodate the rubber elasticity type of temperature dependence mentioned in footnote 1 on page 96.

The stress relations (3.85)–(3.88) are the relevant constitutive relations to be used in solving typical problems. These forms are not convolution integrals and not convenient to use since time is involved both explicity and implicitly through ξ. For this reason, a reduced form for these stress relations involving only ξ is obtained. Now ξ from (3.86) is a monotone increasing function of time t and therefore it may be inverted to have

$$t = g(x_i, \xi) \tag{3.89}$$

where we take $T = T_0$, $t < 0$ and also assume conditions of rest for $t < 0$. Using (3.89), a different strain function $\hat{\epsilon}_{ij}(x_i, \xi)$ is defined through

$$\epsilon_{ij}(x_i, t) = \epsilon_{ij}(x_i, g(x_i, \xi)) = \hat{\epsilon}_{ij}(x_i, \xi). \tag{3.90}$$

We define stress $\hat{\sigma}_{ij}(x_i, \xi)$ and $\hat{\alpha}(x_i, \xi)$ similarly through

$$\sigma_{ij}(x_i, t) = \sigma_{ij}(x_i, g(x_i, \xi)) = \hat{\sigma}_{ij}(x_i, \xi) \tag{3.91}$$

and

$$\alpha(x_i, t) = \alpha(x_i, g(x_i, \xi)) = \hat{\alpha}(x_i, \xi). \tag{3.92}$$

Through the use of (3.90)–(3.92), (3.87) and (3.88) become

$$\hat{s}_{ij}(x_i, \xi) = \int_0^{\xi} G_1(\xi - \xi') \frac{\partial}{\partial \xi'} \hat{e}_{ij}(x_i, \xi')\, d\xi' \tag{3.93}$$

and

$$\hat{\sigma}_{kk}(x_i, \xi) = \int_0^{\xi} G_2(\xi - \xi') \frac{\partial}{\partial \xi'} [\hat{\epsilon}_{kk}(x_i, \xi') - \hat{\alpha}(x_i, \xi')]\, d\xi'. \tag{3.94}$$

Relations (3.93) and (3.94) now involve convolution integrals, and because of these, it would seem that the Laplace transform might conveniently be used to solve boundary value problems involving thermorheologically simple materials. However, in accordance with the notation of (3.93) and (3.94), the equilibrium equations must also be expressed in terms of the $\hat{\sigma}_{ij}(x_i, \xi)$ variables. Thus, $\sigma_{ij,j}(x_i, t) + F_i(x_i, t) = 0$ has the form

$$\hat{\sigma}_{ij,j}(x_i, \xi) + \frac{\partial \hat{\sigma}_{ij}(x_i, \xi)}{\partial \xi} \frac{\partial \xi}{\partial x_j} + \hat{F}_i(x_i, \xi) = 0 \tag{3.95}$$

where (3.91) has been used. This more complicated form for the equilibrium equations renders the direct application of the Laplace transform with respect to ξ inapplicable. A more complicated form for the strain displacement relations also results when the variables $\hat{\epsilon}_{ij}$ and $\hat{u}_i$ are used.

It follows from the considerations just given that, for thermorheologically simple materials, there is, in general, no correspondence principle or relation between elastic solutions and viscoelastic solutions of boundary value problems. However, there are two special cases which merit attention. These two cases are those in which the given temperature field is spatially uniform but time dependent, and in which the temperature field has a spatial dependence but is independent of time. In the latter case, the related boundary value problem is similar to that under isothermal conditions, but the mechanical properties have a spatial dependency. This type of problem is therefore of the type appropriate to nonhomogeneous media, still a necessarily complicated situation. In the other case, involving time dependent but spatially independent temperature fields, the second term in the equilibrium equations (3.95) is identically zero, and integral transform methods, with respect to the reduced time variable, can be profitably employed in a manner similar to that used in the isothermal case.

An Example

An example is now worked out to illustrate the use of (3.93) and (3.94). This example, as well as some of the previous forms, are from Muki and Sternberg [3.19]. The problem is that of an infinite isotropic viscoelastic slab of thickness $2a$ subject to heating conditions which produce a known temperature variation through the thickness. Rectangular Cartesian coordinates are used such that the stressfree boundaries are given by $z = \pm a$. The temperature distribution $T(z, t)$ is considered to be known whether from experimental observations by some analytical means. The problem is to find the resulting stress state.

The displacement field will be taken as

$$u_x = u_y = 0, \qquad u_z = u_z(z, t). \tag{3.96}$$

All of the resulting strains are zero except $\epsilon_{zz} = \partial u_z / \partial z$, and it follows that

$$\begin{aligned} \epsilon_{kk} &= \epsilon_{zz}, \\ e_{xx} &= e_{yy} = -\tfrac{1}{3}\epsilon_{zz}, \\ e_{zz} &= \tfrac{2}{3}\epsilon_{zz} \end{aligned} \tag{3.97}$$

and

with all other strain components vanishing. The only nonzero stress components are correspondingly

$$\begin{aligned} \sigma_{xx} &= \sigma_{yy}, & s_{xx} &= s_{yy} = -\tfrac{1}{3}(\sigma_{zz} - \sigma_{xx}), \\ \sigma_{kk} &= 2\sigma_{xx} + \sigma_{zz}, & s_{zz} &= \tfrac{2}{3}(\sigma_{zz} - \sigma_{xx}). \end{aligned} \tag{3.98}$$

The deviatoric stress strain relation (3.93) for the zz component in view of (3.97) and (3.98) becomes

$$\hat{\sigma}_{zz}(z, \xi) - \hat{\sigma}_{xx}(z, \xi) = \int_0^\xi G_1(\xi - \xi') \frac{\partial}{\partial \xi'} \hat{\epsilon}_{kk}(z, \xi')\, d\xi'. \tag{3.99}$$

Similarly, (3.94), using (3.97) and (3.98), assumes the form

$$\hat{\sigma}_{zz}(z, \xi) + 2\hat{\sigma}_{xx}(z, \xi) = \int_0^\xi G_2(\xi - \xi') \frac{\partial}{\partial \xi'} [\hat{\epsilon}_{kk}(z, \xi') - \hat{\alpha}(z, \xi')]\, d\xi'. \tag{3.100}$$

The only nontrivial equation of equilibrium is

$$\frac{\partial}{\partial z}\sigma_{zz}(z, t) = 0. \tag{3.101}$$

The boundary conditions are

$$\sigma_{zz}(z, t) = 0 \qquad \text{for} \quad z = \pm a. \tag{3.102}$$

From (3.101) and (3.102) it is required that

$$\sigma_{zz}(z, t) = \hat{\sigma}_{zz}(z, \xi) = 0. \tag{3.103}$$

Now we take the Laplace transform with respect to ξ of (3.99) and (3.100). There results

$$\bar{\hat{\sigma}}_{xx}(z, s) = -s\bar{G}_1\bar{\hat{\epsilon}}_{kk}(z, s) \tag{3.104}$$

and

$$2\bar{\hat{\sigma}}_{xx}(z, s) = s\bar{G}_2[\bar{\hat{\epsilon}}_{kk}(z, s) - \bar{\hat{\alpha}}(z, s)] \tag{3.105}$$

where use has been made of (3.103), and s is the transform parameter. We eliminate $\bar{\hat{\epsilon}}_{kk}$ between (3.104) and (3.105) to arrive at

$$\bar{\hat{\sigma}}_{xx}(z, s) = \frac{-s\bar{G}_1\bar{G}_2\bar{\hat{\alpha}}(z, s)}{2\bar{G}_1 + \bar{G}_2}. \tag{3.106}$$

We define the function $R(\xi)$ through the inverse Laplace transform as

$$R(\xi) = \mathscr{L}^{-1}\left[\frac{\bar{G}_1(s)\,\bar{G}_2(s)}{2\bar{G}_1(s) + \bar{G}_2(s)}\right]. \tag{3.107}$$

By using (3.107), we invert (3.106) to get

$$\hat{\sigma}_{xx}(z, \xi) = -\int_0^{\xi} R(\xi - \xi')\frac{\partial}{\partial \xi'}\hat{\alpha}(z, \xi')\, d\xi'. \tag{3.108}$$

Relation (3.108) is now changed to a variation with physical time t through (3.91) and (3.92). We then have

$$\sigma_{xx}(z, t) = \sigma_{yy}(z, t) = -\int_0^{t} R[\xi(z, t) - \xi'(z, \tau)]\frac{\partial}{\partial \tau}\alpha(z, \tau)\, d\tau. \tag{3.109}$$

This completes the solution for the resulting stresses. When the temperature field is specified, consistent with the initial conditions assumed, ξ is found from (3.86), R from (3.107), and thence the stresses from (3.109). A more general treatment of this problem was given by Muki and Sternberg [3.19] with detailed numerical results. Lee and Rogers [3.20] pointed out that this problem may be solved without using the Laplace transform as is used here, and the problem was further discussed by Sternberg [3.21]. A related but more difficult problem, involving thermal stresses in a cylinder, has been studied by Lockett and Morland [3.22].

It has been mentioned earlier that a coupled theory of thermoviscoelasticity, which includes the dependence of the mechanical property functions upon the temperature field variable, must necessarily be nonlinear. This coupling between thermal effects and mechanical effects is due to the temperature history effects in the stress constitutive relations and the effect of the volume change history in the heat conduction equation. But the example just presented was linear in all respects, even though temperature dependent mechanical properties were included. This is due to the fact that the temperature field was considered to be known independently, and it was not necessary to solve a coupled problem. However, problems of coupled thermoviscoelasticity, which include this temperature dependence of mechanical properties upon the unknown temperature, must be expected to be very difficult to solve analytically. A numerical approach to thermoviscoelastic problems is given in Section 5.5.

3.7. GLASS TRANSITION CRITERION

In Section 3.5 we gave several operational definitions of the glass transition temperature T_g. In all of these definitions the method for evaluating T_g involved either a time or frequency of observation. Thus, the methods are not unique, the numerical value of T_g depending on the rate of the corresponding evaluation process. This is a rather unsatisfactory state of affairs. The temperature at which materials crystallize is uniquely definable, yet the temperature at which polymers go through a transition from rubbery to glassy states is not clear cut, according to the methods described in Section 3.5. In this section we remedy this situation by deriving a criterion that does serve to uniquely define T_g.

Transitions are commonplace in materials. Crystallization is of course a normal occurrence, and transitions of this type are referred to as first order transitions, in the sense that they represent a discontinuous change with temperature of the first derivative of the energy, which is physically manifested as a volume change. Thermodynamically higher order transitions are possible, and they represent discontinuous changes with temperature of higher order derivatives of energy. These were first considered by Ehrenfest [3.23] and were termed by him second and higher order transitions. In particular, for a second order transition Ehrenfest showed that the changes in the compressibility κ, the coefficient of thermal expansion α, and the specific heat at constant pressure C_p must satisfy

$$\frac{\Delta\kappa}{\Delta\alpha} = \frac{TV\,\Delta\alpha}{\Delta C_p} \tag{3.110}$$

where T is the absolute temperature and V is the specific volume, with

$$\Delta\kappa = \kappa_{T_{g(+)}} - \kappa_{T_{g(-)}}$$

etc., being the changes in properties at the transition temperature T_g. The underlying basis of Eq. (3.110) is that it is thermodynamically possible for the properties to change with continuity of energy. This change takes place under equilibrium conditions insofar as the variables (volume, pressure, and temperature) are concerned. It is a separate matter to consider whether the material is in an equilibrium state on the molecular scale, and in general it is not. That is, equilibrium is presumed on the molecular scale in the rubbery state, $T \gg T_g$, but not in the glassy state, $T \ll T_g$, and nonequilibrium configurations of molecules are said to be "frozen in." Considerations such as these have met with some success in describing changes in properties in some low-molecular-weight materials. Attempts to use simple second order transitions to characterize the glass transition behavior of polymers have yielded disappointing results.

Classical second order transitions cannot be used as a model for the glass transition behavior of polymers because the underlying reasoning is based on equilibrium thermodynamics, whereas polymers are inherently dissipative and their behavior is necessarily governed by irreversible thermodynamics. The time dependent nature of the mechanical properties of polymers is due to dissipative processes in the material; such effects cannot be modeled by equilibrium changes as in classical second order transitions. Thus, any treatment which fails to account for irreversible thermodynamical processes cannot be expected to succeed for polymers. The approach followed here arrives at a criterion for the glass transition temperature in polymers from the standpoint of irreversible thermodynamics. The present derivation is taken from the work of Christensen [3.24].

Thermodynamical Derivation

For later purposes it will be necessary to have a complete, linear theory of the coupled mechanical and thermal behaviour of viscoelastic materials. Such a theory based upon the Helmholtz free energy was given in Section 3.1. For reasons which will be made apparent later, this derivation is not suitable for application here, and it is necessary to have a theory based upon the use of the Gibbs free energy. With elastic materials the difference between these two approaches is of little account, but with viscoelastic materials it is extremely significant. Such a thermodynamical viscoelastic derivation based on the use of the Gibbs free energy is provided here. The results will be applied to the glass transition problem.

The local balance of energy for the linear theory is given by

$$\rho r - \rho \dot{U} + \sigma_{ij}\dot{\epsilon}_{ij} - Q_{i,i} = 0 \tag{3.111}$$

where U is the internal energy per unit mass and the notation is as in Section 3.1. The related local entropy production inequality is given by (3.2).

The Gibbs free energy per unit mass G, is defined by the transformation

$$\rho U = \sigma_{ij}\epsilon_{ij} + T\rho S + \rho G. \tag{3.112}$$

Combining Eqs. (3.111), (3.2), and (3.112) results in

$$-\rho S\dot{\theta} - \rho\dot{G} - \dot{\sigma}_{ij}\epsilon_{ij} - Q_i(\theta_{,i}/T_0) \geqslant 0 \tag{3.113}$$

where θ is the infinitesimal temperature deviation from any given base temperature T_0 as

$$T = T_0 + \theta. \tag{3.114}$$

If constitutive equations for G and Q_i are postulated in appropriate form, then all other constitutive equations and governing conditions follow from the process of satisfying relations (3.111) and (3.113). The starting point is then to find the proper constitutive equation for the Gibbs free energy. Following the procedure of Section 3.1 involves expressing the energy as a polynomial expansion in linear Stieltjes integrals, with the result

$$\begin{aligned}
\rho G = &-\frac{1}{2}\int_0^t\int_0^t J_1(2t-\tau-\eta)\frac{ds_{ij}}{d\tau}\frac{ds_{ij}}{d\eta}\,d\tau\,d\eta \\
&-\frac{1}{2}\int_0^t\int_0^t J_2(2t-\tau-\eta)\frac{d\sigma_{jj}}{d\tau}\frac{d\sigma_{kk}}{d\eta}\,d\tau\,d\eta \\
&-\int_0^t\int_0^t \alpha(2t-\tau-\eta)\frac{d\sigma_{kk}}{d\tau}\frac{d\theta}{d\eta}\,d\tau\,d\eta \\
&-\frac{1}{2}\int_0^t\int_0^t C(2t-\tau-\eta)\frac{d\theta}{d\tau}\frac{d\theta}{d\eta}\,d\tau\,d\eta + O(\epsilon^3)
\end{aligned} \tag{3.115}$$

where $\sigma_{ij}/J_1(0)$ and θ/T_0 are infinitesimals of $O(\epsilon)$ and where the deviatoric components of stress and strain are employed. As seen from Eq. (3.115), consideration has been restricted to the case of isotropic materials. The linear terms in the polynomial expansion used to find Eq. (3.115) have been dropped since they can be shown to contribute only the addition of certain constants to the later constitutive equations for strain and entropy. The functions $J_1(t)$,

$J_2(t)$, $\alpha(t)$, and $C(t)$ are creep function generalizations of the usual shear compliance, the volumetric compliance, the coefficient of thermal expansion, and the specific heat. Actually, in the most general form of the polynomial expansion procedure, these creep functions have two independent arguments, as $J_1(t-\tau, t-\eta)$; those here have been taken in the special additive form $J_1(2t-\tau-\eta)$ so that the same creep function appears in all constitutive equations as $J_1(t)$ rather than as $J_1(t, 0)$ and $J_1(0, t)$. This point is discussed in detail in Section 3.2. Henceforth, $O(\epsilon^3)$ terms in Eq. (3.115) will be neglected since interest here centers on the related linear theory.

Substituting Eq. (3.115) into Eq. (3.113) and using Leibnitz's rule, we have

$$\begin{aligned}\Big[-\rho S + \int_0^t \alpha(t-\tau)\frac{d\sigma_{kk}}{d\tau}\,d\tau + \int_0^t C(t-\tau)\frac{d\theta}{d\tau}\,d\tau\Big]\,\dot\theta(t)\\ + \Big[-\frac{\epsilon_{kk}}{3} + \int_0^t J_2(t-\tau)\frac{d\sigma_{kk}}{d\tau}\,d\tau + \int_0^t \alpha(t-\tau)\frac{d\theta}{d\tau}\,d\tau\Big]\,\dot\sigma_{kk}(t)\\ + \Big[\int_0^t J_1(t-\tau)\frac{ds_{ij}}{d\tau}\,d\tau - e_{ij}\Big]\,\dot s_{ij}(t) + \Lambda - Q_i\left(\frac{\theta_{,i}}{T_0}\right) \geqslant 0 \end{aligned}\tag{3.116}$$

where

$$\begin{aligned}\Lambda = \frac{1}{2}\int_0^t\int_0^t \frac{\partial}{\partial t} J_1(2t-\tau-\eta)\frac{ds_{ij}}{d\tau}\frac{ds_{ij}}{d\eta}\,d\tau\,d\eta\\ + \frac{1}{2}\int_0^t\int_0^t \frac{\partial}{\partial t} J_2(2t-\tau-\eta)\frac{d\sigma_{jj}}{d\tau}\frac{d\sigma_{kk}}{d\eta}\,d\tau\,d\eta\\ + \int_0^t\int_0^t \frac{\partial}{\partial t} \alpha(2t-\tau-\eta)\frac{d\sigma_{kk}}{d\tau}\frac{d\theta}{d\eta}\,d\tau\,d\eta\\ + \frac{1}{2}\int_0^t\int_0^t \frac{\partial}{\partial t} C(2t-\tau-\eta)\frac{d\theta}{d\tau}\frac{d\theta}{d\eta}\,d\tau\,d\eta. \end{aligned}\tag{3.117}$$

For relation (3.116) to be satisfied for all values of $\dot\sigma_{ij}(t)$ and $\dot\theta(t)$, it is necessary that the coefficients in brackets vanish; thus,

$$e_{ij}(t) = \int_0^t J_1(t-\tau)\frac{ds_{ij}}{d\tau}\,d\tau$$

$$\epsilon_{kk}(t) = 3\int_0^t J_2(t-\tau)\frac{d\sigma_{kk}}{d\tau}\,d\tau + 3\int_0^t \alpha(t-\tau)\frac{d\theta}{d\tau}\,d\tau$$

and

$$\rho S(t) = \int_0^t C(t-\tau)\frac{d\theta}{d\tau}\,d\tau + \int_0^t \alpha(t-\tau)\frac{d\sigma_{kk}}{d\tau}\,d\tau \tag{3.118}$$

which leaves Eq. (3.116) as

$$\Lambda - Q_i(\theta_{,i}/T_0) \geqslant 0. \tag{3.119}$$

Substituting relations (3.112), (3.115), and (3.118) back into the conservation of energy equation (3.111) results in

$$\rho r - T_0 \frac{\partial}{\partial t}\left[\int_0^t C(t-\tau)\frac{d\theta}{d\tau}\,d\tau + \int_0^t \alpha(t-\tau)\frac{d\sigma_{kk}}{d\tau}\,d\tau\right] - Q_{i,i} = 0 \quad (3.120)$$

where terms of $O(\epsilon^2)$ have been neglected. For uniform temperature distributions Eq. (3.119) has the form

$$\Lambda \geqslant 0 \quad (3.121)$$

where Λ is defined by Eq. (3.117). This expression is conventionally called the dissipation inequality, and it must be satisfied for all possible processes.

The derivation has now revealed the appropriate creep integral stress-strain relations, with coupled thermal–mechanical effects, and the associated entropy constitutive relation (3.118). The associated heat conduction equation (3.120) reveals the generalization of the classical heat conduction equation, which is necessary when dealing with memory dependent materials. It is important to note that the coupling between thermal and mechanical effects occurs only through volumetric terms for isotropic materials. As shown by Eq. (3.118), the linear shear stress strain relation is uninfluenced by temperature coupling.

A particular form of the dissipation inequality (3.121) will be useful. This is the one which occurs for the process

$$\begin{aligned} s_{ij}(t) &= 0 \\ \sigma_{kk}(t) &= \sigma_0 h(t) \\ \theta(t) &= \theta_0 h(t) \end{aligned} \quad (3.122)$$

where $h(t)$ is the unit step function.

Insertion of Eq. (3.122) into Eq. (3.121) gives

$$\frac{dJ_2(t)}{dt}\left(\frac{\sigma_0}{\theta_0}\right)^2 + 2\frac{d\alpha(t)}{dt}\left(\frac{\sigma_0}{\theta_0}\right) + \frac{dC(t)}{dt} \geqslant 0. \quad (3.123)$$

From Eq. (3.123) it immediately follows that

$$\begin{aligned} \frac{dJ_2(t)}{dt} &\geqslant 0 \\ \frac{dC(t)}{dt} &\geqslant 0. \end{aligned} \quad (3.124)$$

By finding the minimum value of the expression in Eq. (3.123) as a function of σ_0/θ_0 and requiring that this value satisfy the inequality, there results the restriction

$$\left[\frac{d\alpha(t)}{dt}\right]^2 \leqslant \left[\frac{dJ_2(t)}{dt}\right]\left[\frac{dC(t)}{dt}\right]. \tag{3.125}$$

The preceding thermodynamical derivation now provides the basic tools with which to study glass transition behavior.

Glass Transition Criterion

The dissipation inequality (3.121) reveals that energy is dissipated during mechanical processes in viscoelastic materials. This mathematical result validates the commonly expressed tenet that materials which show a time dependence in their response are necessarily dissipative. There are some notable exceptions, however. For example, viscoelastic solids that are deformed either instantaneously or infinitely slowly are elastic in their response. It is reasonable to inquire if there are any other conditions under which viscoelastic materials could undergo mechanical or thermal processes with no dissipation of energy at an arbitrary base temperature T_0. Such conditions will be shown to occur and their consequences will be shown to relate to the glass transition behavior.

The state to be investigated is that of volumetric changes as influenced by temperature and pressure. As shown in the preceding section, for the linear theory, states of shear deformation are uncoupled from temperature effects and, accordingly, are of no further interest here.

Let us begin by investigating the condition under which the dissipation inequality (3.121) vanishes identically for the process specified by Eq. (3.122). Thus, with the equality form of Eq. (3.123), the solution is given by

$$\frac{\sigma_0}{\theta_0} = \frac{-\alpha' \pm \sqrt{(\alpha')^2 - J_2'C'}}{J_2'} \tag{3.126}$$

where

$$\alpha' = \frac{d\alpha(t)}{dt}$$
$$J_2' = \frac{dJ_2(t)}{dt} \tag{3.127}$$
$$C' = \frac{dC(t)}{dt}.$$

For a solution of Eq. (3.126) to be real, and for the inequality (3.125) to be satisfied, it is necessary that

$$(\alpha')^2 = J_2'C'. \tag{3.128}$$

Now examine the corresponding strain and entropy constitutive relations. Insertion of Eq. (3.122) into Eq. (3.118) gives

$$\begin{aligned} \epsilon_{kk}(t) &= 3J_2(t)\sigma_0 + 3\alpha(t)\theta_0 \\ \rho S &= C(t)\theta_0 + \alpha(t)\sigma_0 . \end{aligned} \tag{3.129}$$

Let us now consider only those processes for which the volume change and the entropy change vanish. Thus, for

$$\epsilon_{kk} = \rho S = 0 \tag{3.130}$$

Eqs. (3.129) give, respectively,

$$\frac{\sigma_0}{\theta_0} = -\frac{\alpha(t)}{J_2(t)} = -c \tag{3.131}$$

$$\frac{\sigma_0}{\theta_0} = -\frac{C(t)}{\alpha(t)} = -c \tag{3.132}$$

where c is a constant. Taking the derivatives of Eqs. (3.131) and (3.132), we obtain the combination

$$\frac{\alpha'(t)}{J_2'(t)} = \frac{C'(t)}{\alpha'(t)} . \tag{3.133}$$

Comparison of Eq. (3.133) with Eq. (3.128) shows complete consistency.

To summarize, for the process represented by Eq. (3.122), the requirement that there be no dissipation of energy (constant Gibbs free energy) and no change in volume or entropy is satisfied if

$$\frac{\alpha(t)}{J_2(t)} = \frac{C(t)}{\alpha(t)} = c \qquad (a \text{ constant}) \tag{3.134}$$

which in turn implies that

$$\left[\frac{d\alpha(t)}{dt}\right]^2 = \left[\frac{dJ_2(t)}{dt}\right]\left[\frac{dC(t)}{dt}\right]. \tag{3.135}$$

In fact, it is now possible to prove that relation (3.134), along with the requirement

$$\frac{\sigma(t)}{\theta(t)} = -c \tag{3.136}$$

where the constant c is given by Eq. (3.134), provides constant Gibbs free energy and constant strain and entropy without restricting the processes to step functions in time. This is proved directly by substituting Eqs. (3.134) and (3.136) into Eqs. (3.118) and (3.121).

The physical meaning of these results can now be interpreted. First, note the similarity between relation (3.135) and the classical second order transition equation (3.110). They are brought into exactly the same form by interpreting the derivatives in Eq. (3.135) as increments and taking the notational agreements

$$\Delta\kappa \sim J_2', \qquad \Delta\alpha \sim \alpha', \quad . \quad \Delta C_p \sim T_0 V C' \tag{3.137}$$

where V is the specific volume. Note that the property $C(t)$ herein thus corresponds to the specific heat at constant pressure. Despite this similarity between forms, relation (3.135) cannot be interpreted as that of a second order transition; rather, it is the analog of such a transition for dissipative materials. It must be emphasized that relation (3.135) follows not from reversible but from irreversible thermodynamics, and its similarity to the Ehrenfest equation should not be given particular significance. The primary observation is that when the ratios of the creep functions are constant, processes are possible for which the Gibbs free energy, the volume, and the entropy are constant, even though the mechanical properties (creep functions) are individually time dependent. In general, of course, relation (3.134) is not satisfied, but the crucial observation to be made here is that if at some base temperature T_0 the creep functions satisfy relation (3.134), then a transition of the type described above is possible. Such a base temperature at which Eq. (3.134) is satisfied is here taken as the glass transition temperature T_g since it is at this temperature that the thermal–mechanical properties may change even though the material is in a state of equilibrium. Clearly, such a behavior as this is a type of transition or critical state which has been shown to be thermodynamically admissible. The interpretation and practical means of employing this criterion for the glass transition temperature will be consideret next.

As a consequence of the preceding derivation, the procedure for establishing T_g involves determining the linear behavior creep functions at a series of base temperatures until the base temperature is found at which relation (3.134) is satisfied. More specifically, the linear volumetric response to step changes in pressure, $J_2(t)$, and that to step changes in temperature, $\alpha(t)$, are found at a series of base temperatures at which the sample is in an equilibrium state prior to being tested. The base temperature is successively reduced until the point is reached at which $\alpha(t)/J_2(t) = \text{constant}$, and this base temperature is taken to be T_g. Note that this procedure does not require any experimental operations in or near the glassy state. It has not been proved that polymers must have some base temperature at which $\alpha(t)/J_2(t) = \text{constant}$. Rather, it has been shown that if such a behavior at a particular temperature occurs, it has the fundamental transition characteristic that the mechanical and thermal properties can change in a time dependent manner under equilibrium conditions. As a practical matter, the procedure described here could be used to define T_g as that base temperature at which the creep functions are the closest to satisfying the proportionality requirement, even though the requirement may not be satisfied exactly. Such a method still gives due account to the inherent rate processes of the material, in contrast to the *ad hoc* procedures in common use, as described in Section 3.5.

It appears that no suitable experimental data are available for incorporation into the procedure outlined here. Superficially, the $J_2(t)$ and $\alpha(t)$ data of Goldbach and Rehage [3.25, 3.26] would seem to be applicable; however, they relate primarily to the glassy state range, whereas the data needed here should relate to the rubbery state. An examination of the data of Goldbach and Rehage for polystyrene reveals the creep function data for $J_2(t)$ and $\alpha(t)$ to be of the same general character, with the relaxation times associated with $J_2(t)$ becoming successively greater than those for $\alpha(t)$ as the base temperature is lowered from near T_g. This is consistent with the situation described here, such that at T_g the retardation spectra for $J_2(t)$ and $\alpha(t)$ are equal, but as the base temperature is changed from T_g, the retardation spectra for $J_2(t)$ and $\alpha(t)$ becomes divergent.

It can now be understood why the present derivation utilizing the Gibbs free energy could not have been arrived at with equal facility with the Helmholtz free energy. As appears from the derivation of Section 3.1, the resulting volumetric relaxation function $G_2(t)$ is a decreasing function of time, whereas the specific heat (at constant volume) $m(t)$ is an increasing function of time. In the present derivation all relevant mechanical properties are time increasing creep functions. Thus, interpretations similar to relations (3.131) and (3.132) are not possible with the Helmholtz free energy.

Within the present theory, other means of evaluating T_g can be ascertained as an alternative to determining $J_2(t)$ and $\alpha(t)$ directly. These procedures are considered in Ref. [3.24]. There are many other approaches to devising criteria for T_g. References cited in [3.24] outline this, principally work by DiMarzio, Gibbs, and Goldstein. In these latter works the frame of reference is with interpretations on a molecular scale, whereas the present approach proceeds from continuum thermodynamics.

3.8. HEAT CONDUCTION

The classical equation of heat conduction is a parabolic differential equation. The fact that this equation is parabolic implies that discontinuous temperature disturbances have instantaneous rates of propagation. This characteristic of the instantaneous propagation of energy is physically anomalous and there have been many attempts to formulate more general theories of heat conduction that remove this inconsistency. For example, a theory by Gurtin and Pipkin [3.27] includes a dependence of the free energy on the history of temperature gradient in addition to the history of temperature, as in Section 3.1. In contrast, Bogy and Naghdi [3.28] developed a nonlinear theory which includes a dependence of the field variables upon temperature rate, in addition to the usual dependence upon temperature and temperature gradient. Both of these theories ensure finite speeds of propagation for thermal pulses. A theory different from these is given next, one based on present results.

A theory of heat conduction will be developed here that does not require any generalizing assumptions over those that have already been employed herein in deriving a consistent thermodynamical theory of materials with memory. We start with the heat conduction equation in the form of (3.50), rewritten here as

$$T_0 \frac{\partial}{\partial t}\left[\int_{-\infty}^{t} m(t-\tau)\,\dot{\theta}(\tau)\,d\tau\right] = k\theta_{,ii} + \rho r. \tag{3.138}$$

The explicit form given in (3.138) differs from that in (3.50) in two respects. First, we have neglected the strain coupling term in (3.50) since here we are investigating uncoupled heat propagation. Second, the heat supply function is explicitly included in (3.138), as it was in (3.20). As noted previously, k is the thermal conductivity, and $m(t)$ is the time dependent thermal property corresponding to the specific heat at constant volume. We could just as well have started with the form of the heat conduction equation in (3.120), neglecting the stress coupling term, taking Q_i as in (3.18), and noting that $C(t)$ corresponds to the specific heat at constant pressure.

Equation (3.138) is not a convenient form to use, nor is it clear in this form whether this theory implies finite or infinite speeds of propagation of thermal disturbances. The heat conduction equation (3.138) is next converted to a more useful form, and it will be shown that a special restriction is needed to ensure that the speeds of propagation of thermal disturbances are always finite.

Equation (3.138) expresses the current value of the heat supply function plus k times the Laplacian of the temperature field in terms of a functional of the history of the rate of the temperature. It is now assumed that this integral equation can be inverted to express the current value of the temperature as a functional of the history of the heat supply function plus k times the Laplacian of the temperature field. Thus

$$\theta(x_i, t) = \frac{1}{T_0}\int_{-\infty}^{t} \psi(t-\tau)[k\theta_{,ii}(x_i, \tau) + \rho r(x_i, \tau)]\,d\tau. \tag{3.139}$$

The mechanical property function $\psi(t)$ in (3.139) is of course not independent of the relaxation function $m(t)$ in (3.138); in fact, one determines the other. The simplest form of the relationship between these two functions is obtained by comparing the Laplace transformed forms of (3.138) and (3.139) using

$\theta(\tau) = 0$, $\tau < 0$. In so doing, it is found that the Laplace transforms $\bar{\psi}(s)$ and $\bar{m}(s)$ of $\psi(t)$ and $m(t)$ are related through

$$\bar{\psi}(s) = 1/s^2\bar{m}(s).$$

This may be inverted to give the integral relation

$$\int_0^t \psi(t-\tau)\, \dot{m}(\tau)\, d\tau = h(t) \tag{3.140}$$

where $h(t)$ is the unit step function. This relation is entirely analogous to that relating relaxation and creep functions in the linear isothermal theory of viscoelasticity [Eq. (1.18)]. Accordingly, $\psi(t)$ is a type of creep function and is here referred to as the specific heat creep function. This function, is assumed to have the characteristics that it is a continuous monotone increasing function of t, $t \geqslant 0$, which asymptotically approaches a finite constant value for large values of t and obeys

$$\psi(t)\Big|_{t=0} = 0. \tag{3.141}$$

The restriction (3.141) is next shown to be that required to ensure that the speeds of propagation of thermal distrubances are always finite.[3] After that, the physical implications of (3.141) will be mentioned.

Consider now the propagation of plane waves of a thermal disturbance in an infinite medium, and take coordinate x to be in the direction of propagation. It is convenient to employ the Laplace transform technique to determine the maximum speed of propagation of the type of thermal disturbance governed by the heat conduction equation (3.139). With initial conditions of the type $\theta = 0$ for $t < 0$ and with $\rho r = 0$, the Laplace transformed equation (3.139) becomes

$$\bar{\theta} = \frac{k\bar{\psi}}{T_0}\frac{\partial^2\bar{\theta}}{\partial x^2}\,. \tag{3.142}$$

The solution of (3.142) corresponding to waves moving in the positive x direction is

$$\bar{\theta} = Ae^{-(\sqrt{T_0/k\bar{\psi}})x} \tag{3.143}$$

where A is some constant, with respect to x. The creep function $\psi(t)$ subject to the restriction (3.141) is now represented by a Taylor expansion, as

$$\psi(t) = t\psi'(0) + \tfrac{1}{2}t^2\psi''(0) + \cdots \tag{3.144}$$

[3] The restriction (3.141) implies that the corresponding function $m(t)$ obtained from (3.140) has a singularity at $t = 0$, as in a Kelvin–Voigt model. This type of behavior is more restrictive than that considered in the preceding section.

where $\psi'(0)$, $\psi''(0)$, etc., are the appropriate derivatives of $\psi(t)$ at $t = 0$. Taking the Laplace transform of (3.144), substituting it into the radical in (3.143), and using the binomial expansion gives

$$\sqrt{\frac{T_0}{k\bar{\psi}}} = \sqrt{\frac{T_0}{k\psi'(0)}}\left[s - \frac{\psi''(0)}{2\psi'(0)} + O\left(\frac{1}{s}\right)\right]. \tag{3.145}$$

Applying the shifting theorem and the initial value theorem of Laplace transform theory to the transformed solution (3.143) and (3.145) reveals that the solution has the form

$$\theta(x, t) = f(x, t)\, h\left(t - \sqrt{\frac{T_0}{k\psi'(0)}}\, x\right) \tag{3.146}$$

where $f(x, t)$ is some undetermined function of x and t and $h(\ )$ is the unit step function. Relation (3.146) displays the maximum speed of propagation of thermal disturbances c as

$$c = \sqrt{k\psi'(0)/T_0} \tag{3.147}$$

where k is the thermal conductivity, T_0 is the base temperature, and $\psi'(0)$ is the slope of the specific heat creep function at $t = 0$. Following a procedure identical with that just outlined but relaxing the requirement (3.141) so that $\psi(t)|_{t=0} = \psi_0$ is now nonzero constant, we find that the maximum speed of propagation in this case is infinitely large. Thus we have found the restriction (3.141) which must be imposed on the specific heat creep function to ensure finite speeds of propagation, and we have determined the maximum speed of propagation of thermal disturbances (3.147) governed by the theory (3.139) and (3.141).

It is instructive to consider the conceptual means by which one might determine the creep function $\psi(t)$ subject to (3.141) and also approaching a long-time asymptotic value $\psi(\infty)$. Take the case of the very rapid uniform deposit of energy in a matarial, as, for example, through the sudden x-ray heating of a thin film. If the mechanism of energy deposit is sufficiently rapid, the heat supply function ρr may be approximated by the Dirac delta function $\delta(t)$ as

$$\rho r = \hat{r}\delta(t) \tag{3.148}$$

where $\hat{r}$ is the total energy per unit volume deposited in the material at time $t = 0$. Specializing (3.139) to the uniform process indicated by (3.148) gives

$$\theta(t) = (\hat{r}/T_0)\psi(t). \tag{3.149}$$

Thus, in this idealized type of experiment, the temperature response to the sudden deposition of energy is proportional to the specific heat creep function, and this conceptually suggests the means of determining the specific heat creep function. The temperature response to the sudden transfer of energy from the source is not instantaneous; rather it is a transient or delayed type of response, consistent with the effects commonly encountered in the mechanical behavior of materials with memory. It is, however, expected that this transfer time, or the time duration of transient effects, is extremely short, which again is consistent with the fact that the molecular or atomic motions which characterize thermal energy are extremely high frequency states. With a specific heat creep function of this type, the heat conduction equation (3.139) would give results effectively the same as those predicted by the classical theory except, roughly speaking, during the time ranges in which extremely rapid heating occurs.

The classical problem of the sudden heating of a half space is chosen as a problem by which to illustrate the theory derived here. Specifically, the region of the half space is specified by $x \geqslant 0$. The boundary conditions are

$$\theta(x, t)\,|_{x=0} = \begin{cases} 0, & t < 0 \\ \Phi, & t \geqslant 0. \end{cases} \tag{3.150}$$

The Laplace transformed heat conduction equation (3.142) has a solution which satisfies the boundary condition (3.150),

$$\bar{\theta}(x, s) = (\Phi/s)\, e^{-[\sqrt{T_0/k\bar{\psi}(s)}]x}. \tag{3.151}$$

The specific heat creep function of the type considered here admits a general representation through the form

$$\psi(t) = \sum_{i=1}^{N} \Psi_i(1 - e^{-t/t_i}) \tag{3.152}$$

where ψ_i and t_i are quantities to be determined from the experimental determination of $\psi(t)$. The Laplace transform of (3.152) has the form

$$\bar{\psi}(s) = B(s)/A(s) \tag{3.153}$$

where $A(s)$ and $B(s)$ are polynomials in s of degree $N + 1$ and $N - 1$, respectively. The inclusion of (3.153) in (3.151) gives the expression which is to be inverted to obtain the solution. It is useful to note the similarity of this procedure to that involved in the study of wave propagation in linear viscoelastic theory. There, forms completely similar to the combination of (3.151) and (3.153) are obtained, but the inversion problem imposes severe difficulties because of the presence of the branch points in the transformed solution. The same situation prevails here with heat conduction. Exact analytical inversions of (3.151), and (3.153) quickly become impractical to obtain as N increases beyond 1 or 2 because of the increasing number of branch points. However, by way of further illustration, the case $N = 1$ will be pursued.

In this case, $N = 1$, the transformed solution is given by

$$\bar{\theta}(x, s) = (\Phi/s)\, e^{-(\sqrt{T_0 t_1 s(s+1/t_1)/k\psi_1})x}. \tag{3.154}$$

Equation (3.154) has exactly the same form as that encountered by Lee and Morrison [3.29] in studying wave propagation relative to a viscoelastic Maxwell model. Following their results, (3.154) can be inverted using standard techniques to obtain

$$\theta(x, t) = \Phi \left\{ e^{-\xi/2} + \frac{\xi}{2} \int_{\xi}^{\tau} e^{-\zeta/2} \frac{I_1[\frac{1}{2}(\zeta^2 - \xi^2)^{1/2}]}{(\zeta^2 - \xi^2)^{1/2}}\, d\zeta \right\} h(\tau - \xi) \tag{3.155}$$

where

$$\tau = t/t_1\,, \qquad \xi = \sqrt{T_0/t_1 k \Psi_1}$$

and where $I_1(\ )$ is the modified Bessel function of the first kind. The form (3.155) displays the finite speed of propagation characteristic which is not present in the classical solution of this problem. The nature of this problem indicates that many of the techniques and procedures which have been developed for the study of wave propagation in the linear theory of viscoelasticity may also have application in the study of heat conduction.

A few general remarks are in order concerning the relationship of solutions obtained by the present theory to those obtained from the classical theory. Of course, the two types of solutions are always different at or near a propagating wave front (representing a discontinuity in temperature), as predicted by the present theory. This is because the classical theory cannot account for such a wave front. However, it would be expected that the predictions of the classical theory would be adequate for sufficiently slowly changing processes, the proof of which is left as an exercise.

Finally, it is appropriate to observe the relationship of the specific heat creep function $\psi(t)$ to the specific heat function $m(t)$ involved in the thermodynamical derivation of Section 3.1. The restriction (3.30) on the derivative of $m(t)$ is incompatible with the behavior considered here for $\psi(t)$. Actually the form of $m(t)$ that corresponds to the type of $\psi(t)$ considered here would necessarily involve a delta function. This type of behavior is completely different from that considered in Section 3.1. The difference relates to the time scale of effects. The creep function $\psi(t)$ with $\psi(0) = 0$ has a range of characteristic times which are governed by the atomic scale vibrations inherent in the initial temperature response to energy input. On the other hand, much larger characteristic times are involved in the mechanical relaxation functions and in $m(t)$, as they relate to molecular scale effects for very high molecular weight materials. These two time scales of effects—atomic and molecular—are so far apart that there is no overlap or inconsistency between the two scales. For practical purposes, in the time range of most thermal mechanical observations, the specific heat creep function should be taken in accordance with the behavior in Section 3.1 and the effects of the instantaneous propagation of heat can be ignored. Nevertheless, the derivation just given shows the power and utility of the memory formalism for describing the thermal mechanical response of materials. In the present case the general theory of materials with memory served to resolve the instantaneous propagation paradox of classical heat conduction theory.

PROBLEMS

3.1. What modifications must be made to account for incompressible materials in the thermodynamical derivation of the linear theory in Section 3.1 ? Assume isothermal conditions and isotropic materials.

3.2. Following the procedure of Section 3.2 and using the dissipation inequality, derive the necessary restrictions upon the isotropic integrating functions in the free energy expression, where these restrictions follow from the consecutive application of two step function changes in strain or temperature.

3.3. Under steady state harmonic oscillation conditions, obtain the following expression for the energy dissipated per cycle:

$$W = \pi E''(\omega)\, \epsilon_0^2$$

where $E''(\omega)$ is the imaginary part of the appropriate complex modulus and ϵ_0 is the corresponding strain amplitude. Proceed by integrating the rate of doing work over a cycle.

3.4. Perform a thermostress analysis for a thermorheologically simple viscoelastic sphere, similar to that given in Section 3.6 for the slab. Assume a point symmetric temperature field which is specified throughout the sphere as $T(r)$. See Ref. [3.19].

3.5. Show that the postulate of thermorheologically simple behavior is equivalent to the specification that the relaxation times τ_i in the relaxation function representation

$$G(t) = G_0 + \sum_{i=1}^{N} G_i e^{-t/\tau_i}$$

have a temperature dependence given by the form

$$\tau_i = \hat{\tau}_i F(T)$$

where $\hat{\tau}_i$ are the relaxation times at some base temperature T_0 and $F(T)$ is a function of temperature, such that

$$F(T_0) = 1.$$

To show this interpretation, begin by taking $F(T)$ in the form

$$F(T) = (10)^{-f(T)}$$

where $f(T) = -\log F(T)$. What are the implications with respect to mechanical models of this interpretation of the thermorheologically simple postulate?

3.6. There are many practical reasons why the glass transition criterion of Section 3.7 may be satisfied only approximately. For example, T_g is known to depend upon molecular weight, and most polymeric systems contain a distribution of molecular weights. Formulate a method by which the creep functions for thermal expansion and mechanical compressibility can be compared at different base temperatures to find that temperature which comes the closest to the definition of T_g.

3.7. Using the concept of retarded processes from Section 1.7 show that behavior according to the classical theory of heat conduction results for the general theory of Section 3.8 when the process is sufficiently slow.

3.8. Using a Maxwell model, verify the correctness of the general form

$$U = \frac{1}{2}\int_{-\infty}^{t}\int_{-\infty}^{t} E(2t - \tau - \eta)\frac{d\epsilon(\tau)}{d\tau}\frac{d\epsilon(\eta)}{d\eta}\,d\tau\,d\eta$$

to predict the energy stored in the spring.

3.9. With regard to complex moduli, it is often claimed that the experimentally observed behaviors $G'(\omega) \geqslant 0$, $G''(\omega) \geqslant 0$, and $dG'(\omega)/d\omega \geqslant 0$ are imposed by the requirements of nonnegative stored energy, nonnegative rate of dissipation of energy, and fading memory, respectively. Attempt to prove these conjectures. (See Christensen [3.30].)

REFERENCES

3.1. Christensen, R. M., and P. M. Naghdi, "Linear Non-Isothermal Viscoelastic Solids," *Acta Mechanica* **3**, 1 (1967).
3.2. Biot, M. A., "Linear Thermodynamics and the Mechanics of Solids," *Proc. 3rd U.S. Nat. Congr. Appl. Mech.* 1 (1958).
3.3. Eringen, A. C., "Irreversible Thermodynamics and Continuum Mechanics," *Phys. Rev.* **117**, 1174 (1960).
3.4. Hunter, S. C., "Tentative Equations for the Propagation of Stress, Strain, and Temperature Fields in Viscœlastic Solids," *J. Mech. Phys. Solids* **9**, 39 (1961).
3.5. Schapery, R. A., "Application of Thermodynamics to Thermomechanical Fracture and Birefringent Phenomena in Viscoelastic Media," *J. Appl. Phys.* **35**, 1451 (1964).

3.6. Truesdell, C., and R. A. Toupin, *in Handbuch der Physik* (S. Flügge, ed.) Vol. 3, No. 1. Springer, Berlin, 1960.

3.7. Breuer, S., and E. T. Onat, "On the Determination of Free Energy in Linear Viscoelastic Solids," *Z. Angew. Math. Phys.* **15**, 184 (1964).

3.8. Gurtin, M. E., and I. Herrera, "On Dissipation Inequalities and Linear Viscoelasticity," *Quart. Appl. Math.* **23**, 235 (1965).

3.9. Kovacs, A. J., "La Contraction Isotherme du Volume des Polymères Amorphes," *J. Polym. Sci.* **30**, 131 (1958).

3.10. Leaderman, H., "Elastic and Creep Properties of Filamentous Materials and Other High Polymers," p. 175. Textile Foundation, Washington, D.C., 1943.

3.11. Ferry, J. D., "Mechanical Properties of Substances of High Molecular Weight; VI. Dispersion in Concentrated Polymer Solutions and Its Dependence on Temperature and Concentration," *J. Amer. Chem. Soc.* **72**, 3746 (1950).

3.12. Schwarzl, F., and A. J. Staverman, "Time-Temperature Dependence of Linear Viscoelastic Behavior," *J. Appl. Phys.* **23**, 838 (1952).

3.13. Ferry, J. D., *Viscoelastic Properties of Polymers*, 3rd ed. Wiley, New York, 1980.

3.14. Crochet, M. J., and P. M. Naghdi, "A Class of Simple Solids with Fading Memory," *Int. J. Eng. Sci.* **7**, 1173 (1969).

3.15. Crochet, M. J., and P. M. Naghdi, "On 'Thermorheologically Simple' Solids," *Proc. IUTAM Symp. Thermoelasticity*, p. 59. Springer, New York, 1970.

3.16. Morland, L. W., and E. H. Lee, "Stress Analysis for Linear Viscoelastic Materials with Temperature Variation," *Trans. Soc. Rheol.* **4**, 233 (1960).

3.17. Crochet, M. J., and P. M. Naghdi, "On a Restricted Non-Isothermal Theory of Simple Materials," *J. Mécanique* **13**, 97 (1974).

3.18. Crochet, M. J., and P. M. Naghdi, "On Thermomechanics of Polymers in the Transition and Rubber Regions," *J. Rheol.* **22**, 73 (1978).

3.19. Muki, R., and E. Sternberg, "On Transient Thermal Stresses in Viscoelastic Materials with Temperature Dependent Properties," *J. Appl. Mech.* **28**, 193 (1961).

3.20. Lee, E. H., and T. G. Rogers, "Solution of Viscoelastic Stress Analysis Problems Using Measured Creep or Relaxation Functions," *J. Appl. Mech.* **30**, 127 (1963).

3.21. Sternberg, E., "On the Analysis of Thermal Stresses in Viscoelastic Solids," *Proc. 3th Symp. Nav. Structural Mech.* 348. Macmillan, New York, 1964.

3.22. Lockett, F. J., and L. W. Morland, "Thermal Stresses in a Viscoelastic, Thin-Walled Tube with Temperature Dependent Properties," *Int. J. Eng. Sci.* **12**, 879 (1967).

3.23. Ehrenfest, P., *Leiden Comm. Suppl.*, 1933, 75b (reprinted in: P. Ehrenfest,*Collected Scientific Papers*. North–Holland, Amsterdam, 1959).

3.24. Christensen, R. M., "A Thermodynamical Criterion for the Glass-Transition Temperature," *Trans. Soc. Rheol.* **21**, 163 (1977).

3.25. Goldbach, G., and G. Rehage, "Die Volumenretardation des Polystyrols nach Druck- und Temperatursprüngen," *Rheol. Acta* **6**, 30 (1967).

3.26. Goldbach, G., and G. Rehage, "Untersuchungen über die lineare Volumennachwirkung bei amorphen Hochpolymeren," *J. Polym. Sci. C* **16**, 2289 (1967).

3.27. Gurtin, M. E., and A. C. Pipkin, "A General Theory of Heat Conduction with Finite Wave Speeds," *Arch. Rational Mech. Anal.* **31**, 113 (1968).

3.28. Bogy, D. B., and P. M. Naghdi, "On Heat Conduction and Wave Propagation in Rigid Solids," *J. Math. Phys.* **11**, 917 (1970).

3.29. Lee, E. H., and J. A. Morrison, "A Comparison of the Propagation of Longitudinal Waves in Rods of Viscoelastic Materials," *J. Polymer Sci.* **19**, 93 (1956).

3.30. Christensen, R. M., "Restrictions upon Viscoelastic Relaxation Functions and Complex Moduli," *Trans. Soc. Rheol.* **16**, 603 (1972).

Chapter IV Mechanical Properties and Approximate Transform Inversion

4.1. INTRODUCTION

This chapter is primarily concerned with the experimental means of determining the mechanical properties which enter the constitutive relations of viscoelasticity theory. To attempt a complete and general treatment of this subject would require a very extensive development, which is beyond the scope intended here. Rather, our purpose here is to merely examine some typical and representative procedures for determining mechanical properties.

The next four sections focus attention upon the mechanical properties of the linear theory, the nonlinear theory is deferred until later. Of the four sections on the linear theory, the first three are restricted to mechanical properties appropriate to the isothermal theory, while the remaining section relaxes this restriction to include some nonisothermal effects. Also, all developments in this chapter are appropriate only to isotropic materials. However, most of the generalizations of isotropic material conditions to those of anisotropic materials are fairly obvious in the present context. The difficulty of determining the mechanical properties of anisotropic materials is mainly that of the actual amount of work involved in the experimental procedures.

With regard to the linear isothermal theory, there are many different but equivalent means of expressing the mechanical properties which enter the stress strain constitutive relations. The most fundamental descriptions of mechanical properties are probably those given by relaxation function, creep function, and complex modulus formulations. Some minor variations of these are the complex compliance and the complex viscosity formulations. All of these can be determined by direct experimental observation. There are even other means of specifying mechanical properties, which are of interest even though they cannot be determined by direct experimental observation. These include relaxation and creep spectra. Our attention first will be directed to the relaxation function, creep function, and complex modulus formulations. The interrelationships between these three types of functions are given by relations (1.18), (1.58), (1.59), (1.68), and (1.69), which are rewritten here for convenience, as

$$\int_0^\infty J_\alpha(t-\tau)\frac{dG_\alpha(\tau)}{d\tau}\,d\tau = h(t), \tag{4.1}$$

$$G_\alpha'(\omega) = \mathring{G}_\alpha + \omega \int_0^\infty \hat{G}_\alpha(\eta) \sin \omega\eta \, d\eta, \tag{4.2}$$

$$G_\alpha''(\omega) = \omega \int_0^\infty \hat{G}_\alpha(\eta) \cos \omega\eta \, d\eta, \tag{4.3}$$

$$\hat{G}_\alpha(t) = \frac{2}{\pi} \int_0^\infty \frac{G_\alpha'(\omega) - \mathring{G}_\alpha}{\omega} \sin \omega t \, d\omega \tag{4.4}$$

and

$$\hat{G}_\alpha(t) = \frac{2}{\pi} \int_0^\infty \frac{G_\alpha''(\omega)}{\omega} \cos \omega t \, d\omega \tag{4.5}$$

where the complex modulus is decomposed into real and imaginary parts through

$$G_\alpha^*(i\omega) = G_\alpha'(\omega) + iG_\alpha''(\omega)$$

and where the relaxation function is decomposed into two parts, as

$$G_\alpha(t) = \mathring{G}_\alpha + \hat{G}_\alpha(t)$$

with

$$\hat{G}_\alpha(t) \to 0 \qquad \text{as} \qquad t \to \infty.$$

The separate cases of $\alpha = 1, 2$ designate mechanical properties appropriate to states of shear deformation and dilatation, respectively. Direct interrelationships between creep functions $J_\alpha(t)$ and the real and imaginary parts of complex moduli may also be obtained. Even though a relaxation function or a creep function over an infinite time range is equivalent to specifying the corresponding complex modulus over an infinite frequency range, the fact remains that these quantities cannot in general be determined over the complete time or frequency spectrum. Accordingly, knowledge of a relaxation function over a partial time range, is not rigorously equivalent to knowledge of a complex modulus over any finite frequency range. For this reason, it may be advisable to directly determine the particular type of mechanical property which is best suited for its intended use. For example, if mechanical properties are needed for use in analyses concerned with a long-time nearly constant load type of problem, obviously it is the creep function which should be determined for use in the analysis. On the other hand, there are circumstances in which the optimal way of proceeding involves the experimental determination of one type of mechanical property, and the conversion, through approximate means, to another type of property for use in a theoretical application. Situations of this type involve the consideration of the interrelations between the time or frequency range of capability of particular experiments for the particular material of interest, and the time or frequency range in which the mechanical properties are needed for use in the analysis. An example of this circumstance is given in Section 4.3. In discussing these typical procedures for determining mechanical properties, we shall attempt to give at least broad guide lines by which the degree and range of usefulness each particular procedure can be judged.

Before proceeding to the discussion of these experimental procedures, it is useful to consider the general characteristics which should be sought in devising a well-founded and reliable method of determining mechanical properties. The typical testing procedure involves an auxillary analysis which is used to relate an experimentally observable quantity to the mechanical property of interest. In the ideal procedure this analysis must yield the exact solution of the field equations of the theory, and it must represent the exact boundary conditions of the experimental specimen. Though this objective may seem obvious, it does not follow that it is easy to attain. In fact, of the following standard procedures to be discussed, only that of Section 4.3 conforms to the above ideal situation. The other procedures either involve an approximate satisfaction of the field equations of the theory or the use of boundary conditions in the analysis which are an approximation to those actually experienced by the experimental specimen. The mention of these matters here should not be construed as an inclusive criticism of most experimental methods. Rather, the intention here is to observe that the theoretical basis of an experiment should be a significant item of consideration, and for reasons of this type some experiments are better conceived than others. With this note of caution in mind, we will discuss some pertinent experimental methods.

We should mention that in the following descriptions of the various testing procedures, we shall be concerned only with the general method and the supporting theoretical basis of the test. Matters of instrumentation techniques and testing apparatus are not explicitly described. For these particulars, reference should be made to the publications cited or to some survey articles on the subject, such as Kolsky [4.1].

After discussing the means of determining mechanical properties, we shall consider some approximate means for interconverting these forms. The integral transform relationships (4.1)–(4.5), of course, are easy to apply only for rather simple analytical forms. We shall find some simple approximations that have utility in practical applications. Specifically, we consider the interrelationships among complex moduli, relaxation and creep functions, and relaxation and creep spectra. Thus knowledge of one of these forms gives quick information upon the other forms. Having considered approximate means of interconverting properties, it is logical to examine the related problem of approximately inverting integral transforms to obtain approximate solutions of boundary value problems. This problem is considered in application of Laplace transform. Fourier transform is amenable to direct numerical evaluation. As a last matter, two approximate means are given for solving steady state, dynamic problems.

4.2. RELAXATION AND CREEP PROCEDURES

Relaxation and creep tests provide the simplest and most direct means of obtaining the mechanical properties involved in the linear theory of viscoelasticity. Such tests are usually applied to uniaxial deformation conditions whereby the appropriate constitutive relations are given by

$$\sigma(t) = \int_{-\infty}^{t} E(t-\tau) \frac{\partial \epsilon(\tau)}{\partial \tau}\, d\tau \tag{4.6}$$

$$\epsilon(t) = \int_{-\infty}^{t} J(t-\tau) \frac{\partial \sigma(\tau)}{\partial \tau}\, d\tau \tag{4.7}$$

where $E(t)$ and $J(t)$ are the respective uniaxial relaxation and creep functions.

For the relaxation test the strain state $\epsilon(t)$ is specified as

$$\epsilon(t) = \epsilon_0 h(t) \tag{4.8}$$

where $h(t)$ is the unit step function, and ϵ_0 is the strain amplitude. The resulting relaxation function is obtained from (4.6) and (4.8) as

$$\begin{aligned} E(t) &= 0, & t &< 0 \\ E(t) &= \sigma(t)/\epsilon_0\,, & t &\geqslant 0 \end{aligned} \tag{4.9}$$

where $\sigma(t)$ is the experimentally observed stress state. Correspondingly, the conditions of the creep test are specified by

$$\sigma(t) = \sigma_0 h(t) \tag{4.10}$$

with σ_0 being the stress amplitude. The resulting creep function is obtained from (4.7) and (4.10) as

$$\begin{aligned} J(t) &= 0, & t &< 0 \\ J(t) &= \epsilon(t)/\sigma_0\,, & t &\geqslant 0 \end{aligned} \tag{4.11}$$

where $\epsilon(t)$ is the experimentally observed strain state.

Relaxation and creep tests can be performed on specimens subject to conditions other than those of uniaxial deformation. However, the uniaxial deformation test is the one normally used, since its execution involves standard procedures and testing machines. Of course, uniaxial test procedures such as these only determine one of the two independent mechanical properties for isotropic materials. The means of determining a second mechanical property will be considered in Section 4.4.

Although creep and relaxation tests have the advantage of simplicity, there are also some associated shortcomings. First, uniaxial creep and relaxation tests procedures assume the stress to be uniformly distributed through the specimen, with the lateral surfaces being free to expand and contract. This condition cannot be satisfied at the ends of a specimen which uses the usual methods of attachment to a testing machine. Of course this effect becomes entirely negligible as the specimen is made to approach the geometry of filaments. A second, more serious, difficulty involves the dynamic effects which are encountered in obtaining data at short times. The relaxation and creep functions which are determined through (4.9) and (4.11) assume the neglect of all transients excited through the dynamic response of the specimen and testing machine. Typically this effect limits relaxation and creep data to times no less than about 0.1–0.01 sec. Mechanical property data in the time range shorter than this must be obtained by some procedure other than the unusual quasi-static creep or relaxation test. Such a procedure for obtaining short time data will be considered in the next section.

There is no limitation, other than patience, on the maximum time at which relaxation or creep data can be obtained. One last condition that should be mentioned is that of the superposition check. As a standard testing procedure, when the prescribed deformation in a relaxation test, or stress in a creep test, is varied by a certain amount, the response must vary in the same proportion. Otherwise the mechanical properties so determined would not be appropriate to the linear theory.

As an example, the data from a relaxation test on polyisobutylene are given in Fig. 4.1 at several different temperatures. As suggested by the data of Fig. 4.1,

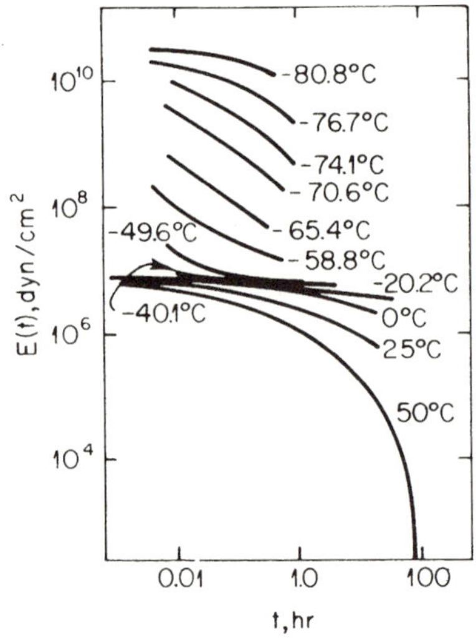

Fig. 4.1. Uniaxial relaxation function of polyisobutylene (after Tobolsky and Catsiff [4.2.]).

this particular polymeric material is not cross-linked and accordingly it is a viscoelastic fluid.

4.3. STEADY STATE HARMONIC OSCILLATION PROCEDURES

In Section 1.6 and the introduction to this chapter we discussed the representation of mechanical properties in forms appropriate to steady state harmonic oscillation conditions. We now consider the means of determining mechanical properties in this form.

As discussed in the preceding section, conventional quasi-static creep and relaxation tests cannot be used to obtain mechanical property data in short time ranges because of complicating dynamic effects. It is theoretically possible to obtain mechanical property data from these types of tests in very short time ranges if an appropriate dynamic analysis is used in the data reduction. Such procedures belong more in the realm of wave propagation effects, and one such method will be considered in the next section. First, however, this section is concerned with relatively simple means of obtaining fairly high frequency mechanical property data and correspondingly fairly short-time data. Specifically, we here describe the means of determining mechanical properties from steady state harmonic oscillation tests, a procedure by which the inclusion of inertia term effects is not an unduly complicating factor. The boundary value problem which is used here was analyzed in Section 2.7. For convenience, we repeat here the appropriate analytical forms.

It is the complex shear modulus $\mu^*(i\omega)$ which the following procedure determines. Consider the torsional oscillation of a right circular cylinder. The relevant equation of motion for an elastic cylinder described by polar cylindrical coordinates r, θ, and z, is

$$\frac{\partial^2 u_\theta}{\partial r^2} + \frac{1}{r}\frac{\partial u_\theta}{\partial r} + \frac{\partial^2 u_\theta}{\partial z^2} - \frac{u_\theta}{r^2} = \frac{\rho}{\mu}\frac{\partial^2 u_\theta}{\partial t^2} \tag{4.12}$$

where $u_\theta = u_\theta(r, z, t)$, and ρ is the mass density. Letting

$$u_\theta = r\Phi(z, t)$$

and

$$\Phi(z, t) = f(z)\, e^{i\omega t} \tag{4.13}$$

where ω is the frequency of harmonic oscillation, then by combining (4.12) and (4.13), $f(z)$ can readily be determined as

$$f(z) = A \sin(\Omega z/h) + B \cos(\Omega z/h) \tag{4.14}$$

where

$$\Omega^2 = \rho\omega^2 h^2/\mu \tag{4.15}$$

and A and B are constants to be determined from the boundary conditions, and

h is the length of the cylinder. The only nonzero component of stress is $\sigma_{\theta z}$, which may be integrated over the cross section to obtain the total twisting moment M as

$$M(z, t) = (\pi/2)(b^4 - a^4)(\rho\omega^2 h/\Omega)[A \cos(\Omega z/h) - B \sin(\Omega z/h)]\, e^{i\omega t} \tag{4.16}$$

where a and b are the inner and outer radii of the cylinder, respectively.

Taking the end $z = h$ as fixed with no displacement while the end $z = 0$ has the applied torque $\hat{M}e^{i\omega t}$, then it is found that the ratio of the twisting moment on the end $z = 0$ to the angle of twist is given by

$$\frac{M(0, t)}{\Phi(0, t)} = \frac{\hat{M}}{f(0)} = \frac{\pi(b^4 - a^4)}{2}\,\rho\omega^2 h\,\frac{\cot\Omega}{\Omega}\,. \tag{4.17}$$

If inertia terms can be neglected the relationship between $M(0, t)$ and $\Phi(0, t)$ is given by the simple form

$$\frac{M(0, t)}{\Phi(0, t)} = \frac{\pi(b^4 - a^4)\mu}{2h}\,. \tag{4.18}$$

Following the method of Section 2.4, the elastic solution (4.17) can be converted to a viscoelastic solution by replacing μ in Ω by $\mu^*(i\omega)$. Experimental measurements of $f(0)$, the amplitude of the angle of twist at $z = 0$, for a specific value of $\hat{M}$ then allows the determination of $\mu^*(i\omega)$ from (4.17) or (4.18). Note that for $\hat{M}$ considered to be real, then $f(0)$ is a complex number which is determined by the magnitude of $\Phi(0, t)$ and the phase angle by which it lags $M(0, t)$.

The experimental results from such a procedure are reported by Gottenberg and Christensen [4.3]. The resulting real and imaginary parts, $\mu'(\omega)$ and $\mu''(\omega)$, of the complex modulus, $\mu^*(i\omega)$, are shown in Figs. 4.2 and 4.3 at different base temperatures for a material having a polyurethane matrix. These data may be converted to a corresponding relaxation function $\mu(t)$ through the use of (4.4). The resulting relaxation function is shown in Fig. 4.4 and is again from Ref. [4.3]. It should be noted that (4.4) involves $G_\alpha'(\omega)$ or correspondingly $\mu'(\omega)$ over the entire frequency domain, whereas the data given in Fig. 4.2 are over about seven decades of frequency. It is nevertheless possible to obtain the relaxation function over a limited time range from complex modulus data over a limited frequency range. This is because the relaxation function at a particular value of time is not equally dependent upon the complex modulus at all values of frequency, but rather has a weighted dependence through (4.4). The means of establishing the convergence of the integral in (4.4) for the data used herein is examined in detail in Ref. [4.3].

The relaxation function obtained in the manner just described provides data for times about three orders of magnitude shorter than those obtainable through direct quasi-static relaxation tests.

This test procedure involving the torsional oscillation of a right circular cylinder is the case mentioned in the introduction to this chapter for which an exact analysis is used along with an exact representation of the experimental boundary conditions. It might therefore seem possible to obtain mechanical property data by this method for as high a frequency, or as short a time as desired. This is, however, not the case. A limitation on high-frequency and short-time data obtainable by this method is provided by unavoidable dynamic resonances in the testing equipment used to support the specimen. In the particular test considered in Ref. [4.3], this upper frequency limitation is about 10^3 Hz. To obtain mechanical property data in the range beyond the capability of this type of test it is necessary to utilize wave propagation type tests, as are considered in the next section.

The testing procedure just described, utilizing the torsional oscillation of a right circular cylinder, is more useful for testing solids than fluids. The types of testing procedures employed for viscoelastic fluids usually involve coaxial rotating cylinders with the fluid in the annulus or a coaxial rotating cone and plate

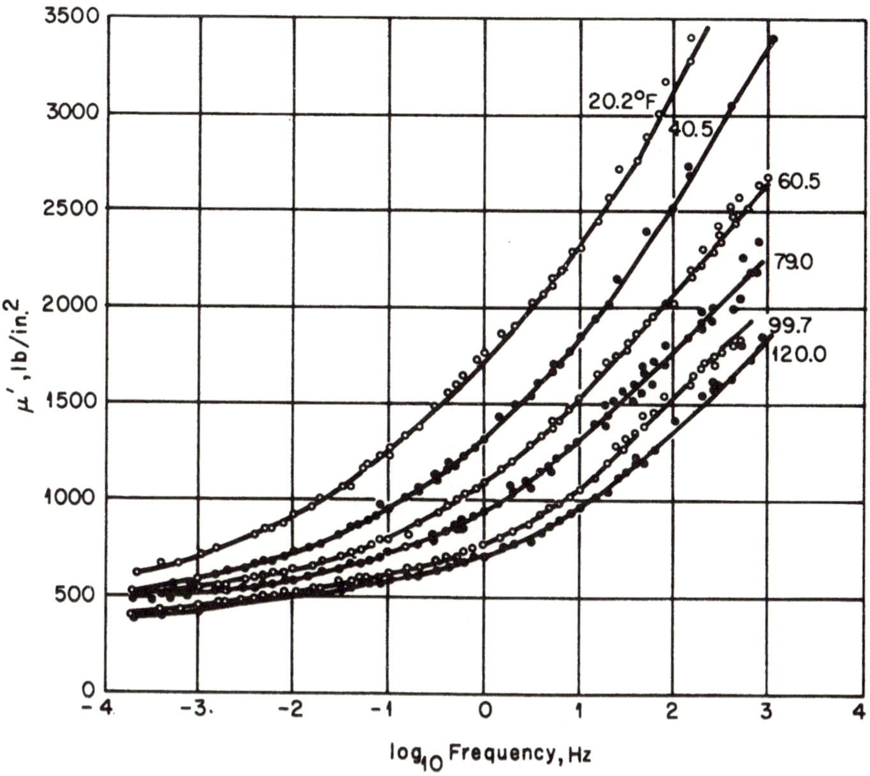

Fig. 4.2. Real part of complex shear modulus, polyurethane matrix containing salt crystals and aluminum powder (after Gottenberg and Christensen [4.3]).

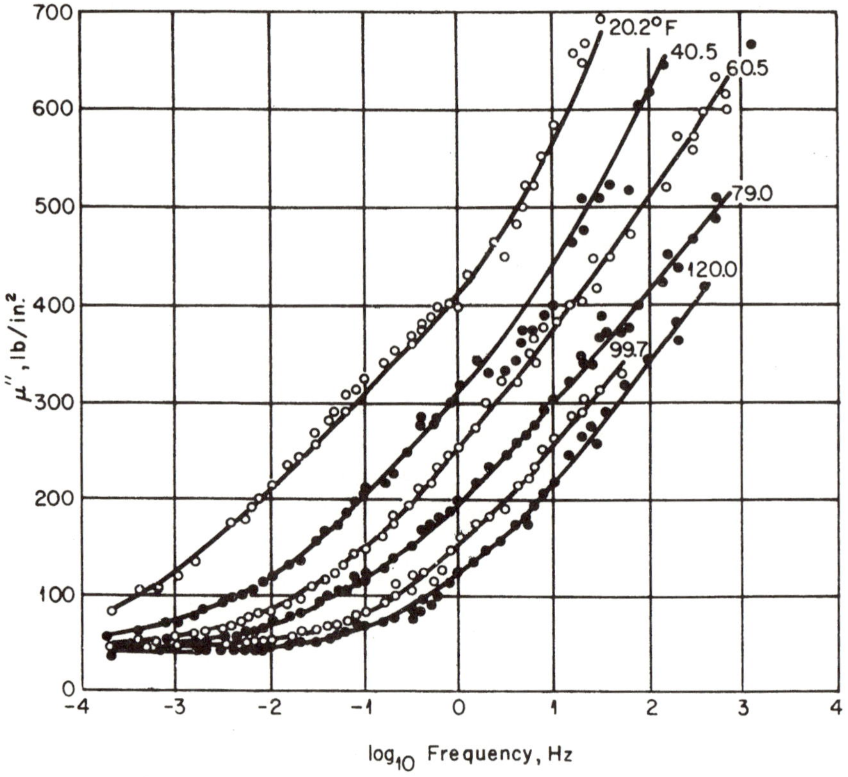

Fig. 4.3. Imaginary part of complex shear modulus, polyurethane matrix containing salt crystals and aluminum powder (after Gottenberg and Christensen [4.3]).

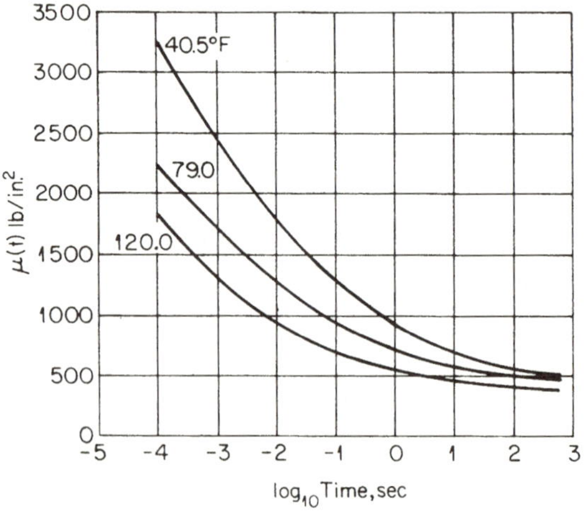

Fig. 4.4. Relaxation function in shear, polyurethane matrix containing salt crystals and aluminum powder (after Gottenberg and Christensen [4.3]).

device with the fluid layer in the spherical wedge between the cone and plate. Of course, these devices may be used to obtain the nonlinear flow properties of the fluid as well as the linear theory mechanical properties, as shown in Section 9.3.

In the preceding section, we discussed certain means of obtaining mechanical properties data appropriate to uniaxial states of deformation. In this section, the mechanical property determined is that corresponding to a state of shear deformation. It therefore follows that for such tests applied to a single isotropic material, the two independent mechanical property functions would thereby be obtained. While this may be a practical reality for some materials, it is a deceptive state of affairs for certain polymeric materials. This is the case if the mechanical properties are to be characterized in the time or frequency range and at the temperature at which they are nearby incompressible. This difficulty may easily be seen even for an almost incompressible elastic material; to obtain a bulk modulus with a given degree of accuracy from a shear modulus and a Young's modulus, these latter two moduli must be specified with much greater accuracy than that allowed for the bulk modulus. To obtain a second independent mechanical property for materials which are almost incompressible under the afore-mentioned conditions, it is necessary to formulate a test involving primarily volumetric change effects. Such a test will be considered in the next section.

Free vibration tests can also be used to determine complex moduli. Such a procedure was considered in Section 2.11 and Problem 2.7. The nature of the data reduction involves certain approximations in this method, and in general, the frequency range in which mechanical property data can be obtained by observing the free vibration of a single specimen is severely limited.

4.4. WAVE PROPAGATION PROCEDURES

Observations from wave propagation type tests can be used to determine viscoelastic mechanical properties. The procedures and techniques which are involved in wave propagation testing are necessarily much more involved than those of the methods described in the preceding sections. However, wave propagation tests offer the means of obtaining data in the high frequency range which is not attainable in the previously mentioned types of tests.

Wave propagation methods can be divided into two general classifications. First, there is the type involving the propagation of single frequency harmonic waves, and second, the type involving more general transient disturbances. We will consider the first type briefly now, and after that we give a detailed study of the second type.

We consider the propagation of disturbances along a filament and accordingly it is the uniaxial complex modulus $E^*(i\omega)$ which is the involved mechanical property. If the filament is very long and one end is subjected to a steady state harmonic oscillation condition, then harmonic waves are propagated along the filament with an attenuation in the direction of propagation. The equation of motion

$$\frac{\partial^2 \bar{u}}{\partial x^2} = \frac{-\rho\omega^2 \bar{u}}{E^*(i\omega)} \tag{4.19}$$

where $\bar{u}(x, \omega)$ is the Fourier transformed displacement, has a solution in the form

$$u(x, t) = Be^{-\alpha x} e^{i\omega[t-(x/c)]} \tag{4.20}$$

where

$$c = | E^* | / \rho^{1/2} E_3 \tag{4.21}$$

and

$$\alpha = \rho^{1/2} \omega E_4 / | E^* | \tag{4.22}$$

with

$$[E^*(i\omega)]^{1/2} = E_3(\omega) + iE_4(\omega). \tag{4.23}$$

The solution (4.20) represents the propagation of a harmonic wave moving in the positive x direction, c and α are the phase velocity and attenuation, respectively, and the complex constant B is to be determined from boundary conditions. Experimental measurements of the phase velocity and attenuation then allow the determination of the real and imaginary parts of the complex modulus $E^*(i\omega)$, at the frequency ω of the test. This method has been widely used to determine high frequency properties, not only with regard to $E^*(i\omega)$ but for other stress states as well. A variation of the method involves the measurement of the characteristics of a wave reflected from the interface between a material of known properties and the viscoelastic material. Typically, the harmonic waves are generated by piezoelectric crystals. Sutherland and Lingle [4.4] and Hunston *et al.* [4.5] provide typical procedures. Ferry [4.6] compiles an extensive list of references of such procedures. The main requirements on the use of the method are that only a single frequency harmonic wave be involved with no unaccounted reflection effects occurring. The single frequency requirement can be relaxed by considering the more general features of the propagation of transient pulses. This situation will now be examined.

Again the propagation of disturbances along a filament is considered. We now are concerned with the propagation of a transient disturbance along the filament, and we use the Fourier integral to synthesize the velocity from the solution given in (4.20). The resulting velocity is then given by

$$\dot{u}(x, t) = \int_{-\infty}^{\infty} F(\omega)\, e^{-\alpha x} e^{i\omega[t-(x/c)]}\, d\omega \tag{4.24}$$

where $F(\omega)$ is a complex function of frequency to be determined from specified end conditions. The phase velocity c and attenuation α in (4.24) are given by (4.21) and (4.22) and through some algebraic reductions, they may be written in the alternate forms

$$c = (| E^* |^{1/2}/\rho^{1/2}) \sec(\delta/2) \tag{4.25}$$

and

$$\alpha = c^{-1}[\omega \tan(\delta/2)] \tag{4.26}$$

where

$$\tan \delta = E''/E' \tag{4.27}$$

and

$$E^*(i\omega) = E'(\omega) + iE''(\omega). \tag{4.28}$$

A reasonable program for determining $E^*(i\omega)$ would involve measuring the velocity at two locations along the filament, then (4.24) could be used to determine $c(\omega)$ and $\alpha(\omega)$ to give the proper dispersion and attenuation of a pulse as it passes from one point of measurement to the other. Specifically, let

$$\dot{u}(x, t) \,|_{x=x_1} = v_1(t) \tag{4.29}$$

and

$$\dot{u}(x, t) \,|_{x=x_2} = v_2(t). \tag{4.30}$$

From (4.24), (4.29) and the Fourier transform inversion formula it follows that

$$F(\omega) = e^{[\alpha+(i\omega/c)]x_1}\bar{v}_1(\omega) \tag{4.31}$$

where

$$\bar{v}_1(\omega) = \frac{1}{2\pi} \int_{-\infty}^{\infty} v_1(t)\, e^{-i\omega t}\, dt. \tag{4.32}$$

Now substitute (4.31) into (4.24), and evaluate the result at $x = x_2$, using (4.30). This procedures gives

$$\int_{-\infty}^{\infty} \bar{v}_1(\omega)\, e^{-\alpha(x_2-x_1)} e^{i\omega[t-(x_2-x_1)/c]}\, d\omega = v_2(t). \tag{4.33}$$

In principle, (4.33) may be used to evaluate $c(\omega)$ and $\alpha(\omega)$ and thereby $E^*(i\omega)$. As a practical matter, however, this is very difficult to accomplish and some simplifying assumptions are needed to render a practical data reduction procedure.

The first simplifying assumption which we shall use is that the duration of the disturbing pulse is sufficiently short that it can be represented by a Fourier series rather than the Fourier integral. This would be expected to be a well-justified procedure in cases where very short duration pulses are initiated by a detonation, and at times before which the dispersion effects have broadened the pulse too much. We shall therefore take a basic time interval which includes the main part of the pulse, and represent the pulse in this interval by a Fourier series. The Fourier series solution corresponding to the Fourier integral solution (4.24) has the form

$$\dot{u}(x, t) = \sum_{n=0}^{\infty} e^{-\alpha_n x}[A_n \cos(npx/c_n) - B_n \sin(npx/c_n)] \cos npt$$

$$+ \sum_{n=1}^{\infty} e^{-\alpha_n x}[A_n \sin(npx/c_n) + B_n \cos(npx/c_n)] \sin npt \qquad (4.34)$$

where A_n and B_n are unknown constants, p is the preselected basic frequency, and c_n and α_n are given by (4.25) and (4.26) respectively for $\omega = np$. In obtaining (4.34) from (4.24), the exponential term in (4.24) has been written in terms of real and imaginary parts, and the formulas for the sine and cosine of the difference of two angles have been used.

For reasons that will be made clear presently it is convenient to associate the $t - (x/c)$ argument in (4.24) in the following form:

$$t - \frac{x}{c} = \left(t - \frac{x}{v}\right) + \left(-\frac{x}{c} + \frac{x}{v}\right).$$

Using this form in (4.34) we may replace the t argument by $t - (x/v)$ and the x/c argument by $(x/c) - (x/v)$. This procedure gives

$$\dot{u}(x, t) = \sum_{n=0}^{\infty} C_n \cos npt' + \sum_{n=1}^{\infty} D_n \sin npt' \qquad (4.35)$$

where

$$t' = t - (x/v), \qquad (4.36)$$

$$C_n = e^{-\alpha_n x}[A_n \cos(npx(c_n^{-1} - v^{-1})) - B_n \sin(npx(c_n^{-1} - v^{-1}))], \qquad (4.37)$$

and

$$D_n = e^{-\alpha_n x}[A_n \sin(npx(c_n^{-1} - v^{-1})) + B_n \cos(npx(c_n^{-1} - v^{-1}))]. \qquad (4.38)$$

The utility of introducing the variable $t' = t - (x/v)$ is now apparent. By properly selecting the velocity v, the pulse will be maintained within the basic interval of the Fourier series representation. In other words the velocity v should be taken to be near an average value of the velocity of propagation of the pulse.

As in the previous Fourier integral representation, we suppose the velocity to be measured at two locations, x_1 and x_2, with the specifications (4.29) and (4.30). We evaluate (4.35) at $x = x_1$, using (4.29), and in the usual way the coefficients are given by

$$C_{n1} = \frac{p}{\pi} \int_{-\pi/p}^{\pi/p} v_1(t') \cos npt' \, dt', \qquad n = 1, 2,... \tag{4.39}$$

$$D_{n1} = \frac{p}{\pi} \int_{-\pi/p}^{\pi/p} v_1(t') \sin npt' \, dt', \qquad n = 1, 2,... . \tag{4.40}$$

The coefficients A_n and B_n are obtained from (4.37) and (4.38) as

$$A_n = e^{\alpha_n x_1}[C_{n1} \cos(npx_1(c_n^{-1} - v^{-1})) + D_{n1} \sin(npx_1(c_n^{-1} - v^{-1}))] \tag{4.41}$$

and

$$B_n = e^{\alpha_n x_1}[-C_{n1} \sin(npx_1(c_n^{-1} - v^{-1})) + D_{n1} \cos(npx_1(c_n^{-1} - v^{-1}))]. \tag{4.42}$$

Thus, relations (4.35)–(4.38) specify the response with A_n and B_n given by (4.39)–(4.42).

At $x = x_2$, we use (4.30) to synthesize the pulse into Fourier series

$$\dot{u}(x, t) |_{x=x_2} = \sum_{n=0}^{\infty} P_n \cos npt' + \sum_{n=1}^{\infty} Q_n \sin npt' \tag{4.43}$$

where

$$P_n = \frac{p}{\pi} \int_{-\pi/p}^{\pi/p} v_2(t') \cos npt' \, dt', \qquad n = 1, 2,... \tag{4.44}$$

and

$$Q_n = \frac{p}{\pi} \int_{-\pi/p}^{\pi/p} v_2(t') \sin npt' \, dt', \qquad n = 1, 2,... . \tag{4.45}$$

We now evaluate the solution (4.35) at $x = x_2$ and equate the result to (4.43). Using (4.37)–(4.42), (4.44), and (4.45), the term by term equivalence of these two forms gives the two general relations

$$\begin{aligned}
&[\cos(npx_2(c_n^{-1} - v^{-1})) \sin(npx_1(c_n^{-1} - v^{-1})) \\
&\quad - \sin(npx_2(c_n^{-1} - v^{-1})) \cos(npx_1(c_n^{-1} - v^{-1}))] \int_{-\pi/p}^{\pi/p} v_1(t') \sin npt' \, dt' \\
&\quad + [\cos(npx_2(c_n^{-1} - v^{-1})) \cos(npx_1(c_n^{-1} - v^{-1})) \\
&\quad + \sin(npx_2(c_n^{-1} - v^{-1})) \sin(npx_1(c_n^{-1} - v^{-1}))] \int_{-\pi/p}^{\pi/p} v_1(t') \cos npt' \, dt' \\
&= e^{-\alpha_n(x_1 - x_2)} \int_{-\pi/p}^{\pi/p} v_2(t') \cos npt' \, dt', \qquad n = 1, 2,...
\end{aligned} \tag{4.46}$$

and

$$\begin{aligned}
&[\sin(npx_2(c_n^{-1} - v^{-1}))\sin(npx_1(c_n^{-1} - v^{-1})) \\
&\quad + \cos(npx_2(c_n^{-1} - v^{-1}))\cos(npx_1(c_n^{-1} - v^{-1}))] \int_{-\pi/p}^{\pi/p} v_1(t') \sin npt'\, dt' \\
&\quad + [\sin(npx_2(c_n^{-1} - v^{-1}))\cos(npx_1(c_n^{-1} - v^{-1})) \\
&\quad - \cos(npx_2(c_n^{-1} - v^{-1}))\sin(npx_1(c_n^{-1} - v^{-1}))] \int_{-\pi/p}^{\pi/p} v_1(t') \cos npt'\, dt' \\
&= e^{-\alpha_n(x_1 - x_2)} \int_{-\pi/p}^{\pi/p} v_2(t') \sin npt'\, dt', \qquad n = 1, 2, \ldots . \qquad (4.47)
\end{aligned}$$

With $v_1(t')$ and $v_2(t')$, the particle velocities at two different locations on the filament, considered as being known from experimental measurements, relations (4.46) and (4.47) comprise two equations in two unknowns, c_n and α_n, for each value of n. Then using relations (4.21) and (4.22) it can be shown that

$$E_{3n} = \rho^{1/2} c_n[(\alpha_n^2 c_n^2/\omega_n^2) + 1]^{-1}$$

and

$$E_{4n} = \rho^{1/2} \alpha_n c_n^2 \omega_n^{-1}[(\alpha_n^2 c_n^2/\omega_n^2) + 1]^{-1} \qquad (4.48)$$

where $\omega_n = np$ and E_{3n} and E_{4n} are values of E_3 and E_4 appropriate to the nth frequency component. The real and imaginary parts of the complex modulus, $E^*(i\omega)$, may then be found directly from (4.23) as

$$E_n' = (E_{3n})^2 - (E_{4n})^2$$

and

$$E_n'' = 2E_{3n}E_{4n} . \qquad (4.49)$$

The procedure just described for determining $E^*(i\omega)$ in the high frequency range has not been explicitly used. Rather, some further simplifying assumptions have usually been employed. These assumptions center around taking particular forms for the phase velocity and attenuation. The simplest assumptions of this type would be to take

$$c = \text{constant}$$

and

$$\tan\delta = \text{constant} \qquad (4.50)$$

where the attenuation would then be found from (4.26). The motivation for this assumption is that if the frequency range of relevance is narrow enough, then $E^*(i\omega)$ is effectively constant in this region and therefore c and $\tan\delta$ can then be found from a slight modification of the procedure just described. Namely the solution is taken as (4.35)–(4.38) with A_n and B_n evaluated in terms of the particle velocity $v_1(t')$ at one location through (4.39)–(4.42). With A_n and B_n so evaluated, the values of c = constant and $\tan \delta$ = constant are so adjusted that the solution (4.35)–(4.38) gives the best possible description of the observed shape of the transmitted pulse at a second location. Hunter [4.7] has discussed this procedure, but he found the results to be unsatisfactory and concluded the forms in (4.50) are too drastic a simplification. A less restrictive assumption was tried earlier by Kolsky [4.8], and it was found to give reasonable results for certain materials.

This somewhat more general assumption has the form

$$\begin{aligned} \tan \delta &= \text{constant} \\ E'(\omega) &= E_0' \left[1 + \frac{\tan \delta}{\pi} \ln \left(\frac{\omega}{p}\right)\right] \end{aligned} \tag{4.51}$$

where E_0' is a constant. The phase velocity c and the attenuation α are evaluated in terms of the quantities in (4.51) by using (4.25) and (4.26). By following a procedure similar to that just described for assumptions (4.50), the values E_0' and $\tan \delta$ are evaluated to give the best possible description of the change in shape of the pulse as it passes between two locations on the filament. As mentioned by Pipkin [4.9], the assumed form given by (4.51) has justification in terms of the power law form of properties given in Problem 4.3. This procedure as applied to polymethylmethacrylate, in Ref. [4.8], gave a very good description of the dispersion and attenuation of the pulse shape for values of c corresponding to E_0' as $c_0 = 2300$ m/sec and $\tan \delta = 0.04$ in the frequency range of about 1000–24,000 Hz. In general, in using this procedure for determining mechanical properties it is necessary that frequency range of excitation and the temperature of the specimen be such that the loss tangent be very small, i.e., $\tan \delta \ll 1$.

A procedure similar to that just described has been applied by Lifshitz and Kolsky [4.10] to obtain a second independent mechanical property which governs the propagation of one-dimensional radially symmetric waves. In fact, the solution given in Section 2.10 could readily be adapted to this purpose. Reference to that solution shows that the relevant mechanical property in complex modulus form is

$$k^*(i\omega) + \tfrac{4}{3}\mu^*(i\omega).$$

Forms corresponding to (4.51) were assumed in Ref. [4.10] for both the complex shear modulus and the complex bulk modulus. With the data for $\mu^*(i\omega)$ known independently, the Fourier synthesis procedure just described for the filament enables the determination of the complex bulk modulus $k^*(i\omega)$ in the present context. For polyethylene this procedure was used in Ref. [4.10] to determine the loss tangent of the complex bulk modulus as

$$\tan \delta_k = 0.023$$

in the frequency range from 5.0×10^4 to 7.5×10^5 Hz. The corresponding loss tangent for the shear modulus, determined independently, and used in this analysis was

$$\tan \delta_\mu = 0.11.$$

A different means of obtaining mechanical property data on the complex bulk modulus, $k^*(i\omega)$, has been given by McKinney *et al.* [4.11]. In their method, a specimen was subjected to a hydrostatic harmonically oscillating pressure transmitted through an oil bath.

4.5. TEMPERATURE DEPENDENT EFFECTS

The mechanical properties appropriate to the linear theory of viscoelasticity generally exhibit a very strong dependence upon the temperature at which the determining test is conducted. This effect is much more pronounced than in the comparable types of tests conducted upon metals, where over a reasonable range the mechanical properties can be taken to be independent of temperature. As we shall see, the nature of the temperature dependence of viscoelastic mechanical properties is interrelated to the time dependence of these materials. It is therefore not too surprising that these temperature dependence effects may appear in polymeric materials, while being negligible in metals.

The simplest situation of this type is that which exists when the temperature dependence effects are to be related to different possible base temperatures of an isothermal theory. In this case, the appropriate mechanical properties are simply obtained in an ambient temperature environment equal to the base temperature at which the theory is to be used, as for example at a particular one of the temperatures in the data of Figs. 4.1–4.4. However, when the mechanical properties are to be used in a nonisothermal theory, the situation is more involved and two further possibilities arise. This nonisothermal case will now be considered.

The general theory of linear coupled thermoviscoelasticity was derived in Chapter 3. It was found to be a result of that derivation that the mechanical properties must be obtained and applied at the fixed base temperature of the intended conditions of application. Thus, although the infinitesimal temperature deviation from the base temperature occurs as an explicit field variable in the theory, any temperature dependence of the mechanical properties upon the infinitesimal temperature deviation from the base temperature is necessarily neglected. If in a given coupled thermoviscoelastic problem the temperature deviations from the base temperature are found to be large enough to cause a significant change in the mechanical properties, then this simply implies that the problem as posed is outside the scope of the linear coupled theory. Thus, the determination of the mechanical properties for use in the linear coupled theory of thermoviscoelasticity follows exactly the same procedure as in the isothermal case: these mechanical properties are simply determined at the constant temperature corresponding to the base temperature of the intended application.

The second possibility under the nonisothermal case is concerned with the mechanical properties appropriate to the uncoupled theory of linear thermo-viscoelasticity. In this case either the history of the temperature field is considered as being given from experimental data, or the coupling term between mechanical and thermal effects can be neglected in the coupled theory such that the temperature variable can be determined through the usual heat conduction equation, independent of the mechanical variables. In either of these situations there is no inconsistency in accounting for the dependence of the mechanical properties upon the total temperature, rather than just upon the fixed base temperature, as in the previous case. This situation is described fully in Sections 3.5 and 3.6, for the particular type of temperature dependence known as that of thermorheologically simple behavior. It is this particular type of behavior which is considered in the remainder of this section.

The basic postulate of thermorheologically simple behavior is that a viscoelastic mechanical property—relaxation functions, creep functions or complex moduli—at a series of different temperatures, when plotted versus the logarithm of time or frequency can be superimposed to form a single curve merely by shifting the various curves at different temperatures along the time or frequency axis. That is, for example, the relaxation data shown in Fig. 4.1 would be tested to see whether or not the various constant temperature curves can be shifted parallel to the log t scale to obtain a single master curve. If this can be done the material conforms to the postulate of thermorheologically simple behavior, if not, it has some more general type of behavior. In fact, the data of Fig. 4.1 can be shifted to obtain the single master curve shown in Fig. 4.5 with the corresponding shift function shown in Fig. 4.6. It therefore follows that this particular material, polyisobutylene, admits the thermorheologically simple characterization for its uniaxial relaxation function. It also follows from the relations (4.1)–(4.3) that the uniaxial creep function and complex modulus corresponding to the relaxation function of Fig. 4.1 also satisfy the thermorheologically simple postulate.

The single master curve description of a mechanical property for a thermorheologically simple material provides a very convenient and concise description of the time (or frequency) and temperature dependence of the mechanical property. However, its usefulness is broader than that of just providing a convenient means of data reduction and description for mechanical properties. As was shown in Section 3.6, this type of temperature dependence of mechanical properties can be conveniently incorporated into the statement of the boundary value problem for the linear uncoupled theory of nonisothermal viscoelasticity.

The question naturally arises as to whether a given material can be expected to exhibit the thermorheologically simple type of behavior. There are no general inclusive guide lines that can be given to aid one in answering this question for a particular material. The only safe and certain answer to this question lies in experimentally verifying or invalidating the shifting procedure for every material of interest. The relaxation data shown in Fig. 4.1 are easily shown to obey the

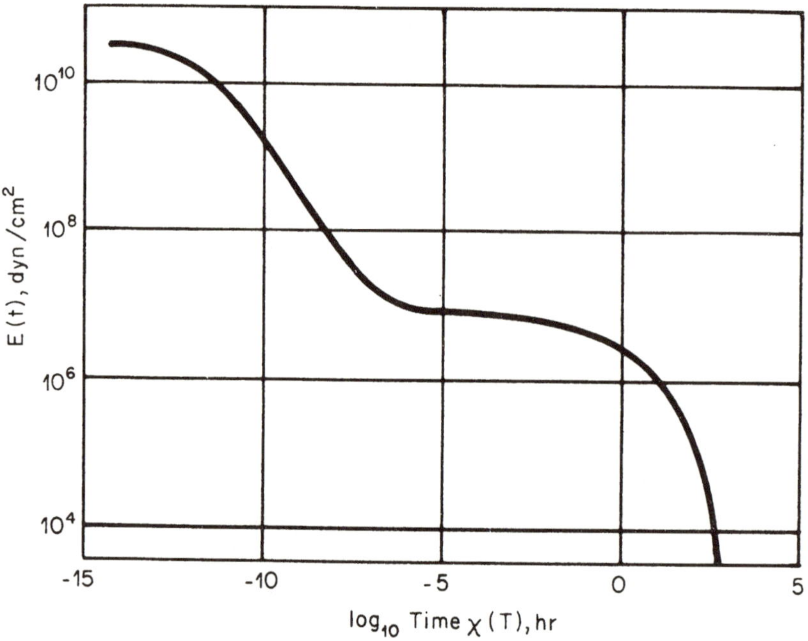

Fig. 4.5. Reduced uniaxial relaxation function of polyisobutylene for data of Fig. 4.1, $\chi(T)$ in Fig. 4.6 (after Tobolsky and Catsiff [4.2]).

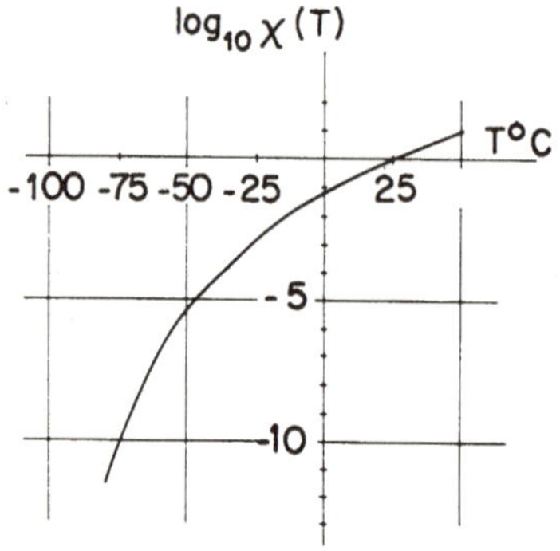

Fig. 4.6. Shift function of polyisobutylene for data of Figs. 4.1 and 4.5 (from the data of Tobolsky and Catsiff [4.2]).

shift hypothesis, whereas as a counterexample, the complex modulus data of Figs. 4.2 and 4.3 can be shifted to form a single curve with only marginal success. Even for materials which are verified to obey the shift hypothesis, such a procedure can only be expected to be valid over a limited time and temperature range, primarily in the rubbery range; see Ferry [4.6] for a detailed description of these types of limitations. An obvious class of materials which would not be expected to obey the thermorheologically simple postulate are those partially crystalline polymers for which the degree of crystallinity undergoes significant changes with temperature.

4.6. APPROXIMATE INTERRELATIONSHIPS AMONG PROPERTIES

We have been considering the means of determining viscoelastic mechanical properties. With one type of mechanical property, the other types can be obtained directly from the interrelationships given in Section 4.1. These interrelationships take the form of infinite integrals that require analytical or numerical evaluation. We are thus motivated to seek approximate interrelationships that circumvent this difficulty for some practical purposes. The approximate forms are useful for estimating properties, but when high accuracy is required, the exact transformations should be employed.

We begin by seeking a simple, reliable approximate relation between the real and imaginary parts of the complex modulus. The exact formula is given by (1.71). Let $E^*(i\omega)$ represent a general form for a complex modulus property. Take the derivative with respect to ω of (1.71), there results,

$$\frac{\pi}{2}\frac{dE'(\omega)}{d\omega} = -2\omega \int_0^\infty \frac{\lambda E''(\lambda)\, d\lambda}{(\omega^2 - \lambda^2)^2}\,. \tag{4.52}$$

Assume that $E''(\omega)$ is changing slowly with respect to ω such that Eq. (4.52) can be approximated by

$$\frac{\pi}{2}\frac{dE'(\omega)}{d\omega} \simeq -2\omega E''(\omega) \int_0^\infty \frac{\lambda\, d\lambda}{(\omega^2 - \lambda^2)^2}\,. \tag{4.53}$$

Carrying out the integration in (4.53) gives

$$E''(\omega) \simeq \frac{\pi}{2}\,\omega\,\frac{dE'(\omega)}{d\omega}\,. \tag{4.54}$$

This very simple formula provides a quick estimate of $E''(\omega)$ if $E'(\omega)$ is known. The formula also reveals the fact that to experimentally determine $E''(\omega)$ requires greater sensitivity than to determine $E'(\omega)$ since $E''(\omega)$ relates to $E'(\omega)$ through the slope of the latter.

Next, we consider the relationship between the relaxation function and the real part of the complex modulus. Specifically, we seek an approximate relation that determines $E(t)$ in terms of $E'(\omega)$. Employing (1.4), or its alternative form (1.70), in the present notation gives

$$E(t) = \frac{2}{\pi}\int_0^\infty \frac{E'(\omega)}{\omega} \sin \omega t\, d\omega. \tag{4.55}$$

Next, a seemingly drastic approximation is made. The term $(\sin \omega t)/\omega$ in the integrand of (4.55) is replaced by the delta function form

$$\frac{\sin \omega t}{\omega} \to \frac{\pi}{2}\,\delta(\omega - \hat{\omega}) \tag{4.56}$$

where the coefficient $\pi/2$ corresponds to the integral of $(\sin \omega t)/\omega$ and $\hat{\omega}$ gives the location of the delta function, which is determined next. The parameter $\hat{\omega}$ is determined by requiring the delta function to have the same first moment about the origin as that of the function $(\sin \omega t)/\omega$. Thus,

$$\hat{\omega} = \int_0^\infty \omega \frac{\sin \omega t}{\omega} \, d\omega \Big/ \int_0^\infty \frac{\sin \omega t}{\omega} \, d\omega. \tag{4.57}$$

To evaluate the upper integral in (4.57) requires the use of an exponential convergence factor, $e^{-a\omega}$. Then, letting $a \to 0$, it is found that

$$\hat{\omega} = 2/\pi t. \tag{4.58}$$

Using the replacement form (4.56) in (4.55) along with (4.58) gives

$$E(t) \simeq E'(\omega) \,|_{\omega \to 2/\pi t} \,. \tag{4.59}$$

The form (4.59) is extremely simple. A test of (4.59) on a Maxwell model shows it to be quite accurate. Furthermore, (4.59) can be reasoned to have increased accuracy on a generalized Maxwell model as the number of terms is increased. It follows that on logarithmic scales (4.59) is expected to be very reliable, in contrast to our expectation from the replacement involved in (4.56). Obviously, a relation comparable to (4.59) applies to creep functions for solids as

$$J(t) \simeq J'(\omega) \,|_{\omega \to 2/\pi t} \,. \tag{4.60}$$

A similar form for fluids can be derived by removing the viscosity term as in Section 1.3.

Now we develop an interrelation between relexation and creep functions. The interrelative "guess" $J(t) \sim 1/E(t)$ is known to be inaccurate, and we therefore seek a rational approximation. Begin with the approximate relation (4.59) and invert it to read

$$E'(\omega) \simeq E(t) \,|_{t \to 2/\pi\omega} \,. \tag{4.61}$$

Substitute the form (4.61) into the approximation (4.54) to get

$$E''(\omega) \simeq -\frac{1}{\omega} \left[\frac{dE(t)}{dt} \right] \Big|_{t \to 2/\pi\omega} . \tag{4.62}$$

Using $J^*(i\omega) = 1/E^*(i\omega)$ gives the real part of $J^*(i\omega)$ as

$$J'(\omega) = \frac{E'(\omega)}{[E'(\omega)]^2 + [E''(\omega)]^2} . \tag{4.63}$$

Substitute (4.61) and (4.62) into (4.63) and then substitute the resulting form into (4.60) to obtain finally

$$J(t) \simeq \frac{E(t)}{E^2(t)+(\pi^2t^2/4)[dE(t)/dt]^2}\,. \tag{4.64}$$

This simple form shows the dependence of $J(t)$ not only on $E(t)$ but also on the slope of the latter. The result (4.64) is consistent with the behavior discussed in Section 1.2 that at $t = 0$ and $t \to \infty$ there results $J(0) = 1/E(0)$ and $J(\infty) = 1/E(\infty)$, respectively. The latter result restricts this approximation to solids, but the procedure can be modified to accommodate fluids. The form (4.64) also applies with $J(t)$ and $E(t)$ interchanged.

To conclude the development of these approximations, we consider the means of determining spectra in terms of the corresponding creep or relaxation functions. Consider the creep spectrum formula (1.90), rewritten here as

$$J(t) = \int_0^\infty L(\tau)(1 - e^{-t/\tau})\,d\tau + J(0). \tag{4.65}$$

Approximate the exponential term $(1 - e^{-t/\tau})$ in (4.65) by the step function

$$(1 - e^{-t/\tau}) \to h(t - \alpha\tau) \tag{4.66}$$

where α is to be determined. The procedure for determining α is to minimize the total square error between the difference of the two terms in (4.66). Thus, with

$$\Gamma = \int_0^\infty [(1 - e^{-t/\tau}) - h(t - \alpha\tau)]^2\,d\tau \tag{4.67}$$

we specify

$$d\Gamma/d\alpha = 0. \tag{4.68}$$

Carrying out this operation it is found that

$$\alpha = \ln 2. \tag{4.69}$$

Substitute the step function from (4.66) into (4.65) to obtain

$$J(t) \simeq \int_0^{t/\alpha} L(\tau)\,d\tau + J(0). \tag{4.70}$$

Now take the derivative of (4.70) to find finally

$$L(\tau) \simeq dJ(\alpha\tau)/d\tau \tag{4.71}$$

where α is given by (4.69). A form similar to (4.71) is readily found for the relaxation function of Section 1.9:

$$H(\tau) \simeq - dE(\alpha\tau)/d\tau \tag{4.72}$$

where α is still given by (4.69).

Note that the approximate form (4.71), although correct for $\tau \to \infty$, incorrectly predicts that $L(0) \neq 0$. $L(0) = 0$ follows from (4.65). To correct this deficiency, we seek an approximate formula for $L(\tau)$ appropriate to short times. Take the derivative of (4.65) to give

$$\frac{dJ(t)}{dt} = \int_0^\infty \frac{L(\tau)}{\tau} e^{-t/\tau} \, d\tau. \tag{4.73}$$

Approximate the exponential in (4.73) by the step function

$$e^{-t/\tau} = h(\alpha\tau - t) \tag{4.74}$$

by minimizing the total square error of the difference between the two terms in (4.74) over the range $0 \leqslant \tau \leqslant \infty$, which determines the parameter α as given in (4.69). Then substituting (4.74) in (4.73) and taking the derivative of the ensuing form gives

$$L(\tau) \simeq - \frac{\tau}{\alpha} \frac{d^2 J(\alpha\tau)}{d\tau^2} \tag{4.75}$$

where α is given by (4.69). The form (4.75) is correct in the limit as $\tau \to 0$. It is simple to reason that formula (4.71) is better than (4.75) at long times, whereas (4.75) is better than (4.71) at short times. It can also be proved that the two expressions (4.71) and (4.75) must cross when plotted against τ, so that there is a transition from one approximation to the other. There is, of course, a formula corresponding to (4.75) for relaxation spectra.

Another approximate form for creep and relaxation spectra can be derived. Specifically, the relaxation spectrum can be related directly to the imaginary part of the complex modulus. By combining relations (4.54), (4.59), and (4.72) it can be readily shown that

$$H(\tau) \simeq \frac{2}{\pi\tau} [E''(\omega)] \big|_{\omega \to 2/\pi\tau \ln 2} .$$

A similar form can be derived using the relaxation spectrum counterpart of (4.75) rather than (4.72) as the starting point. More involved, and presumably more accurate, approximate means of determining spectra have been given by Wiff [4.12] and Yakobson [4.13]. Many other approximate interrelations are referenced in Ferry [4.6], especially those of Schwarzl and Staverman and those of Tschoegl. Many of these forms are much more complicated than those derived here and require graphical or numerical evaluation of high order derivatives. The point of view followed here is that if simple first order forms such as those given here are not adequate, then recourse should be made to iterative, digital computer solutions of the exact relations. The first order approximations provide useful starter solutions for such iterative routines.

This concludes the derivation of approximate interrelations between the various property forms. The form (4.72), but with α arbitrarily taken as 1 rather than derived as here, is known as the Alfrey [4.14] rule. Various forms of (4.54) relating $E''(\omega)$ to $E'(\omega)$ are well known in the literature. The other approximations are developed here with the aim of practical utility. It is important to observe that these approximations are not approximations in the mathematical sense of the term, rather they would more properpy be called estimates. There are no error bounds or other devices by which the "goodness" of the approximation can be assessed to justify the use of that term. Nevertheless, these forms have considerable utility in practical assessments.

4.7. APPROXIMATE INVERSION OF THE LAPLACE TRANSFORM

Having found approximate interrelationships among the mechanical properties, we are motivated to find approximate means to invert the integral transforms that are so useful in viscoelasticity theory. There is a vast body of work on approximate transform inversion, and necessarily we here cite only typical procedures rather than provide a comprehensive survey. In this section we consider the Laplace transform usage, appropriate to quasi-static problems.

We here follow the derivation of Schapery [4.15] in deriving two methods of approximately inverting the Laplace transform. Let the function to be inverted be given by $\bar{\psi}(s)$; by the definition of the Laplace transform it is related to the function $\psi(t)$ being sought through

$$s\bar{\psi}(s) = s\int_0^\infty \psi(t)\, e^{-st}\, dt. \tag{4.76}$$

Take a change of variable specified by

$$u = \log s \tag{4.77}$$

where log to the base 10 is used. Consistent with the change (4.77) let

$$s\bar{\psi}(s) = \hat{f}(u). \tag{4.78}$$

Also, take another change of variable

$$v = \log t \tag{4.79}$$

along with the change

$$\psi(t) = f(v). \tag{4.80}$$

Using (4.78) and (4.80) in (4.76) gives

$$\hat{f}(u) = 10^u \int_{-\infty}^{\infty} f(v)\, e^{-s10^v} 10^v \ln 10\, dv \tag{4.81}$$

where (4.79) has also been employed.

Introduce another notational change whereby

$$w = u + v. \tag{4.82}$$

Inserting (4.82) into (4.81), noting that $dv = dw$, and using (4.77) gives

$$\hat{f}(u) = \ln 10 \int_{-\infty}^{\infty} f(v)\, e^{-10^w} 10^w\, dw. \tag{4.83}$$

At this point expand $f(v)$ as a power series to obtain

$$f(v) = f(v_0) + f'(v_0)(v - v_0) + \tfrac{1}{2} f''(v_0)(v - v_0)^2 + \cdots. \tag{4.84}$$

Substituting (4.84) into (4.83) gives

$$\hat{f}(u) = f(v_0) + f'(v_0) \ln 10 \int_{-\infty}^{\infty} (w - w_0)\, 10^w e^{-10^w} dw + \cdots \tag{4.85}$$

where $w - w_0$ in (4.85) corresponds to $v - v_0$ in (4.84) and the following identity, obtained from (4.83) with $f(v) = \psi(t) = h(t)$ and $\hat{f}(u) = s\bar{\psi}(s) = 1$, has been used:

$$\ln 10 \int_{-\infty}^{\infty} 10^w e^{-10^w}\, dw = 1. \tag{4.86}$$

Note that $w_0 = u + v_0$ in (4.85).

Now introduce an assumption that the terms $f''(v_0)$ and the higher derivatives in (4.85) are sufficiently small that they can be neglected. This assumption implies that $\psi(t)$ versus $\log t$ is nearly a straight line in the approximately two decade interval for which the weighting function $e^{-10^w} 10^w$ in (4.83) is sufficiently different from zero. Thus the higher order terms in (4.85) are neglected, enabling us to write

$$\hat{f}(u) \simeq f(v_0) + f'(v_0) \ln 10 \int_{-\infty}^{\infty} (w - w_0)\, 10^w e^{-10^w} dw. \tag{4.87}$$

At this point select w_0 such that the second term in (4.87) vanishes. Thus,

$$\int_{-\infty}^{\infty} (w - w_0)\, 10^w e^{-10^w} dw = 0. \tag{4.88}$$

Relation (4.88) is satisfied by w_0 being the centroid of the area under $10^w e^{-10^w}$, specified by

$$w_0 = \int_{-\infty}^{\infty} w 10^w e^{-10^w}\, dw \Big/ \int_{-\infty}^{\infty} 10^w e^{-10^w}\, dw. \tag{4.89}$$

Using (4.86) in (4.89) it can be shown that

$$w_0 = \frac{1}{\ln 10} \int_0^{\infty} \ln(\tau)\, e^{-\tau}\, d\tau \tag{4.90}$$

where

$$w = \log \tau. \tag{4.91}$$

Now the integral in (4.90) is just the negative of Euler's constant, and thus we write

$$w_0 = -\, 0.58/\ln 10 \tag{4.92}$$

With w_0 evaluated in the manner described, (4.87) becomes

$$\hat{f}(u) \simeq f(v_0) \tag{4.93}$$

or

$$\hat{f}(u) \simeq f(w_0 - u). \tag{4.94}$$

Using (4.78) and (4.80) in (4.94) gives

$$\psi(t) \simeq s\bar{\psi}(s) \tag{4.95}$$

where

$$w_0 - u = v. \tag{4.96}$$

Using (4.77) and (4.79), (4.96) becomes

$$w_0 = \log s + \log t. \tag{4.97}$$

Solving (4.97) for s gives

$$s = 10^{w_0}/t. \tag{4.98}$$

Thus (4.95) can be written as

$$\psi(t) \simeq s\bar{\psi}(s)\,\big|_{s=e^{w_0 \ln 10}/t} \tag{4.99}$$

where $10^{w_0} = e^{w_0 \ln 10}$ has been used. Combining w_0 from (4.92) with (4.99) gives

$$\psi(t) \simeq s\bar{\psi}(s)\,\big|_{s=0.56/t}\,. \tag{4.100}$$

For practical application, (4.100) can be used as it is, or it can be simply rewritten as

$$\psi(t) \simeq s\bar{\psi}(s)\,\big|_{s=1/2t}\,. \tag{4.101}$$

The final result, (4.101), gives an immediate estimate of the time function in terms of its Laplace transform. However, the approximation in going from (4.85) to (4.87) is difficult to assess. It is certain, however, that it can apply only to quasi-static problems in viscoelasticity, for which the history is rather smooth. This is a situation in which the exact inversion can usually be accomplished using residue theory, as shown in Section 2.8. Schapery [4.15] calls the method just derived the direct method, in contrast to the collocation method, shown next.

For the collocation method, let $\bar{f}(s)$ designate the known Laplace transform of $f(t)$, and assume that $f(t)$ can be expressed as

$$f_A(t) = \sum_{i=1}^{N} A_i e^{-t/t_i} \tag{4.102}$$

where the A_i are undetermined coefficients and the t_i are prescribed positive constants. Thus, $f_A(t)$ is an approximate representation of the variable $f(t)$.

The total square error of the difference between $f(t)$ and $f_A(t)$ is given by

$$E^2 = \int_0^\infty [f(t) - f_A(t)]^2 \, dt. \tag{4.103}$$

The coefficients A_i in (4.102) are found by minimizing the total square error. Thus, from (4.103),

$$\frac{\partial(E)^2}{\partial A_i} = -2\int_0^\infty [f(t) - f_A(t)] \, e^{-t/t_i} \, dt = 0 \tag{4.104}$$

which reduces to

$$\int_0^\infty [f(t) - f_A(t)] \, e^{-t/t_i} \, dt = 0. \tag{4.105}$$

This form represents the Laplace transform of $[f(t) - f_A(t)]$ if $1/t_i$ is replaced by s. Therefore, (4.105) can be written as

$$[\bar{f}(s) - \bar{f}_A(s)]_{s \to 1/t_i} = 0. \tag{4.106}$$

Using (4.102) this can be put into the form

$$\left[\sum_{j=1}^{N} \frac{A_j}{s + (1/t_j)}\right]_{s \to 1/t_i} = [\bar{f}(s)]_{s \to 1/t_i}. \tag{4.107}$$

Relations (4.107) represent a system of N linear algebraic equations to be used to determine the N unknown coefficients A_j in the assumed form of the solution (4.102).

The collocation methld (4.107) is subject to the same smoothness restrictions as discussed for the direct method, (4.101). Certainly neither method could be expected to be usefully applied in dynamic problems with oscillatory conditions.

4.8. APPROXIMATE SOLUTIONS FOR DYNAMIC PROBLEMS

We now consider two methods for obtaining approximate solutions for steady state harmonic problems that include inertial effects. Specifically, we pose the following problem. Suppose we have a numerical solution for an elastic harmonic oscillation problem. How do we use the Correspondence Principle to convert the elastic solution to a steady state harmonic viscoelastic solution? Two different approaches to this problem are given here.

Following the work of Dasgupta and Sackman [4.16], let the elastic response function be of the form

$$f(x_i, t) = F(x_i \,.\, \omega, \mu, \nu, \rho)e^{i\omega t} \tag{4.108}$$

where $F(\ )$ is a complex function of the arguments shown. Symbolically, the corresponding viscoelastic solution is given by

$$f(x_i, t) = F(x_i, \omega, \mu^*, \nu^*, \rho)e^{i\omega t}. \tag{4.109}$$

However, if the elastic solution $F(\ )$ is only known numerically, then we cannot simply replace μ and ν by μ^* and ν^* in the analytical solution. First, we begin by assuming Poisson's ratio in the viscoelastic problem to be a real constant. This restriction will be relaxed later.

Employ the nondimensional frequency

$$\xi = \omega L/c_s \tag{4.110}$$

where

$$c_s = \sqrt{\mu/\rho}$$

and L is the characteristic dimension of the problem. Also define

$$\xi^* = \omega L/c_s^* \tag{4.111}$$

where

$$c_s^* = \sqrt{\mu^*/\rho}.$$

The elastic value of $F(\xi)$ is known. The analytic property of $F(\xi^*)$ will be used to construct its solution. Using the superposition principle, the elastic response to the harmonic excitation can be found from the response to a delta function as

$$F(\xi) = \int_{-\infty}^{\infty} \hat{\Delta}(\tau)\, e^{-i\xi\tau}\, d\tau \tag{4.112}$$

where $\hat{\Delta}(\tau)$ is the response to the unit delta function. Thus the viscoelastic solution is given by

$$F(\xi^*) = \int_{-\infty}^{\infty} \hat{\Delta}(\tau)\, e^{-i\xi^*\tau}\, d\tau. \tag{4.113}$$

Now write

$$\xi = a - ib \tag{4.114}$$

where $b > 0$ follows from (4.111) and $\mu' > 0$, $\operatorname{sgn} \mu'' = \operatorname{sgn} \omega$. Using (4.114) in (4.113) gives

$$F(a - ib) = \int_{-\infty}^{\infty} \hat{\Delta}(\tau)\, e^{-b\tau} e^{-ia\tau}\, d\tau. \tag{4.115}$$

The function $F(a - ib)$ can be shown to satisfy the Cauchy–Riemann equations, and thus it is analytic. Therefore, $F(\xi^*)$ is harmonic and can be found as the solution of Laplace's equation. Furthermore, we must have

$$F(\xi^*) = F(\xi) \qquad \text{for} \quad \xi^* \text{ real.}$$

This problem is the classical Dirichlet problem, whose solution in the lower half plane can be expressed as

$$F(a - ib) = \frac{b}{\pi} \int_{-\infty}^{\infty} \frac{F(\xi')\, d\xi'}{b^2 + (\xi' - a)^2} \tag{4.116}$$

where ξ' is real. The procedure then is very simple. $F(\xi')$ is known from the elastic solution, and a and b follow from (4.111) written in terms of real and imaginary parts, as in (4.114). A numerical integration of (4.116) then gives the viscoelastic solution.

An example will illustrate the procedure. Consider a spherical cavity of radius L in an infinite medium. Take the elastic solution as the ratio of the cavity displacement to the applied stress upon it, as

$$F(\xi) = \frac{(\mu_r\,|_{r=L})/L}{(\sigma_{rr}\,|_{r=L})/4\mu} = \frac{\gamma + i\xi}{\gamma + i\xi - \xi^2/4\gamma} \tag{4.117}$$

where

$$\gamma = \sqrt{2(1 - \nu)/(1 - 2\nu)}$$

and ξ is given by (4.110). The form (4.117) when substituted into (4.116) is directly amenable to numerical integration. As shown by Dasgupta and Sackman [4.16], a reasonable truncation and integration of (4.116) compares well with the exact solution of the problem.

In a sequel to this work, Dasgupta and Sackman [4.17] show that it is possible to treat the case of complex viscoelastic Poisson's ratio. This is accomplished by applying the integration formula (4.116) twice sequentially, once for each property.

The example shows this procedure as applied to an infinite media problem. For bounded media the elastic response gives rise to singularities in $F(\xi)$ at the resonances. In principle the present method can be employed in such cases by considering the Cauchy principal value of the integration at the singularities; however, there might be numerical complications. Next, we derive a method that can treat the resonances directly.

The restriction in the following method, due to Hashin [4.18], is that the loss tangents of the governing shear and bulk moduli be small; thus,

$$\begin{aligned} \tan \delta_\mu &= \mu''/\mu' \ll 1 \\ \tan \delta_k &= k''/k' \ll 1. \end{aligned} \tag{4.118}$$

Write the harmonic viscoelastic solution as

$$u_i = \hat{u}_i(x_j, \mu^*, k^*)e^{i\omega t}$$

and

$$\sigma_{ij} = \hat{\sigma}_{ij}(x_j, \mu^*, k^*)e^{i\omega t} \tag{4.119}$$

for displacement and stress, respectively. Now, with $\hat{u}_i$ and $\hat{\sigma}_{ij}$ as analytic functions, expand them in a Taylor's series to obtain

$$\hat{u}_i(x_j, \mu^*, k^*) = \hat{u}_i(x_j, \mu', k') + i\mu'' \frac{\partial \hat{u}_i}{\partial \mu'} + \cdots + ik'' \frac{\partial \hat{u}_i}{\partial k'} + \cdots \tag{4.120}$$

with a similar form for $\hat{\sigma}_{ij}$. With the restriction to small loss tangents, truncate the form (4.120) just after the explicit terms shown. The solution (4.120) can be used either with an analytical or a numerical solution. The derivatives with respect to μ' and k' could easily be evaluated numerically. However, the form (4.120) is not suitable for use at resonances, and a special method applicable to such cases is given next.

For the elastic system, let the solution be written as

$$u_i(x_j, t) = \frac{N_i(x_i, \omega, G_j)}{\Delta(\omega, G_j)} e^{i\omega t} \tag{4.121}$$

where the G_j are the elastic moduli, and the frequency equation for the problem is given by

$$\Delta(\omega, G_j) = 0. \tag{4.122}$$

Let ω_n denote the spectrum of natural frequencies. The viscoelastic solution is then

$$u_i(x_j, t) = \frac{N_i(x_i, \omega, G_j^*)}{\Delta(\omega, G_j^*)} e^{i\omega t}. \tag{4.123}$$

For analytic functions and small loss tangents, expand N_i and Δ in a Taylor's series to obtain

$$u_i(x_j, t) \simeq \left\{ \left[N_i(x_i, \omega, G'_j) + i \sum_{j=1}^{2} G''_j \frac{\partial N_i}{\partial G'_j} \right] \Big/ \left[\Delta(\omega, G'_j) + i \sum_{j=1}^{2} G''_j \frac{\partial \Delta}{\partial G'_j} \right] \right\} e^{i\omega t}. \tag{4.124}$$

The magnitude of the coefficient of $e^{i\omega t}$ in (4.124) can be written as

$$\left| \frac{N_i + iN''_i}{\Delta + i\Delta''} \right| \simeq \frac{[(N_i \Delta)^2 + (N''_i \Delta - N_i \Delta'')^2]^{1/2}}{\Delta^2 + (\Delta'')^2} \tag{4.125}$$

where N''_i and Δ'' identify with the corresponding terms in (4.124). For small loss tangents the viscoelastic frequency equation is

$$\Delta(\omega, G'_j(\omega)) = 0. \tag{4.126}$$

Using (4.126) in (4.125) gives

$$\left| \frac{N_i + iN''_i}{\Delta + i\Delta''} \right| \simeq \frac{N_i}{\Delta''}. \tag{4.127}$$

Relation (4.127) when evaluated at the resonant frequencies ω_n obtained from (4.126) gives the corresponding amplitudes of the response.

The procedure described in evaluating the response amplitude from (4.127) necessarily applies to analytical solutions. In the case of a numerical solution for the elastic problem, N_i and Δ would not be known separately and the procedure must be modified. Write the elastic solution in inverted form as

$$\eta_i = \frac{1}{\hat{u}_i} = \frac{\Delta(\omega, G_j)}{N_i(x_i, \omega, G_j)}. \tag{4.128}$$

For small loss tangents, the real part of the viscoelastic response η_i^* is

$$\eta'_i \simeq \frac{\Delta(\omega, G'_j)}{N_i(x_i, \omega, G'_j)} = \frac{\Delta}{N_i}. \tag{4.129}$$

Take the derivative of (4.129) to obtain

$$\frac{\partial \eta'_i}{\partial G'_j} \simeq \frac{1}{N_i} \frac{\partial \Delta}{\partial G'_j} - \frac{\Delta}{(N_i)^2} \frac{\partial N_i}{\partial G'_j} \qquad \text{(no sum)}. \tag{4.130}$$

At a resonance ω_n, for $\Delta = 0$, the second term in (4.130) vanishes leaving

$$\frac{\partial \eta'_i}{\partial G'_j} \simeq \frac{1}{N_i} \frac{\partial \Delta}{\partial G'_j}. \tag{4.131}$$

Comparing (4.124) and (4.125), we see Δ'' is given by

$$\Delta'' \simeq \sum_{j=1}^{2} G_j'' \frac{\partial \Delta(\omega, G_j')}{\partial G_j'} . \tag{4.132}$$

Substitute $\partial\Delta/\partial G'$ from (4.131) into (4.132) to get

$$\Delta'' \simeq N_i \sum_{j=1}^{2} G_j'' \frac{\partial \eta_i'}{\partial G_j'} . \tag{4.133}$$

Finally, putting (4.133) into (4.127) gives the amplitude of response at the resonance as

$$\left| \frac{N_i + iN_i''}{\Delta + i\Delta''} \right| \simeq \left(\sum_{j=1}^{2} G_j'' \frac{\partial \eta_i'}{\partial G_j'} \right)^{-1} . \tag{4.134}$$

With η' defined by (4.129), the procedure specified by (1.234) comprises a practical evaluation method to be used with numerical elastic solutions.

The two methods for obtaining approximate viscoelastic solutions just derived apply only to steady state harmonic problems under dynamic conditions. By no means, however, are they the only possible approaches to the problem. In fact, one could take the direct approach and simply convert a digital-computer-type numerical elastic solution to viscoelastic by replacing elastic properties by their complex viscoelastic counterparts. Using complex arithmetic in the solution algorithm will yield the complex viscoelastic solution. The method applies equally well to quasi-static and dynamic problems. In applications to general transient problems, Fourier synthesis could be applied to build up such solutions from those solutions under steady state harmonic conditions. Alternatively, problems that can be formulated in the Laplace transform domain are amenable to approximate inversion. The two methods given in the preceding section for inverting the Laplace transform require a high degree of smoothness in the time domain character of the solution. Obviously, such methods are not applicable to dynamic conditions. A more general means of inverting Laplace transform, due to Bellman, has been applied by Swanson [4.19] to obtain dynamic solutions.

PROBLEMS

4.1. It is intended to obtain short-time mechanical property data directly from relaxation-type testing. Discuss the effects and limitations of testing machine inertia in obtaining reliable short-time data. Can a correction be applied for testing machine inertia?

4.2. In Section 4.3 complex modulus data were converted to relaxation function data through the Fourier transform relation. Discuss a rational procedure by which the infinite frequency range integration can be truncated in a numerical evaluation of the integral.

4.3. In Section 2.8 and Problem 2.10 a method was described for fitting relaxation function data by a series of decaying exponentials. A similar procedure can be followed for creep functions. Alternatively, relaxation and creep function data can be fit by power law forms. Specifically, consider fitting a creep function to the form

$$J(t) = J(0) + \tilde{J}t^n, \qquad 0 \leqslant n \leqslant 1.$$

From the literature, obtain a set of creep function data over several decades. Fit the power law form, above, to the data. Discuss the validity of the power law representation in the short-, medium-, and long-time ranges. Obtain the corresponding power law form for the relaxation function. Show that more general power law forms can be deduced that fit relaxation and creep functions at both the initial value and the long-time asymptote.

4.4. Suppose it is necessary to have high frequency mechanical properties data upon a particular material. It is proposed to obtain the data either by wave propagation testing or by using reduced temperature shifting along with frequency testing in an easily accessible frequency range. Discuss the advantages and disadvantages of each method.

4.5. Use the Maxwell model forms of $E(t)$ and $E'(\omega)$ to evaluate the approximate interconversion formula

$$E(t) \simeq E'(\omega)|_{\omega \to 2/\pi t} \,.$$

Reason the reliability of this interconversion for a generalized Maxwell model.

4.6. Carry out the analytical derivation necessary to evaluate the parameter α of (4.69) in the creep spectrum derivation.

4.7. Derive the relaxation spectrum result, (4.72).

4.8. As discussed in Section 1.9, creep and relaxation spectra are often specified in terms of logarithmic variables. Using a definition similar to (1.91) of $\tilde{L}(\tau)$ derive the approximate result

$$\tilde{L}\left(\frac{t}{\alpha}\right) = \frac{dJ(t)}{d(\log t)}$$

where α is given by (4.69). Derive the counterpart of formula (4.75) for the logarithmic spectrum $\tilde{L}(\tau)$.

4.9. Reason why the formulas for the creep spectrum (4.71) and (4.75) must cross when plotted versus τ.

4.10. Obtain the formula corresponding to (4.75) for the relaxation spectrum.

4.11. Apply the approximate Laplace transform inversion formula (4.101) successively to the Laplace transforms of a step function, an exponential, a linearly increasing function, and a sinusoid.

4.12. Apply the method of formula (4.134) to obtain the approximate resonant response of a Bernoulli–Euler beam of viscoelastic material. Specifically, take the ends to be simply supported and find the displacement amplitude at the center of the beam, where a concentrated harmonic force is applied.

REFERENCES

4.1. Kolsky, H., "Experimental Studies of the Mechanical Behavior of Linear Viscoelastic Solids," *Proc. 4th Symp. Nav. Structural Mech.* 357. Pergamon Press, Oxford, 1967.

4.2. Tobolsky, A. V., and E. Catsiff, "Elastoviscous Properties of Polyisobutylene (and Other Amorphous Polymers) from Stress-Relaxation Studies; IX. A Summary of Results," *J. Polym. Sci.* **19**, 111 (1956).

4.3. Gottenberg, W. G., and R. M. Christensen, "An Experiment for Determination of the Mechanical Property in Shear for a Linear, Isotropic Viscoelastic Solid," *Int. J. Eng. Sci.* **2**, 45 (1964).

4.4. Sutherland, H. J., and R. Lingle, "An Acoustic Characterization of a Polymethyl Methacrylate and Three Epoxy Formulations," *J. Appl. Phys.* **43**, 4022 (1972).

4.5. Huntson, D. L., R. R. Myers, and M. B. Palmer, "A Layered Waveguide Technique for Determination of the Viscoelastic Properties of Liquids and Deformable Solids," *Trans. Soc. Rheol.* **16**, 33 (1972).

4.6. Ferry, J. D., *Viscoelastic Properties of Polymers*, 3rd ed. Wiley, New York, 1980.

4.7. Hunter, S. C., "The Solution of Boundary Value Problems in Linear Viscoelasticity," *Proc. 4th Symp. Nav. Structural Mech.* 257. Pergamon Press, Oxford, 1967.

4.8. Kolsky, H., "The Propagation of Stress Pulses in Viscoelastic Solids," *Phil. Mag. Ser. 8*, **1**, 693 (1956).

4.9. Pipkin, A. C., *Lectures on Viscoelasticity Theory*, Springer–Verlag, Berlin and New York, 1972.

4.10. Lifshitz, J. M., and H. Kolsky, "The Propagation of Spherically Divergent Stress Pulses in Linear Viscoelastic Solids," *J. Mech. Phys. Solids* **13**, 361 (1956).

4.11. McKinney, J. E., S. Edelman, and R. S. Marvin, "Apparatus for the Direct Determination of the Dynamic Bulk Modulus," *J. Appl. Phys.* **27**, 425 (1956).

4.12. Wiff, D. R., "RQP Method of Inferring a Mechanical Relaxation Spectrum," *J. Rheol.* **22**, 589 (1978).

4.13. Yakobson, E. E., "Iterative Method of Determining the Relaxation-Time Spectrum from the Components of the Complex Shear Modulus," *Mekhanika Polimerov* **6**, 1069 (1975).

4.14. Alfrey, T., and P. Doty, "The Methods of Specifying the Properties of Viscoelastic Materials," *J. Appl. Phys.* **16**, 700 (1945).

4.15. Schapery, R. A., "Approximate Methods of Transform Inversion for Viscoelastic Stress Analysis," *Proc. 4th U.S. Nat. Cong. Appl. Mech.* 1075 (1962).

4.16. Dasgupta, G., and J. L. Sackman, "An Alternative Representation of the Elastic-

Viscoelastic Corsespondence Principle for Harmonic Oscillations," *J. Appl. Mech.* **44**, 57 (1977).

4.17. Dasgupta, G., and J. L. Sackman, "A Quadrature Representation of the Viscoelastic Analogy in the Frequency Domain," *J. Appl. Mech.* **45**, 955 (1978).

4.18. Hashin, Z., "Vibration Analysis of Viscoelastic Bodies with Small Loss Tangents," *Int. J. Solids Struct.* **13**, 549 (1977).

4.19. Swanson, S. R., "Approximate Laplace Transform Inversion in Dynamic Viscoelasticity," *J. Appl. Mech.* **47**, 769 (1980).

Chapter V Problems of a Nontransform Type

In Section 2.5 we outlined a very simple method by which elastic solutions can be converted to viscoelastic solutions through the use of integral transforms. The method is often called the elastic-viscoelastic correspondence principle, or simply the correspondence principle. Not all problems, however, fit into this classification. Boundary value problems not allowed include those whereby at a point on the surface of the body the type of boundary condition changes with time. For example, in contact problems the boundary condition changes from traction (traction free) to displacement specification at some points as the indentor is pushed into the body. These types of problems cannot be attacked through the direct use of the correspondence principle. We consider these types of problems in this chapter.

We begin in the first section by studying the contact problem of a rigid indentor being pushed into and withdrawn from a viscoelastic half space. We find it to be completely practical to solve such difficult problems. Next we develop a general approach for solving a specific class of problems for which the correspondence principle does not apply. We shall actually employ integral transforms in this method, known as the extended correspondence principle. However, in this context, the use of integral transforms is much more subtle than in the direct correspondence principle of Section 2.5. Thus, the title of this chapter refers only to the use of integral transform methods in the direct application of the correspondence principle.

Sections 5.3 and 5.4 are given over the solutions of growing crack problems. Probably crack problems are the most important class of problems for which the correspondence principle does not apply. The last section deals with a complicated problem of thermal–mechanical interaction. Integral transform methods are not employed in this problem; rather a practical, numerical approach is formulated.

5.1. CONTACT PROBLEM

As an example of a viscoelastic contact problem, the identation of a viscoelastic half space by a rigid spherical indentor is considered. Early studies on this problem were given by Lee and Radok [5.1] and by Hunter [5.2]. The outline given here is taken from the more recent study of Graham [5.3].

In the present analysis the indentor is taken as being applied at the origin of a system of rectangular Cartesian coordinates x, y, and z, and it moves in the z direction. In this problem, the shear stresses are taken to be identically equal to zero over the entire boundary of the half space, while in the contact region, the normal component of the displacement of the boundary is taken to conform to the shape of the indentor. This problem involves the mixed-type boundary conditions mentioned in Section 2.1.

The statement of the boundary conditions for this problem are given

$$\begin{aligned} &\text{at} \quad z = 0 \quad \text{and} \quad r \leqslant a(t), \qquad u_z = \alpha(t) - (1/2R)(x^2 + y^2)\, h(t), \\ &\text{at} \quad z = 0 \quad \text{and} \quad r \geqslant a(t), \qquad \sigma_{zz} = 0, \end{aligned} \tag{5.1}$$

and

$$\text{at} \quad z = 0, \qquad \sigma_{yz} = \sigma_{zx} = 0$$

where $r = (x^2 + y^2)^{1/2}$ is the radius coordinate, R is the radius of the spherical indentor, $\alpha(t)$ is the displacement of the indentor, and $a(t)$ is the radius of the contact area. A solution is now developed which provides the relationship among the normal stresses under the indentor, the displacement $\alpha(t)$ of the indentor, and the radius of the contact region $a(t)$.

It is convenient to start with the elastic solution of the Boussinesq problem. This involves the formula for the normal displacement of the point, with coordinates x, y on the surface of the half space, due to the concentrated normal load P applied at coordinates ξ, η; thus,

$$u_z(x, y, 0) = [(1 - \nu)P/2\pi\mu][(x - \xi)^2 + (y - \eta)^2]^{-1/2} \tag{5.2}$$

where μ is the elastic shear modulus. This Boussinesq problem, involving the displacement due to concentrated forces, is amenable to the application of the elastic–viscoelastic correspondence principle. Correspondingly, the Laplace transformed viscoelastic relationship between the history of the concentrated force and the history of the surface displacement is given by

$$\bar{u}_z(x, y, 0, s) = \frac{(1 - \nu)\, s\bar{J}(s)\, \bar{P}(s)}{2\pi} [(x - \xi)^2 + (y - \eta)^2]^{-1/2} \tag{5.3}$$

where Poisson's ratio ν has been taken to be a constant, and $J(t)$ is the creep function in shear with the relationship between $\mu(t)$ and $J(t)$ being given by (1.18). Using the convolution theorem to invert (5.3) and generalizing the result to include a distribution of forces, we get

$$u_z(x, y, 0, t) = \frac{(1 - \nu)}{2\pi} J * d \iint_{\Omega_m} \frac{P(\xi, \eta, \tau)\, d\xi\, d\eta}{\rho} \tag{5.4}$$

where

$$\rho = [(x - \xi)^2 + (y - \eta)^2]^{1/2} \tag{5.5}$$

where Ω_m designates the maximum contact area and where the Stieltjes convolution notation of Section 1.2 is used. The orders of space and time integrations in (5.4) may be interchanged to give

$$u_z(x, y, 0, t) = \frac{(1-\nu)}{2\pi} \iint_{\Omega_m} \frac{q(x, y\ a,)}{\rho}\, d\xi\, d\eta \tag{5.6}$$

where

$$q(x, y, a) = J * dP. \tag{5.7}$$

For monotonically increasing contact radius $a(t)$ the maximum area of contact Ω_m in (5.6) may be replaced by the current contact area $\Omega(t)$ to give

$$\alpha(t) - \frac{1}{2R}(x^2 + y^2)\, h(t) = \frac{(1-\nu)}{2\pi} \iint_{\Omega(t)} \frac{q(x, y, a)}{\rho}\, d\xi\, d\eta \tag{5.8}$$

where the boundary condition specification (5.1) also has been used. Relation (5.8) has exactly the same form as results for the corresponding elasticity problem; therefore, the solution of (5.8) is identical to the elasticity solution which is given by

$$q = \frac{4}{\pi(1-\nu)R} \operatorname{Re}(a^2 - r^2)^{1/2} \tag{5.9}$$

and

$$\alpha(t) = \frac{a^2(t)}{R} \tag{5.10}$$

where Re designates the real part. By using (1.18) to invert (5.7), we get

$$P(x, y, t) = \frac{4}{\pi(1-\nu)R} \int_0^t \mu(t-\tau)\, d[\operatorname{Re}\,(a^2(\tau) - r^2)]^{1/2}. \tag{5.11}$$

Relations (5.10) and (5.11) thus give the viscoelastic solution in the case where $a(t)$ is a monotonically increasing function of time. This solution is obtained in a rather straightforward manner, even though it has not been found from the direct application of the correspondence principle to the related contact problem in elasticity. The key step in this procedure is the replacement of Ω_m by $\Omega(t)$ in going from (5.6) to (5.8). This replacement is not possible in the case where the contact radius $a(t)$ decreases after reaching a maximum value.

A more general case is now considered, whereby the contact radius $a(t)$ increases monotonically to a maximum value at time $t = t_m$ and then decreases monotonically to zero. In this situation, we introduce a new time variable $t_1(t)$ which is defined through

and

$$
\begin{aligned}
&t_1(t) = t && \text{for} \quad t \leqslant t_m \\
&a(t_1) = a(t), \qquad t_1 < t_m && \text{for} \quad t \geqslant t_m .
\end{aligned}
\tag{5.12}
$$

Thus, t_1 is the time prior to t_m, whereby the contact radius $a(t)$ is equal to the prior contact radius $a(t_1)$.

Using the identity

$$\beta(t) = J * d(\mu * d\beta)$$

the governing relation (5.6) may be written as

$$u_z(x, y, 0, t) = \frac{(1 - \nu)}{2\pi} J * d\left[\iint_{\Omega_m} \rho^{-1}(\mu * dq)\, d\xi\, d\eta\right] \tag{5.13}$$

where the order of integrations has been interchanged. Writing (5.13) out in operational form we get

$$u_z(x, y, 0, t) = \frac{(1 - \nu)}{2\pi} \int_0^t J(t - \theta)\, d\left[\iint_{\Omega_m} \rho^{-1} \int_0^\theta \mu(\theta - \tau)\, dq(\xi, \eta, \tau)\, d\xi\, d\eta\right]. \tag{5.14}$$

The integral involving the limits 0 to θ in (5.14) will be broken into the two intervals 0 to $t_1(\theta)$ and $t_1(\theta)$ to θ. Doing this, and rearranging the orders of integration, we then have

$$
\begin{aligned}
u_z(x, y, 0, t) &- \frac{(1 - \nu)}{2\pi} \int_0^t J(t - \theta) \int_{t_1(\theta)}^\theta \mu(\theta - \tau)\, d\left[\iint_{\Omega_m} \rho^{-1} q(\xi, \eta, \tau)\, d\xi\, d\eta\right] \\
&= \frac{(1 - \nu)}{2\pi} \int_0^t J(t - \theta)\, d\left[\iint_{\Omega_m} \rho^{-1} \int_0^{t_1(\theta)} \mu(\theta - \tau)\, dq(\xi, \eta, \tau)\, d\xi\, d\eta\right].
\end{aligned}
\tag{5.15}
$$

But now the last space integration in (5.15) can be changed to an integration over the area $\Omega(\theta)$ rather than over Ω_m, since, for $\tau \leqslant t_1$, $\Omega(\tau)$ is a monotonically increasing function of τ. This is the reason for introducing the variable $t_1(t)$. By making this change in (5.15) and using (5.6), we find

$$
\begin{aligned}
u_z(x, y, 0, t) &- \int_0^t J(t - \theta) \int_{t_1(\theta)}^\theta \mu(\theta - \tau)\, du_z(x, y, 0, \tau) \\
&= \frac{(1 - \nu)}{2\pi} \int_0^t J(t - \theta)\, d\left[\iint_{\Omega(\theta)} \rho^{-1} \int_0^{t_1(\theta)} \mu(\theta - \tau)\, dq(\xi, \eta, \tau)\, d\xi\, d\eta\right].
\end{aligned}
\tag{5.16}
$$

The lower limit of 0 in the first integral in (5.16) may be changed to t_m, since, for times less than this, $t_1(\theta) = \theta$ and the integral $t_1(\theta)$ to θ necessarily vanishes. We make this change and substitute for $u_z(x, y, 0, t)$ from (5.1) to get

$$\alpha(t) - \frac{1}{2R}(x^2 + y^2)\,h(t) - \int_{t_m}^{t} J(t-\theta) \int_{t_1(\theta)}^{\theta} \mu(\theta-\tau)\,d\alpha(\tau)$$

$$= \frac{(1-\nu)}{2\pi} \int_0^t J(t-\theta)\,d\left[\iint_{\Omega(\theta)} \rho^{-1} \int_0^{t_1(\theta)} \mu(\theta-\tau)\,dq(\xi,\eta,\tau)\,d\xi\,d\eta\right]. \quad (5.17)$$

Now this form has been obtained from manipulations of the general relation (5.6), but with $\Omega_{\max}$ in (5.6) replaced by $\Omega(\theta)$ in the resulting form (5.17). This replacement of $\Omega_{\max}$ by $\Omega(\theta)$ corresponds to the specialization of the form (5.6) to the elastic case, as we already saw in connection with the case of a monotonically increasing contact area. Thus, the elastic solution for q and α represents the solution of (5.17). Substituting for q and α from (5.9) and (5.10) into (5.17), we get the identity

$$\frac{a^2}{R} - \frac{1}{2R}(x^2 + y^2)\,h(t) - \frac{1}{R}\int_{t_m}^{t} J(t-\theta) \int_{t_1(\theta)}^{\theta} \mu(\theta-\tau)\,d[a^2(\tau)]$$

$$= \frac{2}{\pi^2 R} \int_0^t J(t-\theta)\,d\left\{\iint_{\Omega(\theta)} \rho^{-1} \int_0^{t_1(\theta)} \mu(\theta-\tau)\,d[\mathrm{Re}\,(a^2(\tau) - r^2)^{1/2}]\,d\xi\,d\eta\right\}. \quad (5.18)$$

The governing boundary condition relations (5.1) and the relation (5.4) for the general viscoelastic problem when combined give

$$\alpha(t) - \frac{1}{2R}(x^2 + y^2)\,h(t) = \frac{(1-\nu)}{2\pi}\,J * d\left[\iint_{\Omega_m} \rho^{-1} P(\xi,\eta,\tau)\,d\xi\,d\eta\right]. \quad (5.19)$$

By comparing (5.19) with the identity (5.18) and remembering that $\Omega(\theta)$ in (5.18) may also be taken as Ω_m, we see that (5.19) will be satisfied if

$$\alpha(t) = \frac{a^2}{R} - \frac{1}{R}\int_{t_m}^{t} J(t-\theta) \int_{t_1(\theta)}^{\theta} \mu(\theta-\tau)\,da^2(\tau) \quad (5.20)$$

and

$$P(x, y, t) = \frac{4}{\pi(1-\nu)R} \int_0^{t_1(t)} \mu(t-\tau)\,d[\mathrm{Re}\,(a^2(\tau) - r^2)^{1/2}]. \quad (5.21)$$

These two relations give the general solution, relating the normal stresses under the indentor, the displacement of the indentor, and the radius of the contact area. If $t \leqslant t_m$, then relations (5.20) and (5.21) reduce directly to the more restrictive solution found previously and given by (5.10) and (5.11).

In some cases the resultant force on the indentor is of interest. This is given by

$$F = 2\pi \int_0^{a(t)} rP(x, y, t)\, dr. \tag{5.22}$$

We substitute for P from (5.21) and interchange the orders of integration to get

$$F = \frac{8}{3R(1-\nu)} \int_0^{t_1(t)} \mu(t-\tau)\, da^3(\tau). \tag{5.23}$$

The procedure presented here is a brief account of the results given by Graham [5.3]. More recently Graham [5.4] has extended his procedure to include the case in which the contact area has several relative maxima, rather than a single one as given here. Also, Ting [5.5] has treated this more general case. The problem has been further reconsidered by Sabin and Graham [5.6]. An experimental study which is correlated with the analysis described here has been given by Calvit [5.7]. Graham [5.8] and Ting [5.9] have outlined restricted classes of viscoelastic contact problems which do admit a direct application of the elastic viscoelastic correspondence principle, in contrast to the general case considered here. This circumstance will be considered further in the next section.

A problem similar to that just considered, but even more complicated, involves the problem posed by a cylinder rolling over a viscoelastic half space. Here again, the elastic viscoelastic correspondence principle does not apply, and other techniques of solution must be developed. Such studies have been given by Hunter [5.10] and Morland [5.11, 5.12]. A related problem is that of a punch sliding over a viscoelastic half space. The solution of a particularly difficult problem of that type has been given by Golden [5.13].

5.2. EXTENDED CORRESPONDENCE PRINCIPLE

In the preceding section we saw that it is indeed a practical matter to find solutions to problems for which the correspondence principle does not apply. In this section we obtain a systematic approach to a class of problems for which the correspondence principle does not apply. This approach, developed by Graham [5.8], is called the extended correspondence principle.

Let **n** and **s** denote vectors normal and tangential to the boundary of a body. Form the matrix

$$\begin{bmatrix} \sigma_s & \sigma_s & \sigma_n & \sigma_n & u_s & u_s & u_n & u_n \\ u_n & u_n & u_s & \sigma_s & \sigma_n & u_n & \sigma_s & u_s \\ \sigma_n & u_n & \sigma_s & u_s & u_n & \sigma_n & u_s & \sigma_s \end{bmatrix}$$

and let the elements of any column be denoted by

$$\begin{bmatrix} a \\ b \\ c \end{bmatrix}.$$

The boundary conditions upon the time independent and time dependent regions of the boundary are taken in the form

$$\begin{aligned} a(x_i, t) &= A(x_i, t) \qquad &\text{on} \quad B \\ b(x_i, t) &= B(x_i, t) \qquad &\text{on} \quad B_1(t) \\ c(x_i, t) &= 0 \qquad &\text{on} \quad B_2(t). \end{aligned} \tag{5.24}$$

Obviously the method of integral transforms with respect to time cannot be applied to this problem.

Consider the one parameter (time) family of static elastic boundary value problems governed by (5.24) and the usual elastic stress strain relations. Denote the elastic solutions by

$$u_i^e(t), \quad \epsilon_{ij}^e(t), \quad \sigma_{ij}^e(t).$$

Consider only those types of problems for which the elastic solution has the form

$$\begin{aligned} b^e(x_i, t) &= B^e(x_i, t) \qquad &\text{on} \quad B \\ c^e(x_i, t) &= \xi(\mu, k)\, C^e(x_i, t) \qquad &\text{on} \quad B \end{aligned} \tag{5.25}$$

where

$$\begin{aligned} B^e(x_i, t) &= B(x, t) \qquad &\text{on} \quad B_1 \\ C^e(x_i, t) &= 0 \qquad &\text{on} \quad B_2 \end{aligned} \tag{5.26}$$

and B^e and C^e are independent of the elastic constants while $\xi(\mu, k)$ is independent of coordinates.

Take a different viscoelastic problem of possible interest as governed by the boundary condition forms

$$\begin{aligned} a(x_i, t) &= A(x_i, t) \qquad &\text{on} \quad B \\ b(x_i, t) &= B^e(x_i, t) \qquad &\text{on} \quad B. \end{aligned} \tag{5.27}$$

For this latter, auxiliary problem the correspondence principle can be applied directly, to get the solution designated by

$$u_i^e \to \bar{u}_i, \qquad \epsilon_{ij}^e \to \bar{\epsilon}_{ij}, \qquad \sigma_{ij}^e \to \bar{\sigma}_{ij} \tag{5.28}$$

where $\bar{\mu}_i$, $\bar{\epsilon}_{ij}$, and $\bar{\sigma}_{ij}$ are the Laplace transformed solution of problem (5.27). Thus (5.27) with (5.28) will be the solution of the problem of interest if we can show that the last of (5.24) is satisfied. This then is our main objective, to determine the conditions under which the last of (5.24) is satisfied by the solution (5.28).

Proceed by taking the Laplace transform of the last of (5.25) to get

$$\bar{c}^e(x_i, s) = \xi(\mu, k)\bar{C}^e(x_i, s) \qquad \text{on} \quad B. \tag{5.29}$$

We can see that the elastic form (5.29) and the correspondence principle give the viscoelastic form

$$\bar{c}(x_i, s) = \xi(s\bar{\mu}(s), s\bar{k}(s))\bar{C}^e(x_i, s) \qquad \text{on} \quad B. \tag{5.30}$$

Invert relation (5.30) to obtain

$$c(x_i, t) = K(t)\, C^e(x_i, 0) + \int_0^t K(t-\tau)\frac{dC^e(x_i, \tau)}{d\tau}\, d\tau \qquad \text{on} \quad B \tag{5.31}$$

where

$$K(t) = \mathscr{L}^{-1}\left[\frac{1}{s}\, \xi(s\bar{\mu}(s), s\bar{k}(s))\right].$$

Let $B_1(t)$ be monotonically increasing with time; thus,

$$B_1(t_1) \subseteq B_1(t_2), \quad t_1 \leqslant t_2 \tag{5.32a}$$

but

$$C^e(x_i, t) = 0 \qquad \text{on} \quad B_2(t). \tag{5.32b}$$

Thus from (5.32) it follows that

$$c(x_i, t) = 0 \qquad \text{on} \quad B_2 \tag{5.33}$$

because $C^e = 0$ on B_2 and B_2 is monotonically decreasing. Relation (5.33) is the proof we sought of the last of (5.24). We see that we could not have proved (5.33) if B_1 were not monotonically increasing. Thus, we now understand (5.28), obtained directly from the correspondence principle, to be the solution to the problem with time dependent boundaries. It must be emphasized that this use of the correspondence principle applies only to the special type of problem considered here with time dependent boundary regions.

Indentor Example

We now reconsider the indentor problem of the preceding section as an example of the present procedure. Take the boundary conditions at $z = 0$ as

$$\begin{aligned} \sigma_{rr}(r, 0, t) &= 0, && r \geqslant 0 \\ u_z(r, 0, t) &= D(t) - \beta(r), && 0 \leqslant r \leqslant a(t) \\ \sigma_{zz}(r, 0, t) &= 0, && r > a(t) \end{aligned} \tag{5.34}$$

where cylindrical coordinates are employed and $\beta(r)$ is the shape of the indentor.

Take $a(t)$ as monotonically increasing. The elastic solution of this problem, from Sneddon [5.14], is given by

$$\sigma_{zz}(r, 0, t) = \frac{2\mu(\mu + 3k)}{(3k + 4\mu)\, r} \frac{d}{dr} \int_r^{a(t)} \frac{yg(y, t)\, dy}{(y^2 - r^2)^{1/2}}, \qquad 0 \leqslant r \leqslant a(t) \quad (5.35)$$

where

$$g(y, t) = \frac{2}{\pi} \left[D(t) - y \int_0^y \frac{(d\beta/dr)\, dr}{(y^2 - r^2)^{1/2}} \right]. \qquad (5.36)$$

The contact radius $a(t)$ and displacement $D(t)$ are related by

$$D(t) = a(t) \int_0^{a(t)} \frac{(d\beta/dr)\, dr}{(a^2(t) - r^2)^{1/2}} . \qquad (5.37)$$

Time is treated as a parameter in this solution. Outside the contact region the elastic displacement is

$$u_z(r, 0, t) = \int_0^{a(t)} \frac{g(y, t)\, dy}{(r^2 - y^2)^{1/2}}, \qquad r > a(t). \qquad (5.38)$$

The problem now involves prescribing $\sigma_{rz} = 0$ on B and u_z, given by (5.34) and (5.38), on B. Also it is seen that the elastic constants appear as a separate factor in (5.35), as is required by the method.

Using the correspondence principle to find the pressure under the indentor gives from (5.31) and (5.35)

$$\sigma_{zz}(r, 0, t) = K(t)\, M(r, 0) + \int_0^t K(t - \tau) \frac{d}{d\tau} M(r, \tau)\, d\tau \qquad (5.39)$$

where

$$K(t) = \mathscr{L}^{-1}[2\bar{\mu}(\bar{\mu} + 3\bar{k})/(3\bar{k} + 4\bar{\mu})]$$

and

$$M(r, t) = \frac{h[a(t) - r]}{r} \frac{d}{dr} \int_r^{a(t)} \frac{yg(y, t)}{(y^2 - r^2)^{1/2}}\, dy.$$

The total pressure on the indentor, obtained by integration over the contact region, is found to be

$$P(t) = K(t)\, p(0) + \int_0^t K(t - \tau) \frac{dp(\tau)}{d\tau}\, d\tau \qquad (5.40)$$

where

$$p(t) = 2\pi \int_0^{a(t)} g(y, t)\, dy.$$

Although (5.40) represents the general solution obtained by the present method, it is important to observe that this result is more restrictive than the indentor solution found in Section 5.1. In the present method the contact area must be monotonically increasing, whereas in the preceding method that limitation was overcome.

Crack Growth Example

Following Graham [5.15] we now find a solution for the stress and displacement state in an infinite medium subjected to a growing crack. Reduce the problem to a two-dimensional problem in coordinates x and y, with x being in the direction of crack growth and y being normal to it. As the related half space problem specify

$$\begin{aligned} \sigma_{zx}(x, 0, t) &= 0, & -\infty < x < \infty \\ \sigma_{zz}(x, 0, t) &= -p(|x|, t), & |x| \leqslant a(t) \\ u_z(x, 0, t) &= 0, & |x| > a(t). \end{aligned} \tag{5.41}$$

The state (5.41) with constant pressure $p(x) = p_0$ when superimposed with a uniform stress state p_0 in the z direction provides the complete solution of the crack problem.

The plane strain elastic solution is given by

$$\begin{aligned} \sigma_{zz}^e(x, 0, t) = \left(\frac{2}{\pi}\right)^{1/2} \Bigg\{ & \frac{xg(a(t), t)}{[x^2 - a^2(t)]^{1/2}} - g(0, t) \\ & - \int_0^{a(t)} \frac{x[dg(r, t)\, dr]\, dr}{(x^2 - r^2)^{1/2}} \Bigg\}, \qquad x > a(t) \end{aligned} \tag{5.42}$$

and

$$u_z^e(x, 0, t) = \frac{1 - \nu}{\mu} \left(\frac{2}{\pi}\right)^{1/2} \int_x^{a(t)} \frac{rg(r, t)\, dr}{(r^2 - x^2)^{1/2}}, \qquad x \leqslant a(t) \tag{5.43}$$

where

$$g(r, t) = \left(\frac{2}{\pi}\right)^{1/2} \int_0^r \frac{p(x, t)\, dx}{(r^2 - x^2)^{1/2}}.$$

The problem, as posed, satisfies the conditions for application of the extended correspondence principle. No properties occur in the elastic stress solution (5.42), which thus also represents the viscoelastic solution. Next we find the normal component of displacement over the region $x < a(t)$ in the viscoelastic problem. Following the procedure of the extended correspondence principle, in the special case where $p(x, t) = p_0$ is independent of x, gives

$$\begin{aligned} u_z(x, 0, t) = {} & K(0)\, p_0(t) \operatorname{Re}\{[a^2(t) - x^2]^{1/2}\} \\ & + \int_0^t \frac{dK(\theta)}{d\theta}\, p_0(t - \theta) \operatorname{Re}\{[a^2(t - \theta) - x^2]^{1/2}\}\, d\theta \end{aligned} \tag{5.44}$$

where

$$K(t) = \mathscr{L}^{-1}[(4\bar{\mu} + 3\bar{k})/2s^2\bar{\mu}(\bar{\mu} + 3\bar{k})].$$

Relations (5.42) and (5.44) are the solution of the viscoelastic problem. It is important to observe that the approach provides no information concerning how fast the crack would grow in a viscoelastic medium. Rather, if the crack growth rate is specified through $a(t)$, then the field variables of the problem follow from the solution. The next two sections focus on the important problem of determining the crack rate of growth as a function of the applied stress level.

5.3. CRACK GROWTH—LOCAL FAILURE MODEL

Crack growth theory is an important example of a problem for which the correspondence principle does not apply. This section and the next are used to develop two different approaches to the crack growth problem. Both approaches appeal to certain aspects of the extended correspondence principle of Section 5.2.

Begin by taking an infinite medium with a finite crack of length $2a$. The crack is taken with geometry such that the problem is two dimensional, plane stress or plane strain. The approach followed here is that due to Schapery [5.16–5.18]. The methodology follows that for elastic cracks whereby a set of failure forces are superimposed on the stresses due to the far field, thereby canceling the singular stress state inherent in the classical linear solution. This procedure will be explained for elastic cracks and then it will be extended to viscoelastic cracks.

In the elastic problem the far field stress is σ normal to the plane of the crack. The stress intensity factor is given by

$$K_I = \sqrt{\pi a}\,\sigma. \tag{5.45}$$

The first terms in a stress expansion at the crack tip are given by the square root singularities

$$\begin{aligned}
\sigma_{xx}|_{y=0} &= \frac{K_I}{\sqrt{2\pi x}}\,h(x) \\
\sigma_{yy}|_{y=0} &= \frac{K_I}{\sqrt{2\pi x}}\,h(x) \\
v|_{y=0} &= \frac{4(1-\nu^2)}{E}\,\frac{K_I\sqrt{-x}}{\sqrt{2\pi}}\,h(-x)
\end{aligned} \tag{5.46}$$

where y is normal to the crack with x along its extension and with the origin being at the crack tip. The surface of the crack is free of all tractions. The $(1 - \nu^2)/E$ term is appropriate to plane strain while $1/E$ is appropriate to plane stress. Postulate the existence of a zone of failure forces (stresses) acting over length α. These failure stresses σ_f in the y direction at $y = 0$ produce the stress and displacement state

$$\sigma_{xx}^{f} = \sigma_{yy}^{f} = -\frac{h(x)}{\pi\sqrt{x}}\int_0^{\alpha}\frac{\sqrt{\xi}\,\sigma_f(\xi)\,d\xi}{\xi + x} \tag{5.47}$$

and

$$v^f = -\frac{4(1-\nu^2)}{2\pi E}\,h(\xi)\int_0^{\alpha}\sigma_f(\xi')\ln\left[\left|\frac{(\xi')^{1/2}+\xi^{1/2}}{(\xi')^{1/2}-\xi^{1/2}}\right|\right]d\xi'$$

where ξ is a coordinate in the negative x direction. For $x \geqslant 0$ and also very small (5.47) can be written as

$$\sigma_{xx}^{f} = \sigma_{yy}^{f} = -\frac{1}{\pi\sqrt{x}}\int_0^{\alpha}\frac{\sigma_f(\xi)\,d\xi}{\sqrt{\xi}} + \sigma_f(0). \tag{5.48}$$

Superimpose the two stress states (5.46) and (5.48) to cancel the singularity. The singularity vanishes if

$$\frac{K_I}{(2\pi x)^{1/2}} - \frac{1}{\pi(x)^{1/2}}\int_0^{\alpha}\frac{\sigma_f(\xi)\,d\xi}{(\xi)^{1/2}} = 0. \tag{5.49}$$

Solving this for K_{I} gives

$$K_I = \sqrt{2\alpha/\pi}\,\sigma_m I_1 \tag{5.50}$$

where

$$I_1 = \int_0^1\frac{f(\alpha,\eta)\,d\eta}{\sqrt{\eta}} \tag{5.51}$$

and

$$f(\alpha,\eta) = \sigma_f(\eta)/\sigma_m, \qquad \eta = \xi/\alpha$$

with σ_m being the maximum value of σ_f. The condition (5.49) or (5.50) does not suffice to determine the failure zone size α, which necessarily depends also on the distribution of failure forces $\sigma_f(\xi)$. Superimposing the displacements due to these two stress states, Schapery [5.16] argues that the leading term is given by

$$v = -\frac{8(1-\nu^2)}{3\pi E}\frac{\sigma_m}{\sqrt{\alpha}}I_2\xi^{3/2}h(\xi) \tag{5.52}$$

where

$$I_2 = \int_0^1\frac{(df/d\eta)\,d\eta}{\sqrt{\eta}}. \tag{5.53}$$

We are now ready to employ the extended correspondence principle of Section 5.2 to obtain the viscoelastic solution. The stress state is unchanged from that of the elastic solution, while the viscoelastic displacement solution follows from the elastic displacement solution as

$$v = -\left(\frac{8}{3\pi}\right)\int_{t_1}^{t} J(t-\tau)\frac{\partial}{\partial\tau}\left(\frac{\sigma_m}{\sqrt{\alpha}}I_2\xi^{3/2}\right)d\tau \tag{5.54}$$

where $J(t)$ is the creep function corresponding to $(1-\nu^2)/E$, t_1 is the time when the crack tip first reaches the fixed point x (coordinate x is fixed in space), and

$$\xi = a(\tau) - x. \tag{5.55}$$

Postulate the following criterion for failure of the material in the failure zone:

$$\Gamma = \int_0^{v_m} \sigma_f\, dv \tag{5.56}$$

where v_m is the maximum displacement in the failure zone, and Γ is the energy content of the failure surface. By substituting v from (5.54) in (5.56) we shall obtain an equation to be used to solve for $a(t)$. First, we simplify the convolution integral in (5.54).

Relation (5.54) can be written as

$$v = -\frac{8}{3\pi}\frac{\sigma_m I_2}{\sqrt{\alpha}}(\Delta t\,\dot{a})^{3/2}\,C(\Delta t) \tag{5.57}$$

where

$$C(\Delta t) = \frac{3}{2\,\Delta t^{3/2}}\int_0^{\Delta t} J(\Delta t-\rho)\sqrt{\rho}\,d\rho \tag{5.58}$$

with

$$\rho = \tau - t_1, \qquad \Delta t = t - t_1 \tag{5.59}$$

and to first order

$$\xi(\tau) = (\tau - t_1)\dot{a} \tag{5.60}$$

where $\dot{a}$ is the crack velocity. Now change to log scale with

$$u = \log_{10}\Delta t \qquad \text{and} \qquad w = \log_{10}\rho. \tag{5.61}$$

Then with (5.61) $C(\Delta t)$ becomes

$$C(\Delta t) = \frac{3\ln 10}{2 10^{3u/2}}\int_{-\infty}^{u} \hat{J}(u-w)\,10^{3w/2}\,dw \tag{5.62}$$

where

$$\hat{J}(u) = J(\Delta t).$$

A plot of $10^{3w/2}$ shows it to vary significantly only over about one decade. Accordingly we follow a procedure similar to that employed in Section 4.6. We replace the weighting function in (5.58) by a delta function located at its centroid. In so doing we find (5.62) to be approximated by

$$C(\Delta t) \simeq J(0.49\Delta t). \tag{5.63}$$

Schapery [5.17] follows a slightly different procedure and argues for the approximation

$$C(\Delta t) \simeq J(\tfrac{1}{3}\,\Delta t). \tag{5.64}$$

From (5.57) and (5.64), then,

$$v = -\frac{8}{3\pi}\frac{\sigma_m I_2}{\sqrt{\alpha}}\,\xi^{3/2} J\left(\frac{\xi}{3\dot{a}}\right) \tag{5.65}$$

where (5.60) has been used.

To proceed further we make a simplyfying assumption whereby $\sigma_f(\eta) = \sigma_m$, a constant. Thus the failure criterion (5.56) becomes

$$\Gamma = \sigma_m v_m\,. \tag{5.66}$$

From (5.65)

$$v_m = -\frac{8}{3\pi}\,\sigma_m \alpha I_2 J\left(\frac{\alpha}{3\dot{a}}\right) \tag{5.67}$$

where I_2 is given by (5.53). Then, solving (5.50) for α,

$$\alpha = \frac{\pi}{2}\left(\frac{K_I}{\sigma_m I_1}\right)^2 \tag{5.68}$$

with I_1 from (5.51). It follows that for constant failure stress

$$f = 1, \qquad I_1 = 2, \qquad \alpha = \pi K_I^2\, 8\sigma_m^2\,, \qquad I_2 = -1.$$

Thus, finally, from (5.67)

$$v_m = \frac{1}{3}\frac{K_I^2}{\sigma_m}\,J\left(\frac{\alpha}{3\dot{a}}\right). \tag{5.69}$$

Using this result in the failure criterion (5.66) gives the criterion for crack growth as

$$\Gamma = \tfrac{1}{3}K_I^2 J(\pi K_I^2\, 24\sigma_m^2 \dot{a}). \tag{5.70}$$

For a given stress level, relation (5.70) provides a means to solve for the crack velocity $\dot{a}$. We note the theory to contain an adjustable parameter σ_m, the failure stress. Equivalently, through (5.68) the failure zone length α can be considered to be the parameter adjusted to fit data for a particular material. The result (5.70) is compared with results from the next section in Problem 5.6.

5.4. CRACK GROWTH—ENERGY BALANCE APPROACH

There are many different approaches to problems in mechanics, and this is particularly true for problems in viscoelasticity for which the correspondence principle does not apply. In Section 5.3 we developed an approach for crack growth in viscoelastic materials. In this section we develop a very different approach for the same problem.

The key step in the preceding derivation was the postulated criterion (5.56) for material failure. The present approach avoids the necessity of postulating a failure criterion by appealing directly to the principle of conservation of energy in the problem formulation. Indeed, the concept of the global energy balance serves as the failure criterion in the present approach. The appeal of this method is that it is a generalization of the classical Griffith approach for cracks in elastic materials.

The specific problem to be studied here is as shown in Figure 5.1. An infinitely long strip of material is held in a configuration of fixed transverse

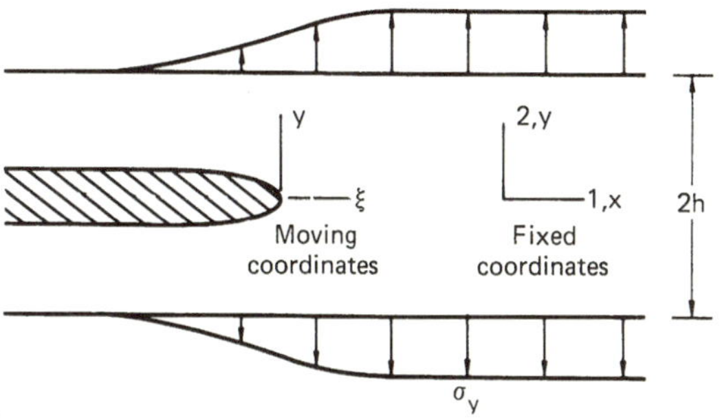

Fig. 5.1. Crack configuration.

strain ϵ_{yy}, which produces a fixed stress of value σ_y far ahead of the crack. The energy stored per unit length of the material far ahead of the crack is given by

$$\tilde{u} = h\sigma_y^2(1 - \nu^2)/2E(\infty) \tag{5.71}$$

where the material is fully relaxed and governed by the asymptote of the relaxation function $E(\infty)$, and Poisson's ratio is taken to be a constant ν. Plane stress conditions are assumed. The problem is to determine the crack velocity as a function of the stress level σ_y or, correspondingly, the strain level $\epsilon = [(1 - \nu^2)/E(\infty)]\, \sigma_\infty$. The approach is taken from the work of Christensen [5.19].

Under isothermal conditions, the appropriate statement of the global conservation of energy is given by

$$c\Gamma + \int_V \Lambda \, dv + \dot{U} = 0 \tag{5.72}$$

where c is the steady state crack velocity, Γ is the surface energy of the crack-generated free surface, V is the volume of the entire infinite strip, Λ is the rate of energy dissipation of the material, and $\dot{U}$ is the release rate of strain energy. To obtain Eq. (5.72) from the usual statement of energy conservation for this particular problem, one assumes that the work done by boundary forces is zero and there are no body forces, and inertial effects are neglected. Note that $\dot{U}$ is just given in terms of (5.71) by

$$\dot{U} = -ch\sigma_y^2(1 - \nu^2)/2E(\infty). \tag{5.73}$$

If the material were perfectly elastic, there would be no dissipation of energy ($\Lambda = 0$) and the criterion (5.72) would simply reduce to the corresponding Griffith criterion. The central problem in our present method involves the rigorous determination of the energy dissipation term Λ.

It is advantageous to work with the viscoelastic formalism that employs creep functions rather than relaxation functions. From Section 3.7 the proper expression for the rate of dissipation of energy is

$$\begin{aligned} \Lambda = {} & \frac{1}{2} \int_{-\infty}^{t} \int_{-\infty}^{t} \frac{\partial}{\partial t} J_1(2t - \tau - \eta) \frac{ds_{ij}(\tau)}{d\tau} \frac{ds_{ij}(\eta)}{d\eta} \, d\tau \, d\eta \\ & + \frac{1}{2} \int_{-\infty}^{t} \int_{-\infty}^{t} \frac{\partial}{\partial t} J_2(2t - \tau - \eta) \frac{d\sigma_{kk}(\tau)}{d\tau} \frac{d\sigma_{jj}(\eta)}{d\eta} \, d\tau \, d\eta \end{aligned} \tag{5.74}$$

where $J_1(t)$ and $J_2(t)$ are the creep functions appropriate to states of shear and hydrostatic stress, respectively.

Because steady state crack propagation conditions are assumed, it is convenient to convert to a coordinate system moving with the crack; the origin of the system is at the crack tip (Fig. 5.1). The relationship between coordinates is then

$$\xi = -ct + x. \tag{5.75}$$

Inserting the coordinate change (5.75) into Eq. (5.74) gives

$$\begin{aligned} \Lambda = & -\frac{c}{2} \int_{\infty}^{\xi} \int_{\infty}^{\xi} \frac{d}{d\xi} J_1\left(\frac{-2\xi + \xi' + \xi''}{c}\right) \frac{ds_{ij}(\xi')}{d\xi'} \frac{ds_{ij}(\xi'')}{d\xi''} \, d\xi' \, d\xi'' \\ & -\frac{c}{2} \int_{\infty}^{\xi} \int_{\infty}^{\xi} \frac{d}{d\xi} J_2\left(\frac{-2\xi + \xi' + \xi''}{c}\right) \frac{d\sigma_{kk}(\xi')}{d\xi'} \frac{d\sigma_{jj}(\xi'')}{d\xi''} \, d\xi' \, d\xi''. \end{aligned} \tag{5.76}$$

For later developments, it is important to have the limiting case form of the dissipation function for situations involving high crack velocity. In such situations, it is more useful to work with the form of Eq. (5.74) rather than with (5.76). Thus, at any fixed value of the x coordinate, let

$$\begin{aligned} s_{ij}(x, y, t) &= f_{ij}(ct - x) \\ \sigma_{kk}(x, y, t) &= F(ct - x) \end{aligned} \tag{5.77}$$

where the right-hand sides of (5.77) have a dependence on y. A change of variables is given by

$$c\tau - x = s_1 \qquad \text{and} \qquad c\eta - x = s_2 \tag{5.78}$$

which, when substituted into Eq. (5.74) and on using Eq. (5.77), give

$$\begin{aligned} \Lambda = \frac{1}{2}\int_{-\infty}^{ct-x}\int_{-\infty}^{ct-x} \frac{\partial}{\partial t} J_1\left(2t - \frac{2x + s_1 + s_2}{c}\right) \frac{df_{ij}(s_1)}{ds_1}\frac{df_{ij}(s_2)}{ds_2}\, ds_1\, ds_2 \\ + \frac{1}{2}\int_{-\infty}^{ct-x}\int_{-\infty}^{ct-x} \frac{\partial}{\partial t} J_2\left(2t - \frac{2x + s_1 + s_2}{c}\right) \frac{dF(s_1)}{ds_1}\frac{dF(s_2)}{ds_2}\, ds_1\, ds_2 \,. \end{aligned} \tag{5.79}$$

For a sufficiently large c, the integration in (5.79) can be carried out, to yield

$$\begin{aligned} \Lambda = \frac{1}{2}\frac{\partial}{\partial t} J_1\left(2t - \frac{2x}{c}\right) [f_{ij}(s_1)\,|_{-\infty}^{ct-x} f_{ij}(s_2)\,|_{-\infty}^{ct-x}] \\ + \frac{1}{2}\frac{\partial}{\partial t} J_2\left(2t - \frac{2x}{c}\right) [F(s_1)\,|_{-\infty}^{ct-x} F(s_2)\,|_{-\infty}^{ct-x}] \qquad (c \to \infty). \end{aligned} \tag{5.80}$$

Noting that

$$f_{ij}(+\infty) = F(+\infty) = 0 \tag{5.81}$$

because the stress state decays to zero after the crack passes by, Eq. (5.80) becomes

$$\begin{aligned} \Lambda = \frac{1}{2}\frac{\partial}{\partial t} J_1\left(2t - \frac{2x}{c}\right) [f_{ij}(-\infty) f_{ij}(-\infty)] \\ + \frac{1}{2}\frac{\partial}{\partial t} J_2\left(2t - \frac{2x}{c}\right) [F(-\infty)]^2 \qquad (c \to \infty). \end{aligned} \tag{5.82}$$

To obtain the volume integral of Λ,

$$\int_V \Lambda\, dv = \int_0^h \int_{-\infty}^{\infty} \Lambda\, d\xi\, dy \tag{5.83}$$

we substitute (5.82) into (5.83), to give

$$\int_V \Lambda\, dv = \frac{-hf_{ij}(-\infty)\, f_{ij}(-\infty)\, c}{2} \int_{-\infty}^{\infty} \frac{\partial}{\partial \xi} J_1\left(\frac{-2\xi}{c}\right) d\xi$$
$$- \frac{hF^2(-\infty)\, c}{2} \int_{-\infty}^{\infty} \frac{\partial}{\partial \xi} J_2\left(\frac{-2\xi}{c}\right) d\xi \qquad (c \to \infty). \tag{5.84}$$

Integrating (5.84), we then obtain

$$\int_V \Lambda\, dv = hf_{ij}(-\infty)\, f_{ij}(-\infty)\, c[J_1(\infty) - J_1(0)]$$
$$+ hF^2(-\infty)\, c[J_2(\infty) - J_2(0)] \qquad (c \to \infty). \tag{5.85}$$

For plane stress conditions we can show that

$$\begin{aligned}
f_{11}(-\infty) &= (-\tfrac{1}{3} + \tfrac{2}{3}\nu)\, \sigma_y \\
f_{22}(-\infty) &= (\tfrac{2}{3} - \tfrac{1}{3}\nu)\, \sigma_y \\
f_{33}(-\infty) &= -\tfrac{1}{3}(1 + \nu)\, \sigma_y \\
f_{12}(-\infty) &= f_{23}(-\infty) = f_{31}(-\infty) = 0 \\
F(-\infty) &= (1 + \nu)\, \sigma_y
\end{aligned}$$

where σ_y is the equilibrium stress due to the applied strain. Then, combining (5.85) with the above forms,

$$\int_V \Lambda\, dv = \tfrac{2}{3}h(1 - \nu + \nu^2)\, \sigma_y^2 c[J_1(\infty) - J_1(0)]$$
$$+ \tfrac{1}{2}h(1 + \nu)^2\, \sigma_y^2 c[J_2(\infty) - J_2(0)] \qquad (c \to \infty). \tag{5.86}$$

This result reveals that for very high rates of crack propagation the energy dissipated by the material depends only on the difference between the initial and final values of the creep functions and is independent of the details of the stress distribution. This result is of importance for later developments in our crack growth criterion.

We now return to the problem of evaluating the dissipation term Λ under more general conditions than were used to evaluate the high speed values in (5.86). To do this we must have an analytical characterization of the stress field in the problem. There is no exact analytical solution for this problem. Consistent with the developments in Section 5.2 concerning the extended correspondence principle, we assume a stress state that is independent of crack velocity and is given in series form. It is necessary to express separate forms in the regions

$\xi \gtrless 0$. First, for the region $\xi < 0$ stress fields satisfying the equilibrium equations are given by

$$\left.\begin{aligned}\sigma_{\xi y} &= \sum_{i=1}^{\infty}\sum_{j=1}^{\infty} \alpha_{ij}\, y^i e^{\beta_j \xi}\\ \sigma_{\xi\xi} &= -\sum_{i=1}^{\infty}\sum_{j=1}^{\infty} \frac{i\alpha_{ij}}{\beta_j}\, y^{i-1} e^{\beta_j \xi}\\ \sigma_{yy} &= -\sum_{i=1}^{\infty}\sum_{j=1}^{\infty} \frac{\beta_j \alpha_{ij}}{i+1}\, y^{i+1} e^{\beta_j \xi}\end{aligned}\right\} \quad (\xi < 0,\ \beta_j > 0). \tag{5.87}$$

These stress expressions satisfy vanishing traction conditions on the surface of the crack. Similarly, in the region $\xi \geqslant 0$ stresses that satisfy the equilibrium equations are given by

$$\left.\begin{aligned}\sigma_{\xi y} &= \sum_{i=1}^{\infty}\sum_{j=1}^{\infty} \gamma_{ij}\, y^i e^{-\lambda_j \xi}\\ \sigma_{\xi\xi} &= \sum_{i=1}^{\infty}\sum_{j=1}^{\infty} \frac{i\gamma_{ij}}{\lambda_j}\, y^{i-1} e^{-\lambda_j \xi} + \sigma_x\\ \sigma_{yy} &= \sum_{i=1}^{\infty}\sum_{j=1}^{\infty} \frac{\lambda_j \gamma_{ij}}{i+1}\, y^{i+1} e^{-\lambda_j \xi} + \sigma_y\end{aligned}\right\} \quad (\xi \geqslant 0,\ \lambda_j > 0) \tag{5.88}$$

where σ_x and σ_y are the stresses at large distances away from the crack tip resulting from the imposed strain state. Imposing continuity of $\sigma_{\xi y}$ and $\sigma_{\xi\xi}$ at $\xi = 0$, we obtain

$$\begin{aligned}\sum_{j=1}^{\infty} \alpha_{ij} &= \sum_{j=1}^{\infty} \gamma_{ij}\\ -\sum_{j=1}^{\infty} \frac{\alpha_{1j}}{\beta_j} &= \sum_{j=1}^{\infty} \frac{\gamma_{1j}}{\lambda_j} + \sigma_x\\ -\sum_{j=1}^{\infty} \frac{\alpha_{ij}}{\beta_j} &= \sum_{j=1}^{\infty} \frac{\alpha_{ij}}{\beta_j} \qquad (i \geqslant 2).\end{aligned} \tag{5.89}$$

Proof of the mathematical completeness of the forms in Eqs. (5.87) and (5.88) is difficult. Unquestionably, a great many terms would be needed to provide an accurate representation near the crack tip. Relations (5.87) and (5.88) are appropriate to the characterization of a stress state without a singularity, as in Section 5.3.

In the following development it is convenient to employ a retardation spectrum characterization for the viscoelastic properties. From Section 1.9, the creep functions are expressed as

$$\begin{aligned} J_1(t) &= \int_0^\infty L_1(\tau)(1 - e^{-t/\tau})\, d\tau + J_1(0) \\ J_2(t) &= \int_0^\infty L_2(\tau)(1 - e^{-t/\tau})\, d\tau + J_2(0) \end{aligned} \tag{5.90}$$

where $L_1(\tau)$ and $L_2(\tau)$ are the retardation spectra in shear and dilatation, respectively. In addition, because the viscoelastic Poisson's ratio is a constant,

$$L_2(\tau) = \frac{1 - 2\nu}{3(1+\nu)} L_1(\tau). \tag{5.91}$$

Using (5.90) and (5.91) in (5.87) and (5.88), the rate of dissipation of energy, given by (5.76), can be calculated. Then Λ can be integrated over the volume to obtain the global rate of dissipation of energy. The details are given by Christensen [5.19]. We find that the general form of the global energy dissipation function can be expressed by

$$\begin{aligned} \int_V \Lambda\, dv = \sigma_y^2 h \sum_{i=1}^\infty \sum_{j=1}^\infty \sum_{k=1}^\infty \sum_{l=1}^\infty \int_0^\infty \frac{L_1(\tau)}{\tau} \\ \times \left[\frac{\Delta_{ijkl}^{(1)}(c\tau)^5 + \Delta_{ijkl}^{(2)}(c\tau)^4 + \Delta_{ijkl}^{(3)}(c\tau)^3 + \Delta_{ijkl}^{(4)}(c\tau)^2}{\delta_{ijkl}^{(1)}(c\tau)^2 + \delta_{ijkl}^{(2)}(c\tau)^3 + \delta_{ijkl}^{(3)}(c\tau)^2 + \delta_{ijkl}^{(4)}(c\tau) + \delta_{ijkl}^{(5)}} \right] d\tau \end{aligned} \tag{5.92}$$

where $\Delta_{ijkl}^{(1)}, \ldots, \Delta_{ijkl}^{(4)}$ and $\delta_{ijkl}^{(1)}, \ldots, \delta_{ijkl}^{(5)}$ are coefficients independent of c and τ; they are restricted by Eq. (5.89) and must be evaluated from the integration process used to obtain (5.92). It is an extremely involved process to determine these coefficients in general. Fortunately, in the following developments it is not necessary to have the explicit values for these coefficients.

Substituting Eq. (5.92) into the crack growth criterion of (5.72) and using Eq. (5.73), we obtain

$$\begin{aligned} \sum_{i,j,k,l} \int_0^\infty L_1(\tau) \left[\frac{\Delta_{ijkl}^{(1)}(c\tau)^5 + \Delta_{ijkl}^{(2)}(c\tau)^4 + \Delta_{ijkl}^{(3)}(c\tau)^3 + \Delta_{ijkl}^{(4)}(c\tau)^2}{\delta_{ijkl}^{(1)}(c\tau)^5 + \delta_{ijkl}^{(2)}(c\tau)^4 + \delta_{ijkl}^{(3)}(c\tau)^3 + \delta_{ijkl}^{(4)}(c\tau)^2 + \delta_{ijkl}^{(5)}(c\tau)} \right] d\tau \\ + \frac{\Gamma}{h\sigma_y^2} - \frac{(1-\nu^2)}{2E(\infty)} = 0 \end{aligned} \tag{5.93}$$

where

$$\sum_{i,j,k,l} \sim \sum_{i=1}^\infty \sum_{j=1}^\infty \sum_{k=1}^\infty \sum_{l=1}^\infty .$$

Equation (5.93) provides the general criterion for time dependent crack growth. In the present context, it must be used to evaluate the crack velocity c. Without the explicit values for the coefficients $\Delta^{(1)}_{ijkl},\ldots,\Delta^{(4)}_{ijkl}$ and $\delta^{(1)}_{ijkl},\ldots,\delta^{(5)}_{ijkl}$, Eq. (5.93) cannot be evaluated explicitly. Even with the values for these coefficients, the solution of (5.93) for c is very complex. However, asymptotic results can be extracted for cases of very rapidly and very slowly growing cracks. These results are unexpectedly simple and clear.

Arranging the integrand in (5.93) as a power series in $(c\tau)$ gives

$$\sum_{i,j,k,l}\int_0^\infty L_1(\tau)\left\{\frac{\Delta^{(4)}_{ijkl}}{\delta^{(5)}_{ijkl}}(c\tau)+\left[\frac{\Delta^{(3)}_{ijkl}\,\delta^{(5)}_{ijkl}-\delta^{(4)}_{ijkl}\,\Delta^{(4)}_{ijkl}}{(\delta^{(5)}_{ijkl})^2}\right](c\tau)^2+\cdots\right\}d\tau$$

$$+\frac{\Gamma}{h\sigma_y^2}-\frac{(1-\nu)^2}{2E(\infty)}=0. \tag{5.94}$$

When the velocity c is sufficiently slow, all other terms in the integrand in (5.94) can be neglected in comparison with the first term. Then

$$c_S=\left[\frac{1-\nu^2}{2E(\infty)}-\frac{\Gamma}{h\sigma_y^2}\right]\Big/\left(\sum_{i,j,k,l}\frac{\Delta^{(4)}_{ijkl}}{\delta^{(5)}_{ijkl}}\right)\int_0^\infty L_1(\tau)\,\tau\,d\tau, \tag{5.95}$$

where c_S is the slow-speed asymptotic solution for the crack velocity. Before this form is interpreted, the corresponding high speed result will be obtained.

To obtain high speed results, the integrand in Eq. (5.93) is expanded as a power series in $1/c\tau$ to give

$$\sum_{i,j,k,l}\int_0^\infty L_1(\tau)\left\{\frac{\Delta^{(1)}_{ijkl}}{\delta^{(1)}_{ijkl}}+\left[\frac{\Delta^{(2)}_{ijkl}\,\delta^{(1)}_{ijkl}-\Delta^{(1)}_{ijkl}\,\delta^{(2)}_{ijkl}}{(\delta^{(1)}_{ijkl})^2}\right]\frac{1}{c\tau}+\cdots\right\}d\tau$$

$$+\frac{\Gamma}{h\sigma_y^2}-\frac{(1-\nu^2)}{2E(\infty)}=0. \tag{5.96}$$

For this form to be in accordance with Eq. (5.86) under high speed conditions it is necessary that

$$\sum_{i,j,k,l}\int_0^\infty L_1(\tau)\frac{\Delta^{(1)}_{ijkl}}{\delta^{(1)}_{ijkl}}\,d\tau-\frac{1-\nu^2}{2E(\infty)}=-\frac{1-\nu^2}{2E(0)}\cdot \tag{5.97}$$

Combining Eqs. (5.96) and (5.97) and neglecting the higher order terms in the integrand in (5.96), we find under high speed conditions that

$$c_F=\left[\sum_{i,j,k,l}\frac{\Delta^{(1)}_{ijkl}\,\delta^{(2)}_{ijkl}-\Delta^{(2)}_{ijkl}\,\delta^{(1)}_{ijkl}}{(\delta^{(1)}_{ijkl})^2}\right]\int_0^\infty\frac{L_1(\tau)}{\tau}\,d\tau\Big/\left(\frac{\Gamma}{h\sigma_y^2}-\frac{1-\nu^2}{2E(0)}\right) \tag{5.98}$$

where c_F is the high speed asymptotic solution for the crack velocity.

The results in (5.95) and (5.98) have a very simple form. It is extremely fortunate that the terms in the numerator of Eq. (5.98) and the denominator of Eq. (5.95) occur as separate factors in product form, where one term is the viscoelastic properties characterization through the retardation spectrum $L_1(\tau)$ and the second factor results only from the characteristics of the stress field given by the summation terms. Only in these asymptotic cases does this separation of effects occur.

Now we can conveniently rewrite Eqs. (5.95) and (5.98) as

$$c_S = \gamma \left\{ \left[\frac{(1-\nu^2)\, h}{2E(\infty)} - \frac{\Gamma}{\sigma_y^2} \right] \Big/ \int_0^\infty L_1(\tau)\, \tau \, d\tau \right\}$$

and (5.99)

$$c_F = \alpha \left\{ \int_0^\infty \frac{L_1(\tau)}{\tau}\, d\tau \Big/ \left[\frac{\Gamma}{h^2 \sigma_y^2} - \frac{(1-\nu^2)}{2hE(0)} \right] \right\}$$

where α and γ are nondimensional parameters dependent only on the characteristics of the fixed stress distribution, the dimension h, and Poisson's ratio. We emphasize that in any particular problem α and γ are independent of the load level and of the viscoelastic material property in shear.

The asymptotic solutions in (5.99) are the principal results of this approach. The parameters α and γ have values for a given ν that are independent of the load level or the material of application, since in this idealization the stress distribution is independent of the crack velocity, consistent with the results of Section 5.2.

It is interesting to observe that in the formula for c_S the integrand $L_1(\tau)\tau$ weights the long-time portion of the retardation spectrum, whereas in the formula for c_F the integrand $L_1(\tau)/\tau$ weights the short-time portion of the retardation spectrum; this is in agreement with physical intuition. Also, we observe that the surface energy term Γ in (5.99) represents the energy content of the new surface due to all processes involved in its creation. Thus, it is a generalization of the classical surface energy term which results only from broken bonds. Perhaps Γ is best viewed as an energy term that is balanced by the work to create the new surface, which necessarily involves very complicated processes on a molecular scale. This interpretation also applies to the derivation of Section 5.3.

In the limit of $c_S \to 0$ and $c_F \to \infty$, the formulas (5.99) are of the Griffith type. In this case the governing mechanical properties are the final asymptotic value of the relaxation function $E(\infty)$ for $c_S \to 0$ and the initial value of the relaxation function $E(0)$ for the case $c_F \to \infty$. The formulas in (5.99) contain more information than merely the limiting case forms, and they are useful for predicting crack velocities within their range of asymptotic applicability.

The results (5.99) serve to establish the validity of the approach to crack growth using the principle of conservation of energy. There is, however, an inherent limitation in these results. Specifically, in using the energy balance approach there is a separate term for the energy content of the crack generated new surface. For consistency with the presence of this surface energy term in the energy balance, a surface tension effect should be included in the solution of the related boundary value problem to determine the stress state. The surface tension effect was not considered here; rather traction-free conditions were imposed on the surface of the crack, consistent with the results of Section 5.2 on the extended correspondence principle. It is beyond the scope intended here to study the full, complicated problem with the surface tension effect included. In a simplified approach, one can take a single term stress distribution, uniform (average) across the width of the strip and exponential in ξ behind the crack tip. The energy balance method can be used to deduce the following result.

$$\sigma_y^2 = \frac{2\Gamma}{h(1-\nu^2)J(1/\beta c)} \tag{5.100}$$

where β determines the decay rate of average stress behind the crack tip, similar to (5.87), and $J(\ )$ is the creep function corresponding to $E(t)$. The significant values of stress applied to the crack surface are due to the curvature of the surface tension layer near the crack tip. Accordingly β is taken as

$$\beta = \frac{1}{\rho} \tag{5.101}$$

with ρ being the radius of curvature at the crack tip. As a practical matter, β, or correspondingly, ρ, would be treated as a parameter to be determined from crack kinetics data, taken as independent of crack velocity in the first approximation.

The crack growth theory approaches of this section and the preceding section compare well with some experimental results [5.18, 5.19]. The general field of crack kinetics is evolving. A lively and energetic discussion of crack growth theory has occurred between McCartney [5.20, 5.21] and Christensen [5.22, 5.23]. The field shows promise in application to the life prediction of load bearing polymeric structures. Certainly the results found here are persuasive arguments that practical viscoelastic problems can be solved even when the straightforward application of integral transforms is not permitted.

5.5. THERMOVISCOELASTIC STRESS ANALYSIS PROBLEM

A complete problem of interactive thermal and mechanical effects is considered here. The viscoelastic material is taken to behave in accordance with the thermorheologically simple postulate of Section 3.6. As mentioned in that section, the elastic-viscoelastic correspondence principle does not apply to such problems. We here seek a practical, numerical approach to the problem, following the work of Frutiger and Woo [5.24].

The problem is suggested by that which is encountered during the cooling phase in the manufacture of plate glass. Consider a circular plate of radius a subjected to a given nonuniform transient temperature distribution during the cooling. The governing thermorheologically simple stress strain relations, from (3.87) and (3.88), with constant thermal expansion are given by

$$s_{ij}(t) = \int_0^t 2\mu(\xi - \xi') \frac{\partial e_{ij}(\tau)}{\partial \tau} \, d\tau$$

and (5.102)

$$\sigma_{kk}(t) = \int_0^t 3k(\xi - \xi') \frac{\partial}{\partial \tau} \{\epsilon_{kk}(\tau) - 3\alpha[T(\tau) - T_0]\} \, d\tau$$

where the reduced time variables are specified through the forms

$$\xi = \int_0^t \chi(T(x_i, \eta)) \, d\eta, \qquad \xi' = \int_0^\tau \chi(T(x_i, \eta)) \, d\eta \tag{5.103}$$

in terms of the shift function $\chi(\ )$. Employing cylindrical coordinates and taking axial symmetry, we assume the plate is free to expand or contract through its thickness. Setting $\sigma_{zz} = 0$ and evaluating ϵ_{zz} then gives

$$\sigma_{rr} = \int_0^t D(\xi - \xi') \frac{\partial}{\partial \tau} [\epsilon_{rr}(\tau) + \epsilon_{\theta\theta}(\tau)] \, d\tau - 2 \int_0^t \mu(\xi - \xi') \frac{\partial \epsilon_{\theta\theta}(\tau)}{\partial \tau} \, d\tau$$
$$- 3 \int_0^t R(\xi - \xi') \frac{\partial}{\partial \tau} [\alpha(T(\tau) - T_0)] \, d\tau \tag{5.104}$$

where $R(t)$ is the same as in (3.107); i.e.,

$$\bar{R}(s) = \frac{6\bar{k}(s)\,\bar{\mu}(s)}{4\bar{\mu}(s) + 3\bar{k}(s)} \tag{5.105}$$

and

$$\bar{D}(s) = \frac{4[\bar{\mu}(s) + 3\bar{k}(s)]\,\bar{\mu}(s)}{4\bar{\mu}(s) + 3\bar{k}(s)} . \tag{5.106}$$

A similar form is found for $\sigma_{\theta\theta}$.

Following classical plate theory, take the strains as

$$\epsilon_{rr} = \frac{\partial u}{\partial r} - z \frac{\partial^2 w}{\partial r^2}$$

(5.107)

$$\epsilon_{\theta\theta} = \frac{u}{r} - \frac{z}{r} \frac{\partial w}{\partial r} .$$

Substitute (5.07) into the stress strain relations, and then integrate these over the thickness to get the force and moment resultants, such as

$$
\begin{aligned}
N_r = \int_{-h}^{h} \int_0^t \Big\{ & D(\xi - \xi') \frac{\partial}{\partial \tau} \Big[\frac{\partial u(\tau)}{\partial r} + \frac{u(\tau)}{r} - \frac{\partial^2 w(\tau)}{\partial r^2} - \frac{1}{r} \frac{\partial w(\tau)}{\partial r} \Big] \\
& + 2\mu(\xi - \xi') \frac{\partial}{\partial \tau} \Big[- \frac{u(\tau)}{r} + \frac{1}{r} \frac{\partial w(\tau)}{\partial r} \Big] \\
& - 3R(\xi - \xi') \frac{\partial}{\partial \tau} [\alpha(T(\tau) - T_0)] \Big\} \, d\tau \, dz
\end{aligned}
\tag{5.108}
$$

with similar forms for the other resultants.

The equilibrium equations give

$$
\frac{\partial}{\partial r} [r N_r(r, t)] - N_\theta(r, t) = 0
$$

and

$$
\frac{\partial}{\partial r} [r M_r(r, t)] - M_\theta(r, t) = 0. \tag{5.109}
$$

The edge conditions are taken as traction free; thus,

$$
N_r(a, t) = M_r(a, t) = 0. \tag{5.110}
$$

Take the relaxation functions in terms of decaying exponentials, as

$$
\mu(t) = G_0 + \sum_{i=1}^{I} G_i e^{-t/\tau_i}
$$

$$
k(t) = k_0 + \sum_{i=1}^{I} k_i e^{-t/\gamma_i}
$$

$$
R(t) = R_0 + \sum_{i=1}^{I} R_i e^{-t/\lambda_i} \tag{5.111}
$$

and

$$
D(t) = D_0 + \sum_{i=1}^{I} D_i e^{-t/\beta_i}.
$$

The solution procedure is as follows: evaluate the terms in $\mu(t)$ and $k(t)$ to fit the data, and then evaluate $R(t)$ and $D(t)$ from (5.105) and (5.106). Next, integrate the resultants with respect to z and t, and solve the differential equations with respect to r considering temperature $T(r, z, t)$ as given.

Now we develop the explicit numerical procedure to effect the solution. Take a typical term from (5.108) as

$$
\int_{-h}^{h} \int_0^t D(\xi(r, z, t) - \xi(r, z, \tau)) \frac{\partial}{\partial \tau} f(r, \tau) \, d\tau
$$

where $f(r, \tau)$ has the form of any of the terms in (5.108). Consider first the time integral portion of this term and write

$$F(t) = \int_0^t D(\xi(t) - \xi(\tau)) \frac{\partial}{\partial \tau} f(\tau)\, d\tau. \tag{5.112}$$

Consider discrete time intervals by taking

$$\Delta t_q = t_q - t_{q-1}, \quad q = 1, \ldots, p, \qquad t_p = t, \qquad t_0 = 0 \tag{5.113}$$

with similar terminology for ξ. Then (5.112) has the form

$$F(t) = D_0 f(t) + \sum_{q=1}^{p} \left\{ \sum_{i=1}^{I} D_i e^{-(\xi_p - \xi_q)/\beta_i} \int_{t_{q-1}}^{t_q} e^{-(\xi_q - \xi')/\beta_i}\, d\tau \right\} \frac{\Delta f(t_q)}{\Delta t_q}. \tag{5.114}$$

Interchange the orders of summation in (5.114) to get

$$F(t) = D_0 f(t) + \sum_{i=1}^{I} D_i F_p^{(i)} \tag{5.115}$$

where

$$F_p^{(i)} = \sum_{q=1}^{p} e^{-(\xi_p - \xi_q)/\beta_i} \left[\int_{t_{q-1}}^{t_q} e^{-(\xi_q - \xi')/\beta_i}\, d\tau \right] \frac{\Delta f(t_q)}{\Delta t_q}. \tag{5.116}$$

The following recursion relation for $F_p^{(i)}$ can be proved by direct substitution:

$$F_p^{(i)} = e^{-(\xi_p - \xi_{p-1})/\beta_i} \left\{ F_{p-1}^{(i)} + \left[\int_{t_{p-1}}^{t_p} e^{-(\xi_p - \xi')/\beta_i}\, d\tau \right] \frac{\Delta f(t_p)}{\Delta t_p} \right\} \tag{5.117}$$

where

$$F_0^{(i)} = 0.$$

The recursion relation (5.117) is the key to the present approach. Although the convolution integrals account for the entire past history, we see that computationally we do not need to store the entire past history. Rather, using (5.117) we only need to update from the previous time level. In computing the time integral in (5.117) we take the temperature to be constant over the interval. Then ξ varies linearly with time, and the integration can be performed to give

$$\int_{t_{p-1}}^{t_p} e^{-(\xi_p - \xi')/\beta_i}\, d\tau = \frac{\beta_i}{\chi_p} [1 - e^{-(\xi_p - \xi_{p-1})/\beta_i}]. \tag{5.118}$$

Combining (5.117) and (5.118) gives the final result

$$F_p^{(i)} = e^{-(\xi_p - \xi_{p-1})/\beta_i} \left\{ F_{p-1}^{(i)} + [1 - e^{-(\xi_p - \xi_{p-1})/\beta_i}] \frac{\beta_i}{\chi_p} \frac{\Delta f(t_p)}{\Delta t_p} \right\}. \tag{5.119}$$

Thus we see that with the recursion relation (5.119) the time dependence is completely expressed in discrete form and one need only solve for the new step of time dependence at each new time increment. All the forms in (5.108) admit time treatment of the type given in (5.119). Thus the complete time dependence of the problem has been reduced to simple algebraic operations that do not require large storage capacity in a computer. To complete the numerical process, following Frutiger and Woo [5.24], the integration over the thickness is accomplished by dividing the thickness into increments and using the trapezoidal rule. The integration with respect to the radial coordinate is done by a finite difference technique. The complete procedure reduces to a very manageable routine for a digital computer.

The float glass cooling examples given by Frutiger and Woo [5.24] reveal a very important physical effect, that of residual stresses. In the examples the shear relaxation function was modeled by the terms in (5.11), the bulk property was taken to be a constant k, and a temperature cooling range 621–30° C was taken for the tempering process. The final, residual stresses result from a competition of effects. On the one hand, the material is relaxing as temperature changes occur, and thus the induced stresses can relax out; on the other hand, the material is cooling and approaching the glassy state, which slows down the relaxation process. Ultimately, when the glassy state is reached, there remains a significant state of permanent stress, the residual stress state. In the examples the residual stresses were found to be of size $\pm$ 70 MPa. This problem illustrates effects in the manufacture of glass. Other related and difficult problems of glass processing have received a very complete treatment by Crochet [5.25]. Comparable types of problems of residual stress in polymeric resins and composites have been considered by Weitsman [5.26] and by Gurtin and Murphy [5.27].

PROBLEMS

5.1. The simple beam contact problem of Section 2.12 included a tensile force at the center of the contact region. Reformulate the problem in the case in which only compressive forces are allowed in the contact region.

5.2. What is the shape of crack tip region in the crack growth theory of Section 5.3? What would be the shape without the distribution of failure forces at the crack tip? What would be the shape of the crack tip region in the formulation mentioned in Section 5.4 which includes a surface tension layer?

5.3. Carry out the analytical procedure necessary to derive (5.63) from (5.62) in the crack growth theory involving failure forces.

5.4. Derive the general form, (5.92), of the global rate of dissipation of energy in the energy balance approach to crack growth.

5.5. The extended correspondence principle of Section 5.2 has been shown to apply to certain types of contact problems and crack growth problems. Are there any other types of problems for which this procedure applies?

5.6. Using the power law form for creep functions given in Problem 4.3, compare the forms of the two crack growth theories given by (5.70) and by (5.100). Specifically, in the creep function form

$$J(t) = J(0) + \tilde{J} t^n$$

neglect the initial value of the creep function $J(0)$, and compare the two theories as a function of exponent n, $0 \leqslant n \leqslant 1$.

5.7. Using the defining form of the creep spectrum (5.90) put the high speed asymptotic formula (5.99) into the following form:

$$c_F = \frac{\alpha d J_1(t)/dt \,|_{t=0}}{\Gamma/h_y^2 \sigma^2 - (1 - \nu^2)/2hE(0)} \,.$$

Because $E(0) = 2(1 + \nu)]J_1(0)$, this formula expresses the high speed crack velocity as a function of the initial value and initial slope of the creep function.

5.8. Consider an idealized two bar linkage subject to cooling.

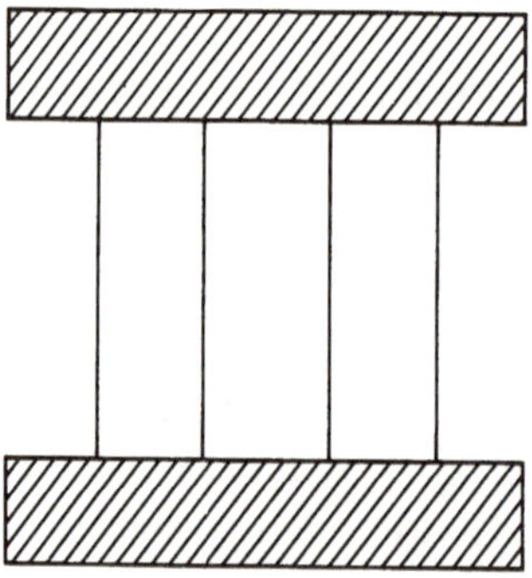

Take the end fixtures to be rigid and remaining parallel but unconstrained by external force. Subject the two viscoelastic bars to uniform temperature cooling at exponential rates that are different for each bar. Assume the material is thermorheologically simple, and formulate an analysis of the type given in Section 5.5 to determine the residual stresses.

5.9. Verify the recursion relation (5.117), used in the residual stress problem of Section 5.5.

REFERENCES

5.1. Lee, E. H., and J. R. M. Radok, "The Contact Problem for Viscoelastic Bodies," *J. Appl. Mech.* **27**, 438 (1960).

5.2. Hunter, S. C., "The Hertz Problem for a Rigid Spherical Indentor and a Viscoelastic Half-Space," *J. Mech. Phys. Solids* **8**, 219 (1960).

5.3. Graham, G. A. C., "The Contact Problem in the Linear Theory of Viscoelasticity," *Int. J. Eng. Sci.* **3**, 27 (1965).

5.4. Graham, G. A. C., "The Contact Problem in the Linear Theory of Viscoelasticity When the Time Dependent Contact Area Has Any Number of Maxima and Minima," *Int. J. Eng. Sci.* **5**, 495 (1967).

5.5. Ting, T. C. T., "Contact Problems in the Linear Theory of Viscoelasticity," *J. Appl. Mech.* **35**, 248 (1968).

5.6. Sabin, G. C. W., and G. A. C. Graham, "The Normal Aging Viscoelastic Contact Problem," *Int. J. Eng. Sci.* **18**, 751 (1980).

5.7. Calvit, H. H., "Numerical Solution of the Problem of Impact of a Rigid Sphere onto a Linear Viscoelastic Half-Space and Comparison with Experiment," *Int. J. Solids Structures* **3**, 951 (1967).

5.8. Graham, G. A. C., "The Correspondence Principle of Linear Viscoelasticity Theory for Mixed Boundary Value Problems Involving Time-Dependent Boundary Regions," *Quart. Appl. Math.* **26**, 167 (1968).

5.9. Ting, T. C. T., "A Mixed Boundary Value Problem in Viscoelasticity with Time-Dependent Boundary Regions, *in Developments in Mechanics* (H. J. Weis, D. F. Young, W. F. Riley, and T. R. Rogge, eds.), Vol. 5, p. 591. Iowa State Univ. Press, Ames, Iowa, 1969.

5.10. Hunter, S. C., "The Rolling Contact of a Rigid Cylinder with a Viscoelastic Half-Space," *J. Appl. Mech.* **28**, 611 (1961).

5.11. Morland, L. W., "A Plane Problem of Rolling Contact in Linear Viscoelasticity Theory," *J. Appl. Mech.* **29**, 345 (1962).

5.12. Morland, L. W., "Exact Solutions for Rolling Contact between Viscoelastic Cylinders," *Quart. J. Mech. Appl. Math.* **20**, 73 (1967).

5.13. Golden, J. M., "The Problem of a Moving Rigid Punch on an Unlubricated Viscoelastic Half-Plane," *Quart. J. Mech. Appl. Math.* **32**, 25 (1980).

5.14. Sneddon, I. N., "The Relation Between Load and Penetration in the Axisymmetric Boussinesq Problem for a Punch of Arbitrary Profile," *Int. J. Eng. Sci.* **3**, 47 (1965).

5.15. Graham, G. A. C., "Two Extending Crack Problems in Linear Viscoelasticity Theory," *Quart. Appl. Math.* **27**, 497 (1970).

5.16. Schapery, R. A., "A Theory of Crack Initiation and Growth in Viscoelastic Media. I. Theoretical Development," *Int. J. Fract.* **11**, 141 (1975).

5.17. Schapery, R. A., "A Theory of Crack Initiation and Growth in Viscoelastic Media. II. Approximate Methods of Analysis," *Int. J. Fract.* **11**, 369 (1975).

5.18. Schapery, R. A., "A Theory of Crack Initiation and Growth in Viscoelastic Media. III. Analysis of Continuous Growth," *Int. J. Fract.* **11**, 549 (1975).

5.19. Christensen, R. M., "A Rate-Dependent Criterion for Crack Growth," *Int. J. Fract.* **15**, 3 (1979).

5.20. McCartney, L. N., Discussion, *Int. J. Fract.* **16**, R229 (1980).

5.21. McCartney, L. N., Discussion, *Int. J. Fract.* **17**, R161 (1981).

5.22. Christensen, R. M., Response, *Int. J. Fract.* **16**, R233 (1980).

5.23. Christensen, R. M., Response, *Int. J. Fract.* **17**, R169 (1981).

5.24. Frutiger, R. L., and T. C. Woo, "A Thermoviscoelastic Analysis for Circular Plates of Thermorheologically Simple Material," *J. Therm. Stress* **2**, 45 (1979).

5.25. Crochet, M. J., and A. Denayer, "Transient and Residual Thermoviscoelastic Stresses in Glass," *J. Appl. Mech.* **47**, 254 (1980).
5.26. Weitsman, Y., "Optimal Cool-Down in Linear Viscoelasticity," *J. Appl. Mech.* **47**, 35 (1980).
5.27. Gurtin, M. E., and L. F. Murphy, "On Optimal Temperature Paths for Thermorheologically Simple Viscoelastic Materials," *Quart. Appl. Math.* **28**, 179 (1980).

Chapter VI Wave Propagation

The propagation of waves in viscoelastic materials has many different and important aspects. Several such aspects are covered here for isotropic media. In the first section the isothermal transient propagation of disturbances in semi-infinite rods is analyzed. Some of the results of this study are generalized to the case of the propagation of plane disturbances in three-dimensional media. Specifically, the two possible speeds of propagation of planes of discontinuous stress and strain in infinite media are found. The study of the characteristics of the propagation of harmonic waves is deferred until Sections 6.3 and 6.4.

The complications involved in solving transient wave propagation problems in viscoelasticity motivates the development of a special method of solving such problems, which is applicable in certain cases. Specifically, in Section 6.2, a rather complete means is given for analyzing dynamic response problems for bodies of finite extent. The method given is more practical to use than the ordinary direct application of integral transform techniques usually used in wave propagation studies.

Another wave propagation problem of importance involves the propagation of plane harmonic waves in unlimited media. This case is analyzed, and, in fact, coupling effects with thermal waves are considered, as an application of the coupled thermoviscoelastic theory developed in Section 3.1. The specialization of the results in this section to isothermal conditions gives the speeds of propagation and rates of attenuation of isothermal harmonic waves in three-dimensional media. In the comparable elasticity theory, pulses can be propagated through the media without change of shape or attenuation. But now the results of this section show that speeds of propagation of harmonic waves in viscoelastic media are frequency dependent. Consequently, pulses composed of a spectrum of harmonic waves experience a dispersive effect, and the shape of the pulse cannot be preserved. This effect, along with the attenuation effect in the viscoelastic waves, reveals the striking distinction between the characteristics of wave propagation in viscoelastic and elastic media.

In the study of Section 6.3, only plane waves are considered, and as such, the directions of propagation and attenuation are restricted to be the same. In Section 6.4, it is shown that this restriction need not always be the case. The problem studied in this section is that of the reflection of harmonic waves from the free surface of a half space. It is shown that one of the two types of reflected waves may indeed have different directions of propagation and attenuation.

In Section 6.5 the technically important but difficult problem of a very rapidly moving load on a viscoelastic half space is studied. Some specific analytical results are found although a complete solution is not possible. In the last section, viscoelastic Rayleigh waves are treated. A solution rich in physical effects is found which shows the very limited form of results obtainable from the corresponding elastic solution. The viscoelastic solution is of obvious relevance in geophysics.

6.1. ISOTHERMAL WAVE PROPAGATION

Uniaxial Case

When an elastic body is subjected to a discontinuous change in stress over all or part of its boundary, this discontinuity in stress is propagated throughout the body. In the elastic case the speed of propagation of the singular surface across which stress undergoes a jump can easily be determined in several situations. We now consider the corresponding viscoelastic problem. Specifically, we begin by seeking to determine the speed at which a discontinuity in stress and strain will be propagated along a viscoelastic bar, assuming plane wave conditions, and neglecting lateral inertia effects.

The equation of motion is not valid at the singular surface across which stress is discontinuous. However, a jump condition which is valid at the singular surface can be derived from the global form of the balance of linear momentum; see for example Thomas [6.1]. Such a jump condition, in the present context, has the form

$$[\![\sigma(x, t)]\!] = -\rho v \left[\!\!\left[\frac{\partial u(x, t)}{\partial t}\right]\!\!\right] \tag{6.1}$$

where σ and u are the uniaxial stress and displacement respectively, ρ is the mass density, v is the speed of propagation of the singular surface, and the notation $[\![h]\!]$ for the jump in the variable $h(x, t)$ across the singular surface is defined by

$$[\![h]\!] = \lim_{\substack{x^+ \to x \\ t^+ \to t}} h(x^+, t^+) - \lim_{\substack{x^- \to x \\ t^- \to t}} h(x^-, t^-)$$

with x^+, t^+ and x^-, t^- being on opposite sides of the singular surface at x, t.

The uniaxial viscoelastic constitutive relation is

$$\sigma(x, t) = \int_{-\infty}^{t} E(t - \tau) \frac{\partial \epsilon(x, \tau)}{\partial \tau} \, d\tau. \tag{6.2}$$

Integrating (6.2) by parts and substituting into (6.1) gives

$$\left[\!\!\left[E(0) \frac{\partial u(x, t)}{\partial x} + \int_{-\infty}^{t} \frac{\partial}{\partial t} E(t - \tau) \frac{\partial u(x, \tau)}{\partial x} \, d\tau\right]\!\!\right] = -\rho v \left[\!\!\left[\frac{\partial u(x, t)}{\partial t}\right]\!\!\right] \tag{6.3}$$

where the strain displacement relation $\epsilon = \partial u/\partial x$ has been used. We are concerned with the propagation of discontinuous stress and strain waves, which implies the propagation of discontinuities in the first derivatives of displacement. These types of waves are commonly called shock waves. Assuming that $dE(t)/dt$ is bounded and continuous for $0 \leqslant t < \infty$, then the term involving the integral in (6.3) contributes nothing to the jump because $\partial u(x, \tau)/\partial x$ is continuous except at $\tau = t$ and the integral is independent of $\partial u(x, t)/\partial x$. Thus, (6.3) reduces to

$$E(0) \left[\!\left[\frac{\partial u(x, t)}{\partial x}\right]\!\right] = -\rho v \left[\!\left[\frac{\partial u(x, t)}{\partial t}\right]\!\right]. \tag{6.4}$$

From Thomas [6.1] the kinematical condition relating the jumps in the first derivatives of displacement is given by

$$\left[\!\left[\frac{\partial u(x, t)}{\partial t}\right]\!\right] = -v \left[\!\left[\frac{\partial u(x, t)}{\partial x}\right]\!\right]. \tag{6.5}$$

Combining (6.4) and (6.5) results in

$$E(0) \left[\!\left[\frac{\partial u(x, t)}{\partial x}\right]\!\right] = \rho v^2 \left[\!\left[\frac{\partial u(x, t)}{\partial x}\right]\!\right].$$

Thus, the speed of propagation is given by

$$v = [E(0)/\rho]^{1/2}. \tag{6.6}$$

Relation (6.6) specifies the speed of propagation of a uniaxial viscoelastic shock wave as being determined by the initial value of the uniaxial relaxation function and the density.

The same wave speed result can be determined through the direct application of integral transform methods to the equation of motion. However, such a procedure is necessarily less rigorous than that just followed, due to the fact that the derivation of the equation of motion, a local equation, assumes conditions which are violated at the singular surface of a shock wave. Nevertheless, such procedures are widely used and, in the present context, give results in accordance with those found through the rigorously established jump conditions.

Either the Fourier or Laplace transform can be used to solve wave propagation problems. If the Fourier transform is used, the procedure will follow the same steps as in the example of Section 2.10 which is concerned with one-dimensional wave propagation in a spherically symmetric condition. Accordingly, the response of the rod, expressed in terms of either stress, strain, velocity, or displacement can be found in terms of a real infinite integral in the frequency domain. In certain special cases this integral can be evaluated through the use of residue theory, but, in general, the evaluation is not practical analytically, for reasons which presently will be explained. The approach to be followed here uses the Laplace transform method, but the difficulties which are encountered are common to all integral transform methods. The governing equation of motion is

$$\frac{\partial \sigma(x, t)}{\partial x} = \rho \frac{\partial^2 u(x, t)}{\partial t^2}. \tag{6.7}$$

The Laplace transforms of (6.2) and (6.7) when combined give

$$s\bar{E}(s)\frac{\partial \bar{\epsilon}(x, s)}{\partial x} = \rho s^2 \bar{u}(x, s) \tag{6.8}$$

where the rod is assumed to be initially at rest. The strain displacement relation $\epsilon = \partial u/\partial x$, when used in (6.8), gives

$$\frac{\partial^2 \bar{u}}{\partial x^2} - \frac{\rho s}{\bar{E}(s)}\bar{u} = 0. \tag{6.9}$$

Through using the transform of (6.2) and the strain displacement relation, we get an alternative form to (6.9):

$$\frac{\partial^2 \bar{\sigma}}{\partial x^2} - \frac{\rho s}{\bar{E}(s)}\bar{\sigma} = 0. \tag{6.10}$$

Exactly the same form of equations as (6.9) and (6.10) can be shown to govern the Laplace transformed strain $\bar{\epsilon}$ and particle velocity $\bar{v} = s\bar{u}$.

The general solution of these equations is given by

$$(\bar{\sigma}, \bar{\epsilon}, \bar{u}, \bar{v}) = A(s)e^{\Omega(s)x} + B(s)e^{-\Omega(s)x} \tag{6.11}$$

where

$$\Omega(s) = [\rho s/\bar{E}(s)]^{1/2} \tag{6.12}$$

where $A(s)$ and $B(s)$, which are to be determined from boundary conditions, are appropriate to any one of the four solution variables in (6.11). A particular type of boundary condition will be considered in the following work.

Semi-Infinite Rod

The specific case of a semi-infinite rod, with the end at $x = 0$, is now considered. The constant A must be taken equal to zero to have waves propagating in the $+x$ direction only. Thus, the solution for the stress is given by

$$\bar{\sigma} = B(s)e^{-\Omega(s)x}. \tag{6.13}$$

Some general information can be obtained from (6.13) before attempting to find specific solutions. We write the inversion of (6.13) which, from Appendix B, is given by

$$\sigma(x, t) = \frac{1}{2\pi i}\int_{\gamma-i\infty}^{\gamma+i\infty} B(s)\, e^{-\Omega(s)x+st}\, ds \tag{6.14}$$

where the real number γ is to the right of all singularities in the complex s plane.

The contour is closed to the right of $s = \gamma$ with an infinite semicircle, which has $|s| \to \infty$. The exponential term in (6.14) is written as

$$\exp[-s(x(\rho/s\bar{E})^{1/2} - t)].$$

But, for $|s| \to \infty$, it follows that $s\bar{E}(s) \to E(0)$ from the initial value theorem of Laplace transform theory. For $|s| \to \infty$, $t < x[\rho/E(0)]^{1/2}$ it further follows that $\exp[-s(x(\rho/s\bar{E})^{1/2} - t)] \to 0$. Thus, under these conditions, the contour integral along the infinite semicircular arc vanishes, and, since there are no singularities inside the contour, Cauchy's integral theorem gives the corresponding stress in (6.14) as being identically equal to zero. That is,

$$\sigma(x, t) = 0 \qquad \text{for} \quad t < x[\rho/E(0)]^{1/2}. \tag{6.15}$$

Relationship (6.15) and the corresponding ones for displacement, velocity, and strain show that disturbances in the viscoelastic rod can propagate no faster than the speed $[E(0)/\rho]^{1/2}$. This is simply the speed of propagation of disturbances in an elastic rod with a modulus $E = E(0)$, the initial value of the relaxation function.

A particular example is now worked out. This example, from Lee and Morrison [6.2], is that of the wave propagation in a material represented by a Maxwell model, which is discussed in Section 1.5. The relaxation function appropriate to the Maxwell model is given by

$$E(t) = Ee^{-t/\tau}, \qquad \tau = \eta/E \tag{6.16}$$

where E and η are the mechanical model parameters. The Laplace transform of (6.16) with undisturbed initial conditions, when combined with (6.12), gives

$$\Omega(s) = (\rho/E)^{1/2}(s^2 + Es/\eta)^{1/2}. \tag{6.17}$$

The end condition is taken as that of a discontinuous change in particle velocity resembling impact, given by

$$v(0, t) = Vh(t). \tag{6.18}$$

We evaluate $B(s)$ in (6.13) from the Laplace transform of (6.18). This gives

$$\bar{\sigma}(x, s) = [-\rho cV/(s^2 + Es/\eta)^{1/2}] \exp[-c^{-1}x(s^2 + Es/\eta)^{1/2}] \tag{6.19}$$

where $c = (E/\rho)^{1/2}$. The problem is now to invert (6.19). Relation (6.19) is noted to have branch points which are involved in the contour integration process of the inversion. Actually, the inverse of (6.19) is obtained from a table of Laplace transforms as

$$\sigma(x, t) = \rho cVe^{-(Et/2\eta)}I_0\left(\frac{E}{2\eta}\left(t^2 - \frac{x^2}{c^2}\right)^{1/2}\right) h\left(t - \frac{x}{c}\right) \tag{6.20}$$

where $I_0(\ )$ is the modified Bessel function of the first kind of order zero. Although the inversion of (6.19) was easily obtained, it can also be seen that more complicated stress strain relations than (6.16) would result in an inversion problem that quickly becomes intractable. The same considerations also apply to the inversion of Fourier transformed solutions. This use of the Maxwell model in solving a rod response problem is of course highly artificial, not only for the reason of the usual limitations of simple mechanical models, but also because the Maxwell model, as used here, implies the characteristics of a fluid.

The transform inversion difficulties just mentioned can be overcome if an asymptotic solution is sought. This approach is now illustrated with an example. In this example, the end condition is taken to be the discontinuous change in stress given by

$$\sigma(0, t) = ph(t). \tag{6.21}$$

By evaluating $B(s)$ in (6.13), from (6.21) we get

$$\bar{\sigma}(x, s) = s^{-1}pe^{-\Omega(s)x} \tag{6.22}$$

where $\Omega(s)$ is given by (6.12). Now we expand $E(t)$ about $E(0)$ in a Taylor series, as

$$E(t) = E(0) + tE'(0) + (t^2/2)\, E''(0) + \cdots \tag{6.23}$$

where $E'(0) = [dE(t)/dt]_{t=0}$ with similar forms for the higher derivatives. The Laplace transform of (6.23) is then

$$\bar{E}(s) = \frac{E(0)}{s} + \frac{E'(0)}{s^2} + \cdots. \tag{6.24}$$

We substitute (6.24) into (6.12) and carry out the indicated division to obtain

$$\Omega(s) = \left[\frac{\rho s^2}{E(0)} - \frac{\rho s E'(0)}{(E(0))^2} + \cdots\right]^{1/2}. \tag{6.25}$$

We apply the binomial expansion to this to get

$$\Omega(s) = s\left[\frac{\rho}{E(0)}\right]^{1/2} - \frac{1}{2}\left[\frac{\rho}{E(0)}\right]^{1/2}\frac{E'(0)}{E(0)} + \cdots \tag{6.26}$$

where the additional terms in (6.26) involve positive powers of $(1/s)$. We already know that the solution has the form

$$\sigma(x, t) = f(x, t)\, h(t - (\rho/E(0))^{1/2}x) \tag{6.27}$$

where $f(x, t)$ is some as yet undetermined function of its arguments. As in (6.15), (6.27) reveals the maximum rate of propagation of disturbances. But more than this can be obtained from (6.26); using the initial value and shifting theorems of Laplace transform theory with (6.22) and (6.26), we get

$$\sigma(x, t)\,|_{t=(\rho/E(0))^{1/2}x} = p \exp[\tfrac{1}{2}(\rho/E(0))^{1/2}(E'(0)/E(0))x]. \tag{6.28}$$

This gives the magnitude of a propagating discontinuity in stress in a viscoelastic rod, with no restriction to a particular mechanical model. Relation (6.28) shows that the decay of the wave front, or discontinuity, is directly determined by the initial slope of the relaxation function $E'(0) \leqslant 0$. This same type of procedure can be used to obtain an asymptotic solution in the region of the propagating front at $x = [E(0)/\rho]^{1/2}t$.

Achenbach and Reddy [6.3] have obtained an asymptotic solution, similar to that described above, which gives the solution as a function of time at a fixed location, after the propagating discontinuity has passed by. Berry [6.4] was the first to determine the maximum speed of propagation of viscoelastic waves. Chu [6.5] studied several different effects in viscoelastic wave propagation and first derived the expression for the decay of a wave front. There have been many different types of manipulations of the equations governing viscoelastic wave propagation in order to put then in a form most suitable for numerical evaluation. Such studies include those of Sackman and Kaya [6.6], Valanis and Chang [6.7], and Knauss [6.8].

Infinite Media

In the rod analysis just given, discontinuities in stress and strain are shown to propagate with the velocity $[E(0)/\rho]^{1/2}$ where the relaxation function $E(t)$ is appropriate to uniaxial conditions. To investigate the similar possibility in unlimited media, the three-dimensional equations of motion must be employed. The Laplace transform of the equations of motion results in

$$s\bar{\mu}(s)\,\bar{u}_{i,jj}(x_i, s) + s[\bar{\lambda}(s) + \bar{\mu}(s)]\,\bar{u}_{k,ki}(x_i, s) = \rho s^2\bar{u}_i(x_i, s) \tag{6.29}$$

where undisturbed initial conditions are assumed.

We consider two special types of motion: First we consider the case where the divergence of the displacement vector vanishes. This is specified by $u_{k,k} = 0$, which implies a shear disturbance, and (6.29) thereby reduces to

$$\bar{u}_{i,jj} - \frac{\rho s}{\bar{\mu}(s)}\bar{u}_i = 0. \tag{6.30}$$

In the second case, we assume that the curl of the displacement vector vanishes, which implies an irrotational motion. As in elasticity theory this condition reduces (6.29) to the form

$$\bar{u}_{i,jj} - \frac{\rho s}{\bar{\lambda}(s) + 2\bar{\mu}(s)}\bar{u}_i = 0. \tag{6.31}$$

Equations (6.30) and (6.31) have exactly the same form as equation (6.9), which applies to the case of wave propagation in a rod. By analogy with the results obtained for the rod, it follows that plane disturbances governed by (6.30) can propagate no faster than at the speed $[\mu(0)/\rho]^{1/2}$ and plane disturbances governed by (6.31) propagate no faster than at the speed $[(\lambda(0) + 2\mu(0))/\rho]^{1/2}$. In other words, shear (transverse) discontinuities propagate at the speed $[\mu(0)/\rho]^{1/2}$ while irrotational (longitudinal) discontinuities propagate at the speed $[(\lambda(0) + 2\mu(0))/\rho]^{1/2}$. The same result can be established in a more rigorous manner through the use of the proper jump conditions of the type which were employed here in the uniaxial case. This is the approach followed by Fisher and Gurtin [6.9] who consider not only shock waves but also waves involving discontinuities in second derivatives of displacement (acceleration waves) and higher order waves. A similar type of study was given by Valanis [6.10].

Both types of disturbances just discussed are involved in propagating waves in the general case of unlimited three-dimensional media. In this case, the Helmholtz resolution is employed to represent the tranform of the displacement vector as

$$\bar{u}_i = \bar{\varphi}_{,i} + (\text{curl } \bar{\psi})_i$$

where $\varphi(x_i, t)$ is a scalar potential and $\psi(x_i, t)$ is a vector potential with $(\text{curl } \bar{\psi})_i$ being the ith component of the curl of $\bar{\psi}$. Using this representation in the transformed equations of motion (6.29), we find that these equations are satisfied if

$$\bar{\varphi}_{,jj} - \frac{\rho s}{\bar{\lambda}(s) + 2\bar{\mu}(s)} \bar{\varphi} = 0$$

and

$$\bar{\psi}_{i,jj} - \frac{\rho s}{\bar{\mu}(s)} \bar{\psi}_i = 0.$$

In comparing these relations with (6.30) and (6.31) or through independent reasoning we see that the scalar potential φ governs the propagation of irrotational waves while the vector potential ψ governs the propagation of shear waves. The maximum speeds of propagation of these two types of waves have the values already mentioned.

The brief description given in this section focuses on the speed of propagation of discontinuous disturbances. Sections 6.3 and 6.4 are, among other things, concerned with the speeds of propagation of harmonic waves in viscoelastic media.

6.2. DYNAMIC RESPONSE PROBLEMS

In addition to problems of the type just discussed, another class of practical problems is that of the propagation of disturbances in bodies of finite extent with regard to the coordinate directions in which the field variables are not constant. We consider for example, the problem of the dynamic response of finite length rods. Certainly the transform solution (6.11) can be evaluated to satisfy both end conditions, and the problem is similar to that of the semi-infinite rod. Some simplification can be achieved by following the argument of Lee and Kanter [6.11], which shows that the solution for a finite length viscoelastic rod can be obtained directly from the solution for the semi-infinite rod through the superposition of waves traveling both left and right. This is similar to the means of solution in elastic rods. This method does not, however, provide any simplification in the integral transform inversion problem, which still must be accomplished for the semi-infinite rod. And, in general, the direct application of integral transform methods in dynamic response problems always leads to a complicated inversion process. The more involved forms of the viscoelastic stress strain relations inevitably lead to the necessity for more branch cuts in the contour integration inversion process. The entire contour integration procedure quickly becomes impractical for reasonable mechanical properties representations. An alternative method of solution is now given, which completely eliminates these difficulties for certain types of dynamic response problems.

The method of solution is an adaptation to viscoelasticity of the method of time dependent boundary conditions in elasticity theory, developed by Mindlin and Goodman [6.12], Herrmann [6.13], and Berry and Naghdi [6.14]. The types of problems to be considered here are those problems in which the viscoelastic Poisson's ratio is a real constant, and for which the boundary conditions have the separable form given by

$$\begin{aligned} \sigma_{ij} n_j &= S_i(x_i)\, p(t) \qquad \text{on} \quad B_\sigma \\ u_i &= U_i(x_i)\, u(t) \qquad \text{on} \quad B_u . \end{aligned} \tag{6.32}$$

It should be noted that the above form for the boundary conditions is not as restrictive as it might at first seem, since a much wider class of problems can be obtained through superposition.

The Laplace transformed equations of motion are given by

$$\bar{u}_{i,jj} + \frac{1}{1-2\nu}\bar{u}_{j,ji} = \frac{\rho s}{\bar{\mu}(s)}\bar{u}_i \tag{6.33}$$

where the initial conditions on displacement and velocity have been taken identically equal to zero. The form of the solution is taken as

$$u_i(x_i\,, t) = \sum_n C_n(t)\, U_i^n(x_i) + v_i(x_i\,, t) \tag{6.34}$$

where $C_n(t)$ are unknown functions of time which are to be determined, and U_i^n and v_i are obtained in the following manner. The functions $U_i^n(x_i)$ are the eigenvectors from the associated elastic eigenvalue problem specified by

$$U_{i,jj} + \frac{1}{1-2\nu} U_{j,ji} + k^2 U_i = 0 \tag{6.35}$$

where

$$\begin{aligned} \sigma_{ij} n_j &= 0 \quad \text{on} \quad B_\sigma \\ u_i &= 0 \quad \text{on} \quad B_u . \end{aligned} \tag{6.36}$$

The corresponding eigenvalues k_n and eigenvectors U_i^n then satisfy the relation

$$U_{i,jj}^n + \frac{1}{1-2\nu} U_{j,ji}^n = -k_n^2 U_i^n. \tag{6.37}$$

The functions $v_i(x_i, t)$ represent the quasi-static solution of the viscoelastic boundary value problem. That is, $v_i(x_i, t)$ satisfy

$$v_{i,jj} + \frac{1}{1-2\nu} v_{j,ji} = 0 \tag{6.38}$$

where

$$\begin{aligned} \sigma_{ij} n_j &= S_i(x_i)\, p(t) \quad \text{on} \quad B_\sigma \\ v_i &= U_i(x_i)\, u(t) \quad \text{on} \quad B_u . \end{aligned} \tag{6.39}$$

The transformed quasi-static solution $\bar{v}_i(x_i, s)$ can be written in the form

$$\bar{v}_i(x_i, s) = \frac{V_i(x_i)\, \bar{p}(s)}{s\bar{\mu}(s)} + W_i(x_i)\, \bar{u}(s) \tag{6.40}$$

where $V_i(x_i)$ and $W_i(x_i)$ are the spatial parts of the quasi-static solution.

With U_i^n and v_i defined in this manner, we see that the form of the solution (6.34) identically satisfies the boundary conditions (6.32) of the dynamic response problem. It remains only to satisfy the equations of motion (6.33). We substitute the displacements (6.34) into the equations of motion (6.33) and use (6.40) to get

$$\sum_n \bar{C}_n \left(U_{i,jj}^n + \frac{1}{1-2\nu} U_{j,ji}^n - \frac{\rho s}{\bar{\mu}(s)} U_i^n \right) = \frac{\rho \bar{p}}{\bar{\mu}^2} V_i + \frac{\rho s \bar{u}}{\bar{\mu}} W_i . \tag{6.41}$$

Using (6.37), this can be written as

$$\sum_n \bar{C}_n (k_n^2 \bar{\mu} + \rho s)\, U_i^n = \frac{-\rho \bar{p}}{\bar{\mu}} V_i - \rho s \bar{u} W_i . \tag{6.42}$$

The orthogonality of the elastic eigenvectors U_i^n, in normalized form, allows $\bar{C}_n$ to be determined from (6.42) as

$$\bar{C}_n = \frac{-\rho\bar{p}}{\bar{\mu}(k_n^2\bar{\mu} + \rho s)} \int_V V_i(x_i)\, U_i^n(x_i)\, dv - \frac{\rho s\bar{u}}{(k_n^2\bar{\mu} + \rho s)} \int_V W_i(x_i)\, U_i^n(x_i)\, dv. \tag{6.43}$$

By taking the relaxation function $\mu(t)$ as a series of decaying exponentials, we see that (6.43) involves only simple poles, and the inversion process may be carried out directly to complete the solution.

To summarize, this means of solving the viscoelastic dynamic response problems, for which Poisson's ratio must be a real constant, involves solving the corresponding quasi-static viscoelastic problem, the corresponding elastic eigenvalue problem, and inverting (6.43). As was shown in Chapter 2, it is entirely practical to solve quasi-static problems with realistic mechanical properties representations. Similarly, the form (6.43) is likewise amenable to inversion for realistic mechanical properties representations. All these steps are straightforward and practical for a number of technically important problems. Such solutions have been discussed and given by Achenbach [6.15] and Christensen [6.16]. Also, Valanis [6.17] uses a method similar but slightly different from that used here and has solved a number of important problems.

Finite Length Rods

Two simple examples are now given to illustrate the method just presented. The first problem is that of determining the response of a viscoelastic rod with one end fixed and the other end subjected to a suddenly applied load.

We assume a displacement solution in the form

$$u(x, t) = \sum_{n=1}^{\infty} C_n(t) \sin \frac{(2n-1)\pi x}{2L} + K(t)\frac{x}{L}. \tag{6.44}$$

The first term on the right-hand side represents the eigenfunctions of an elastic fixed-free bar of length L with coordinate x originating from the fixed end. The last term in (6.44) corresponds to the quasi-static solution of a fixed-free bar acted upon by an end stress. In this term $K(t)$ is a function of time, which will be evaluated to satisfy the time dependent boundary condition. This boundary condition is taken as

$$\sigma(x, t) = ph(t), \qquad \text{at} \quad x = L. \tag{6.45}$$

It is found that the Laplace transform of (6.45) is satisfied by

$$\bar{K}(s) = pL/s^2\bar{E}(s) \tag{6.46}$$

where (6.2), (6.44), and $\epsilon = \partial u/\partial x$ have been used.

We substitute (6.44) and (6.46) into the equation of motion (6.7) with the result

$$\sum_{n=1}^{\infty}\left[\frac{(2n-1)^2\pi^2}{4L^2}-\frac{\rho s}{\bar{E}(s)}\right]\bar{C}_n(s)\sin\frac{(2n-1)\,\pi x}{2L}=-\frac{\rho p x}{s(\bar{E}(s))^2}. \quad (6.47)$$

The coefficients $\bar{C}_n$ are evaluated by multiplying (6.47) by $\sin[(2m-1)\pi x/(2L)]$, $m=1, 2, 3,\ldots$, and integrating over the interval 0 to L, which with the utilization of the appropriate orthogonality conditions gives

$$\bar{C}_n(s)=\frac{16\rho p L^4(-1)^n}{(2n-1)^2\pi^2 s\bar{E}(s)[(2n-1)^2\pi^2\bar{E}(s)+4L^2\rho s]}. \quad (6.48)$$

With (6.48) the transform of the stress solution is then given by

$$\bar{\sigma}(x,s)=\frac{p}{s}+\sum_{n=1}^{\infty}\frac{8\rho p L^3(-1)^n\cos[(2n-1)\pi x/(2L)]}{(2n-1)\,\pi[(2n-1)^2\pi^2\bar{E}(s)+4L^2\rho s]}. \quad (6.49)$$

A general viscoelastic material is accounted for by representing the relaxation function in the form

$$E(t)=\sum_{j=0}^{N}E_j e^{-t/\tau_j}. \quad (6.50)$$

The transform of (6.50) is written as

$$\bar{E}(s)=A(s)/B(s) \quad (6.51)$$

where $A(s)$ is a polynomial of degree N in s, and $B(s)$ is a polynomial of degree $N+1$ in s. We let the polynomial in s, $P_n(s)$, be defined by

$$P_n(s)=(2n-1)^2\pi^2A(s)+4L^2\rho sB(s) \quad (6.52)$$

and write it in factored form as

$$P_n(s)=D_n\prod_{j=1}^{N+2}(s-a_{j,n}) \quad (6.53)$$

where $a_{j,n}$ are the complex roots. With the use of (6.51)–(6.53), (6.49) is inverted directly to obtain

$$\frac{\sigma(x,t)}{p}=h(t)\left\{1+\frac{8\rho L^3}{\pi}\sum_{n=1}^{\infty}\frac{(-1)^n\cos[(2n-1)\pi x/(2L)]}{D_n(2n-1)}\right.$$

$$\left.\times\left[\sum_{k=1}^{N+2}\frac{B(a_{k,n})\,e^{a_{k,n}t}}{\lim\limits_{s\to a_{k,n}}\dfrac{\prod_{j=1}^{N+2}(s-a_{j,n})}{(s-a_{k,n})}}\right]\right\}. \quad (6.54)$$

This represents the complete dynamic solution for the stress response of the finite length rod to a suddenly applied end stress. The inversion process involved here is the same as is involved in the quasi-static problems of Chapter 2, and there are no restrictions upon the mechanical properties other than are implied by the representation (6.50). The roots $a_{j,n}$ of the polynomial (6.53) are most conveniently found through the use of standard digital computer programs.

A Shear Flow Response Problem

In the second example, the general method of analysis for dynamic response problems is applied to a flow problem. Specifically the problem posed is that which occurs when an infinite layer of a viscoelastic fluid, with thickness h, initially at rest, has one face subjected to a discontinuous change in velocity

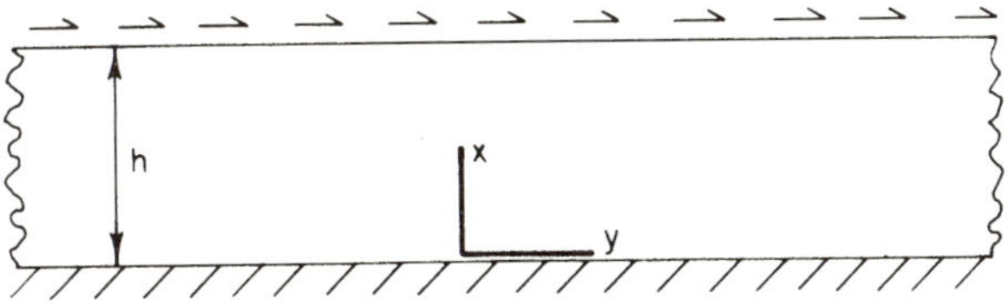

Fig. 6.1. Shear flow problem.

while the other face is held at rest. Such a system is shown in Fig. 6.1. The appropriate boundary conditions are

$$\begin{aligned} u_y(x, y, t) &= 0, \qquad \text{at} \quad x = 0 \\ u_y(x, y, t) &= vth(t), \qquad \text{at} \quad x = h \end{aligned} \tag{6.55}$$

with $u_x = u_z \equiv 0$. Symbol v represents the velocity of the moving face. This case is that of the shearing flow of a viscoelastic fluid, and we seek the transient dynamic solution of this problem. In this problem we are limited by the assumptions of infinitesimal deformation which we have employed in formulating the linear theory. Following the discussion of Section 1.3, the solution so obtained is either limited to small values of time from that of the initial state or the shear rate must be infinitesimally small.

The governing equation for shear deformation, corresponding to (6.7) for the uniaxial case, is given by

$$\frac{\partial^2 \bar{u}_y}{\partial x^2} = [\rho s/\bar{\mu}(s)]\, \bar{u}_y \tag{6.56}$$

where the relaxation function in shear $\mu(t)$ necessarily must satisfy the condition $\mu(t) \to 0$ as $t \to \infty$ to have an unlimited flow capability; see Section 1.3.

The displacement field is taken in the form

$$u_y(x, t) = \sum_{n=1}^{\infty} C_n(t) \sin \frac{n\pi x}{h} + vt \frac{x}{h} h(t). \tag{6.57}$$

This form for the displacement solution satisfies the boundary conditions, (6.55), of the problem. The unknown amplitudes $C_n(t)$ are evaluated to satisfy the equation of motion (6.56) by a procedure similar to that followed for the finite length rod case. The solution for the stress may be shown to be given by

$$\sigma_{xy}(x, t) = \frac{v}{h} \sum_{i=0}^{K} G_i \tau_i (1 - e^{-t/\tau_i})$$

$$+ 2\rho v h \sum_{n=1}^{\infty} \frac{(-1)^n \cos(n\pi x/h)}{D_n} \left[\sum_{k=1}^{K+2} \frac{A(a_{k,n})\, e^{a_{k,n} t}}{\lim\limits_{s \to a_{k,n}} \dfrac{\prod_{j=1}^{K+2} (s - a_{j,n})}{(s - a_{k,n})}} \right] \tag{6.58}$$

where

$$n^2\pi^2 A(s) + h^2 \rho s B(s) = D_n \prod_{j=1}^{K+2} (s - a_{j,n}) \tag{6.59}$$

with

$$\mu(t) = \sum_{i=0}^{K} G_i e^{-t/\tau_i} \qquad \text{and} \qquad \bar{\mu}(s) = \frac{A(s)}{B(s)}. \tag{6.60}$$

$A(s)$ is a polynomial of degree K in s, and $B(s)$ is of degree $K + 1$ in s.

Relation (6.58) gives solution of this problem for a general viscoelastic material. In using (6.60) to represent a particular material, K, G_i, and τ_i must be given particular values after which the roots of $n^2\pi^2 A(s) + h^2\rho s B(s)$ can be found as indicated in (6.59). These roots are, in general, complex, and when substituted into (6.58) the resulting stress is given numerical values. We see that this procedure is exactly the same as is outlined in Chapter 2 for solving quasi-static problems. In the quasi-static response problem of Section 2.8 it is shown to be entirely practical to solve such problems with the relaxation function of the type (6.60) represented by eight or more terms.

As an illustration of the type of behavior which the solution (6.58) represents, we evaluate the case where $\mu(t)$ in (6.60) has a single term with a single relaxation time τ_0 and amplitude G_0. This represents the Maxwell model case, with coefficient of viscosity $\eta = G_0 \tau_0$, which, of course, is the simplest type of viscoelastic fluid. The solution (6.58), evaluated at $x = h$ for this type of material, has the form

$$\frac{\sigma_{xy}(x, t)\,|_{x=h}}{\sigma_{xy}\,|_{\text{steady state}}} = (1 - e^{-t/\tau_0}) + \sum_{n=1}^{\hat{N}} \frac{e^{[(-1/2)+\gamma_n](t/\tau_0)} - e^{[(-1/2)-\gamma_n](t/\tau_0)}}{\gamma_n}$$

$$+ 2 \sum_{n=\hat{N}+1}^{N} \frac{e^{(-1/2)(t/\tau_0)} \sin(\mathscr{L}_n t/\tau_0)}{\mathscr{L}_n} \tag{6.61}$$

where

$$\gamma_n = (\tfrac{1}{4} - n^2\lambda)^{1/2}, \qquad \mathscr{L}_n = (n^2\lambda - \tfrac{1}{4})^{1/2}, \qquad \lambda = \pi^2 G_0 \tau_0^2/h^2\rho,$$

$$\sigma_{xy}\,|_{\text{steady state}} = vG_0\tau_0/h$$

and $\hat{N}$ is $(1/4\lambda)^{1/2}$, rounded off to the next lower integer. In specializing (6.58) to (6.61), the series has been truncated at the level $n = N$. The value shown above for $\sigma_{xy}\,|_{\text{steady state}}$ is simply the stress necessary to deform the fluid in a steady shearing flow, after the effects of all starting transients have died out.

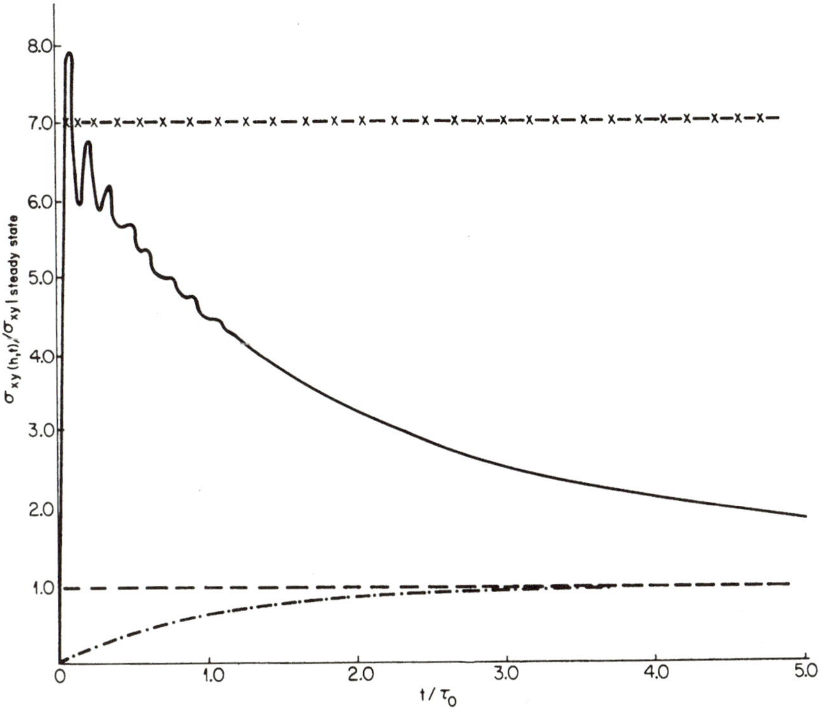

Fig. 6.2. Shear flow boundary stress solution. —— denotes dynamic viscoelastic solution, $N = 100$, $\lambda = \pi^2/(49)$. —×— denotes equivalent dynamic elastic problem solution, $0 < t/\tau_0 < 14$. – – – denotes steady state flow solution. —·— denotes quasi-static viscoelastic solution.

Relation (6.61) is evaluated for $N = 100$ and the result is shown in Fig. 6.2. We see that the dynamic solution rapidly decays to that of a steady state flow condition. It is also of interest to compare this solution with that of a corresponding dynamic elastic problem. For an elastic medium, subjected to the boundary conditions (6.55), the stress σ_{xy} at $x = h$ as a function of time can be shown to be given by an increasing staircase type function. The time duration of each step corresponds to the time needed for an elastic wave front to propagate across the thickness from $x = h$ to $x = 0$ and be reflected back to $x = h$ again. For the elastic shear modulus taken as $\mu = G_0$, the initial value of the viscoelastic relaxation function, the first step of the elastic solution has a value of exactly seven for the ordinate of Fig. 6.2. In terms of the data of Fig. 6.2, the time for the wave front to propagate a distance $2h$ corresponds to $t/\tau_0 = 14$. In as much as the wave front of a viscoelastic shear disturbance propagates with velocity $[\mu(0)/\rho]^{1/2}$, the time for the wave front to propagate a distance $2h$ in the corresponding elastic problem applies also to the viscoelastic problem of Fig. 6.2. If a smooth monotonically decreasing curve were drawn through the oscillating dynamic solution of Fig. 6.2, it would have an initial value of approximately seven on the ordinate. This value is the solution of the corresponding elastic problem. The oscillation in the viscoelastic solution is not due to dynamic effects, but rather is caused by the truncation of the series representation used in this problem, analogously to the behavior of Fourier series. By retaining more terms in the series, the oscillatory character of the solution is minimized and more nearly confined to the very short time region. Keeping this in mind, we can now see that the dynamic viscoelastic solution has an initial response that is elastic in character, but eventually decays into the type of response which is that of a viscous fluid. This example illustrates the variety of effects which can be manifested by viscoelastic materials.

In the case of the Maxwell model fluid, the closed form solution of this problem is given by relation (6.20), for times before reflection effects have occurred, with suitably modified notation. However, more complicated models cannot be treated by the method used to deduce (6.20). In contrast to this, it is a simple matter to perform calculations of the type given in this example whereby more terms are included in the relaxation function representation and more terms are included in the series representation of the solution in order to obtain a more accurate solution.

6.3. HARMONIC THERMOVISCOELASTIC WAVES IN UNLIMITED MEDIA

The propagation of harmonic waves in unlimited isotropic media, under both isothermal and nonisothermal conditions, is the subject of interest here. In the nonisothermal case, this study represents an application of the linear coupled thermoviscoelastic theory developed in Chapter 3. The isothermal results are merely deduced as a special case of the nonisothermal results.

In using the linear thermoviscoelastic theory developed in Chapter 3, the temperature deviations from a given base temperature are necessarily of infinitesimal order and the mechanical properties are independent of these temperature deviations. The mechanical properties could be different for different base temperatures. Some of the implications of these restrictions in the present application are discussed at the end of this section. However, one of the immediate consequences of these restrictions is that, insofar as the mechanical waves are concerned, the shear waves are uncoupled from the thermal disturbance, while the irrotational waves have a coupling effect with the thermal waves. These two cases are examined separately.

First, the case of the propagation of plane irrotational waves is considered. The analysis to be given here uses the equations developed in Chapter 3, while the method follows that presented by Hunter [6.18]. We take the x coordinate in the direction of propagation, with the only nonzero displacement component being $u(x, t)$ in the x direction. The equation of motion is

$$\frac{\partial \sigma_{xx}(x, t)}{\partial x} = \rho \frac{\partial^2 u(x, t)}{\partial t^2} . \tag{6.62}$$

The stress constitutive relation to be used is obtained from (3.48) which, using (3.46), then becomes

$$\sigma_{xx}(x, t) = \int_{-\infty}^{t} [\lambda(t-\tau) + 2\mu(t-\tau)] \frac{\partial^2 u(x, \tau)}{\partial \tau\, \partial x} d\tau - \int_{-\infty}^{t} \varphi(t-\tau) \frac{\partial \theta(x, \tau)}{\partial \tau} d\tau \tag{6.63}$$

where θ is the infinitesimal temperature deviation from the base temperature T_0. The heat conduction equation (3.50) becomes

$$T_0 \frac{\partial}{\partial t} \int_{-\infty}^{t} m(t-\tau) \frac{\partial \theta(x, \tau)}{\partial \tau} d\tau + T_0 \frac{\partial}{\partial t} \int_{-\infty}^{t} \varphi(t-\tau) \frac{\partial^2 u(x, \tau)}{\partial \tau\, \partial x} d\tau = k \frac{\partial^2 \theta(x, t)}{\partial x^2} . \tag{6.64}$$

By combining (6.62) and (6.63) we get

$$\int_{-\infty}^{t} G_3(t-\tau) \frac{\partial^3 u(x, \tau)}{\partial \tau\, \partial x^2} d\tau - \int_{-\infty}^{t} \varphi(t-\tau) \frac{\partial^2 \theta(x, \tau)}{\partial \tau\, \partial x} d\tau = \rho \frac{\partial^2 u(x, t)}{\partial t^2} \tag{6.65}$$

where

$$G_3(t) = \lambda(t) + 2\mu(t).$$

Equations (6.64) and (6.65) are coupled partial differential equations to be solved for $u(x, t)$ and $\theta(x, t)$.

Only the case of harmonic waves are studied here; therefore, we let

$$u(x, t) = \hat{u}e^{i(\eta x+\omega t)}$$
$$\theta(x, t) = \hat{\theta}e^{i(\eta x+\omega t)} \tag{6.66}$$

where $\hat{u}$ and $\hat{\theta}$ are amplitudes, ω is the frequency, and η is a parameter which will later be related to the phase velocity and attenuation. By substituting (6.66) into (6.64) and (6.65), they assume the form

$$iT_0\omega m^*\hat{\theta} - T_0\varphi^*\omega\eta\hat{u} = -k\eta^2\hat{\theta} \tag{6.67}$$

and

$$-G_3{}^*\eta^2\hat{u} - i\varphi^*\eta\hat{\theta} = -\omega^2\rho\hat{u} \tag{6.68}$$

where $G_3{}^*(i\omega)$, $\varphi^*(i\omega)$, and $m^*(i\omega)$ are the complex moduli which are given in terms of the respective relaxation functions $G_3(t)$, $\varphi(t)$, and $m(t)$ through the usual Fourier transform relations, such as (1.58) and (1.59). For a solution of (6.67) and (6.68) to exist it is necessary that the determinant of the coefficients vanish, which gives

$$\left(\omega^2 - \frac{G_3{}^*\eta^2}{\rho}\right)(-k\eta^2 - iT_0\omega m^*) + \frac{i(\varphi^*)^2\eta^2T_0\omega}{\rho} = 0. \tag{6.69}$$

We consider the frequency ω as being a given real positive constant. Then (6.69) is written as

$$a\eta^4 + b\eta^2 + c = 0 \tag{6.70}$$

where

$$a = \frac{G_3{}^*}{\rho}, \quad b = \frac{i\omega m^*G_3{}^*T_0}{k\rho} - \omega^2 + \frac{i(\varphi^*)^2T_0\omega}{k\rho}, \quad \text{and} \quad c = \frac{-i\omega^3m^*T_0}{k}. \tag{6.71}$$

The quartic equation (6.70) determines η as

$$\eta = \pm\left[-\frac{b}{2a} \pm \frac{1}{2a}(b^2 - 4ac)^{1/2}\right]^{1/2} \tag{6.72}$$

where a, b, and c are from (6.71). Parameter η is complex and its real and imaginary parts can be directly related to the phase velocity and attenuation, with respect to distance traveled of the waves. It would be possible to take η as being a given real quantity. Then ω would be found as a complex form. This procedure corresponds to the situation considered in Section 2.11 on free vibration problems, where time damped waves are found.

For convenience of interpretation, an asymptotic solution for η now is found. For simplicity, it is assumed henceforth that the complex moduli φ^* and m^* are real constants, with

$$\varphi^*(i\omega) = \varphi \quad \text{and} \quad T_0m^*(i\omega) = m. \tag{6.73}$$

We let η_M and η_T designate the asymptotic expansion values of η appropriate to mechanical waves and thermal waves, respectively. We first consider the mechanical modes and let

$$\eta_M = \pm[(\rho\omega^2/G_3{}^*(i\omega))^{1/2} + c_1\epsilon + c_2\epsilon^2 + \cdots] \tag{6.74}$$

where

$$\epsilon = \varphi^2/m \mid G_3{}^*(i\omega)| \tag{6.75}$$

and c_i are to be determined. Relation (6.74) is an expansion in the nondimensional coupling parameter ϵ which, for most materials, is expected to be small; if it were negligible, there would be no coupling between thermal and mechanical effects.

We take the second and fourth power of η_M in (6.74) and substitute into (6.69), using (6.73). We next collect coefficients of powers of ϵ and set them equal to zero to satisfy (6.69). The zeroth order term vanishes identically. The vanishing of the coefficient of ϵ gives

$$c_1 = \frac{(-im/2k)[\mid G_3{}^* \mid/(\rho G_3{}^*)^{1/2}]}{1 + (imG_3{}^*/\omega k\rho)} . \tag{6.76}$$

When $G_3{}^*(i\omega) = G_3{}'(\omega) + iG_3''(\omega)$ is used in (6.76) it can be separated into real and imaginary parts. This gives

$$c_1 = c_1{}' + ic_1'' \tag{6.77}$$

where

$$c_1{}' = -\frac{m \mid G_3{}^* \mid}{2k\rho^{1/2}} \times \left\{ \frac{G_3^{\text{IV}}\left(1 - \dfrac{mG_3''}{\omega k\rho}\right) + G_3^{\text{III}}\dfrac{mG_3{}'}{\omega k\rho}}{\left[G_3^{\text{III}}\left(1 - \dfrac{mG_3''}{\omega k\rho}\right) - G_3^{\text{IV}}\dfrac{mG_3{}'}{\omega k\rho}\right]^2 + \left[G_3^{\text{IV}}\left(1 - \dfrac{mG_3''}{\omega k\rho}\right) + G_3^{\text{III}}\dfrac{mG_3{}'}{\omega k\rho}\right]^2} \right\} \tag{6.78}$$

and

$$c_1'' = -\frac{m \mid G_3{}^* \mid}{2k\rho^{1/2}} \times \left\{ \frac{G_3^{\text{III}}\left(1 - \dfrac{mG_3''}{\omega k\rho}\right) - G_3^{\text{IV}}\dfrac{mG_3{}'}{\omega k\rho}}{\left[G_3^{\text{III}}\left(1 - \dfrac{mG_3''}{\omega k\rho}\right) - G_3^{\text{IV}}\dfrac{mG_3{}'}{\omega k\rho}\right]^2 + \left[G_3^{\text{IV}}\left(1 - \dfrac{mG_3''}{\omega k\rho}\right) + G_3^{\text{III}}\dfrac{mG_3{}'}{\omega k\rho}\right]^2} \right\} \tag{6.79}$$

and where

$$[G_3'(\omega) + iG_3''(\omega)]^{1/2} = G_3^{\text{III}}(\omega) + iG_3^{\text{IV}}(\omega). \tag{6.80}$$

Using (6.77)–(6.80), we can separate (6.74) into real and imaginary parts as

$$\eta_M = \left[\frac{\omega\rho^{1/2}G_3^{\text{III}}}{(G_3^{\text{III}})^2 + (G_3^{\text{IV}})^2} + c_1'\epsilon + \cdots\right] + i\left[\frac{-\omega\rho^{1/2}G_3^{\text{IV}}}{(G_3^{\text{III}})^2 + (G_3^{\text{IV}})^2} + c_1''\epsilon + \cdots\right] \tag{6.81}$$

where the higher order terms in ϵ are implied.

From (6.66) we see that $u(x, t)$ can be written as

$$u(x, t) = \hat{u}e^{\eta_2 x}e^{i\eta_1(x+(\omega/\eta_1)t)} \tag{6.82}$$

where $\eta_M = \eta_1 - i\eta_2$. The situation indicated by (6.82) is that of a wave moving in the minus x direction, with phase velocity $v = \omega/\eta_1$, and attenuation η_2. Using the real and imaginary parts of (6.81), we find that the phase velocity and attenuation are given respectively by

$$v = \frac{|G_3^*|}{G_3^{\text{III}}\rho^{1/2}} + \epsilon\frac{m|G_3^*|^3[G_3^{\text{IV}} - (m/\omega k\rho)(-G_3'G_3^{\text{III}} + G_3''G_3^{\text{IV}})]}{2k\omega(\rho)^{3/2}(G^{\text{III}})^2\Delta_1} + \cdots \tag{6.83}$$

and

$$\eta_2 = \frac{\omega\rho^{1/2}G_3^{\text{IV}}}{|G_3^*|} + \epsilon\frac{m|G_3^*|[G_3^{\text{III}} - (m/\omega k\rho)(G_3''G_3^{\text{III}} + G_3'G_3^{\text{IV}})]}{2k\rho^{1/2}\Delta_1} + \cdots \tag{6.84}$$

where

$$\Delta_1 = [G_3^{\text{III}} - (m/\omega k\rho)(G_3''G_3^{\text{III}} + G_3'G_3^{\text{IV}})]^2 + [G_3^{\text{IV}} - (m/\omega k\rho)(-G_3'G_3^{\text{III}} + G_3''G_3^{\text{IV}})]^2$$

and where higher order terms in ϵ are omitted, and $G_3^*(i\omega) = \lambda^*(i\omega) + 2\mu^*(i\omega)$. For no coupling $\epsilon = 0$ the first terms of (6.83) and (6.84) give the phase velocity and attenuation of isothermal irrotational viscoelastic waves. The next terms in (6.83) and (6.84) give the first order correction to the isothermal propagation characteristics due to thermomechanical coupling. Contrary to the situation in coupled thermoelasticity, these correction terms can either increase or decrease both the isothermal phase velocity and the attenuation, depending upon the value of the ratio G_3''/G_3'.

For the thermal waves, the parameter η_T is taken in the form

$$\eta_T = \pm[(-i\omega m/k)^{1/2} + d_1\epsilon + d_2\epsilon^2 + \cdots] \tag{6.85}$$

where ϵ is again given by (6.75). We follow a similar procedure to that just outlined, in order to get the phase velocity and attenuation for the thermal waves. These are found to be

$$v = \omega\left(\frac{2k}{\omega m}\right)^{1/2} - \epsilon\left(\frac{2\omega m}{k}\right)^{1/2} \frac{\omega \mid G_3{}^* \mid [(G_3{}' - G_3'')(\omega m/k\rho) + \omega^2]}{\rho \Delta_2} + \cdots \tag{6.86}$$

and

$$\eta_2 = \left(\frac{\omega m}{2k}\right)^{1/2} + \epsilon\left(\frac{\omega m}{2k}\right)^{1/2} \frac{\omega m \mid G_3{}^* \mid [(G_3{}' + G_3'')(\omega m/k) - \omega^2]}{k\rho \Delta_2} + \cdots \tag{6.87}$$

where

$$\Delta_2 = [(G_3{}' - G_3'')(\omega m/k\rho) + \omega^2]^2 + [(G_3{}' + G_3'')(\omega m/k\rho) - \omega^2]^2.$$

The second terms in (6.86) and (6.87) give the first order corrections, due to thermomechanical coupling of the phase velocity and attenuation of uncoupled thermal waves.

This completes the analysis of the propagation of harmonic coupled thermoviscoelastic waves. Although the type of waves considered are irrotational in terms of the mechanical response characteristics, we readily see that the propagation of harmonic shear waves, that is, waves for which $u_{i,i} = 0$, does not have any coupling with thermal effects. In fact, by an analysis similar to that given here for irrotational waves, it may be shown that the phase velocity and attenuation of plane harmonic shear waves are given by the first terms in (6.83) and (6.84) except that $G_3{}^*(i\omega)$ is replaced by $\mu^*(i\omega)$. Thus, the phase velocity and attenuation of harmonic shear waves are given by

$$v - \mid \mu^* \mid / \mu^{\mathrm{III}} \rho^{1/2} \tag{6.88}$$

and

$$\eta_2 = \omega \rho^{1/2} \mu^{\mathrm{IV}} / \mid \mu^* \mid \tag{6.89}$$

where

$$[\mu^*(i\omega)]^{1/2} = \mu^{\mathrm{III}}(\omega) + i\mu^{\mathrm{IV}}(\omega). \tag{6.90}$$

These analyses show that not only is there attenuation in the propagation of viscoelastic waves, but also there is a dispersion effect due to the fact that the phase velocity is frequency dependent. It is, therefore, not possible to transmit a pulse in a viscoelastic media without a change in shape of the pulse; see also Section 4.4.

The analysis of irrotational waves just given, which used the linear theory equations of Section 3.1, necessarily neglects the effects of energy dissipation and the temperature dependence of the viscoelastic mechanical properties (other than a dependence upon the fixed base temperature T_0). In a problem of the type considered here these effects would be expected to be significant, and for this reason the preceding analysis is highly idealized. Studies, which include these nonlinear effects of energy dissipation and temperature dependent mechanical properties, are those of Petrof and Gratch [6.19], Wolosewik and Gratch [6.20], and Hunter [6.21].

Another restriction of the analysis of this section which should be recognized is that of the assumption of plane waves, whereby a single space coordinate suffices to describe both the direction of propagation and the direction of attenuation. This is certainly a legitimate special case; but, as we will see in the next section, it is not necessarily true. In solving the problem of the reflection from a free boundary of harmonic viscoelastic waves we encounter waves which have a direction of attenuation that is different from the direction of propagation.

6.4. REFLECTION OF HARMONIC WAVES

In the problems discussed up to this point in this chapter there has been no direct consideration of the effect of boundaries on traveling trains of waves. Such effects have been included only indirectly in the examples of Section 6.2. Now, a study is made of the general problem of the reflection of harmonic waves from the free surface of a viscoelastic half space. The treatment here follows that of Lockett [6.22].

The Fourier transformed equations of motion are

$$\mu^*(i\omega)\,\bar{u}_{k,jj} + [\lambda^*(i\omega) + \mu^*(i\omega)]\,\bar{u}_{j,jk} = -\rho\omega^2\bar{u}_k\,. \tag{6.91}$$

These equations of motion are uncoupled, as was done in arriving at (6.30) and (6.31), by alternately taking $u_{i,i}(x_i\,,\,t) = 0$ and curl $\mathbf{u}(x_i\,,\,t) = 0$, where $\mathbf{u}$ is the displacement vector. The resulting equations of motion are given by

$$\mu^*\bar{u}_{k,jj} = -\rho\omega^2\bar{u}_k \tag{6.92}$$

$$(\lambda^* + 2\mu^*)\bar{u}_{k,jj} = -\rho\omega^2\bar{u}_k\,. \tag{6.93}$$

Relations (6.92) and (6.93) govern the propagation of shear and irrotational waves, respectively. These waves are sometimes referred to as S waves and P waves, and (6.92) and (6.93) can be written in the common form

$$(c^*_{\mathrm{P\,or\,S}})^2\bar{u}_{k,jj} = -\omega^2\bar{u}_k \tag{6.94}$$

where

$$\begin{aligned}(c_\mathrm{P}^*)^2 &= \rho^{-1}(\lambda^* + 2\mu^*)\\ (c_\mathrm{S}^*)^2 &= \mu^*/\rho.\end{aligned} \tag{6.95}$$

For harmonic waves, the displacement vector $\mathbf{u}$ is given by

$$\mathbf{u} = \mathbf{U}e^{i(Lx+My+\omega_0 t)} \tag{6.96}$$

where $\mathbf{U}$ is a complex amplitude vector, L and M are complex parameters, and $\omega = \omega_0$ is the particular frequency of interest. In writing (6.96) it has been assumed that the wave motion is independent of the z coordinate, relative to a rectangular Cartesian coordinate system x, y, z. The Fourier transform of (6.96) gives

$$\bar{\mathbf{u}} = \mathbf{U}e^{i(Lx+My)}[2\pi\,\delta(\omega - \omega_0)] \tag{6.97}$$

where $\delta(\ )$ is the delta function. We substitute (6.97) into (6.94) and get

$$(c_0^*)^2(L^2 + M^2) = \omega_0^2 \tag{6.98}$$

where c_0^* can be identified with either c_S^* or c_P^* at $\omega = \omega_0$. Actually, (6.98) could have been obtained without recourse to the Fourier transform since only harmonic waves are involved here, but its employment here is a matter of convenience.

The parameters L and M in (6.98) are in general complex, but before considering the most general case, the special case whereby the ratio L/M is real is studied.

L/M Real

When the ratio of complex numbers, L/M is real, the solution of (6.98) gives

$$L = (\omega_0/c_0^*)\cos\alpha$$

and

$$M = (\omega_0/c_0^*)\sin\alpha. \tag{6.99}$$

The ratio ω_0/c_0^* is written as

$$\omega_0/c_0^* = a(1 + i\varphi) \tag{6.100}$$

which determines a and φ in terms of ω_0 and c_0^*. From (6.99) and (6.100) there results

$$L = l(1 + i\varphi)$$

and

$$M = m(1 + i\varphi) \tag{6.101}$$

where

$$l = a\cos\alpha$$

and

$$m = a\sin\alpha. \tag{6.102}$$

With (6.101), (6.96) becomes

$$\mathbf{u} = \mathbf{U}e^{-\varphi(lx+my)}e^{i(lx+my+\omega_0 t)}. \tag{6.103}$$

From (6.103) we see that, in this special case, the wave is attenuated according to the distance traveled, and the direction of attenuation is the same as the direction of propagation. A wave which has this characteristic is called a simple wave, using the terminology of Ref. [6.22]. From (6.102) we see that $\tan\alpha = m/l$, where angle α, measured from the x axis, gives the direction of propagation. The parameter φ determines the attenuation characteristics. Now we will consider the general case whereby the ratio L/M is not real.

L/M Not Real

In this case the ratio of the complex constants, L and M, is complex. For this situation the attenuation is not in the direction of propagation of the wave, and the form corresponding to (6.103) is

$$\mathbf{u} = \mathbf{U}e^{\beta x+\gamma y}e^{i(lx+my+\omega_0 t)} \tag{6.104}$$

where now

$$\beta/\gamma \neq l/m.$$

This type of wave is more general than the harmonic waves considered in Section 6.3 where the directions of propagation and attenuation were taken to be the same.

From this point on, it will be convenient to utilize component rather than vector notation. Thus, the transform of the displacement components u and v in the x and y directions, respectively, are taken as

$$(\bar{u}, \bar{v}) = 2\pi\,\delta(\omega - \omega_0)(U, V)\,e^{i(Lx+My)} \tag{6.105}$$

where U and V are complex constants. The two equations of motion from (6.91) and (6.105) are then

$$\mu^*(L^2 + M^2)U + (\lambda^* + \mu^*)L(LU + MV) = \rho\omega_0^2 U \tag{6.106}$$

and

$$\mu^*(L^2 + M^2)V + (\lambda^* + \mu^*)M(LU + MV) = \rho\omega_0^2 V. \tag{6.107}$$

For S waves, (6.98) combined with (6.106) gives

$$V = -(L/M)U, \qquad \text{S waves.} \tag{6.108}$$

For P waves (6.98) combined with (6.107) gives, after some algebraic reduction,

$$V = (M/L)U, \qquad \text{P waves.} \tag{6.109}$$

Now, using (6.105), (6.108), and (6.109), the displacement components are written as

$$(u, v)|_{\text{S wave}} = (MU, -LU)\, e^{i(Lx+My+\omega_0 t)}$$

and (6.110)

$$(u, v)|_{\text{P wave}} = (LU, MU)\, e^{i(Lx+My+\omega_0 t)}.$$

The solutions (6.110) will be further specialized in the following case.

Reflection of Incident S Wave at Free Boundary

We let $x = 0$ define the stress free boundary of the half space. The reflection characteristics of an incident S wave polarized in the xy plane with angle of incidence α_i from the x axis are now found.

We let u_{Si}, v_{Si} be the displacement components of the incident S wave and we let u_{S}, v_{S} and u_{P}, v_{P} be the displacement components of the reflected S and P waves, respectively. Then, following (6.110), these are expressed as

$$(\bar{u}_{\text{Si}}, \bar{v}_{\text{Si}}) = 2\pi\, \delta(\omega - \omega_0)(M_{\text{Si}} U_{\text{Si}}, -L_{\text{Si}} U_{\text{Si}})\, e^{i(L_{\text{Si}}x+M_{\text{Si}}y)}, \tag{6.111}$$

$$(\bar{u}_{\text{S}}, \bar{v}_{\text{S}}) = 2\pi\, \delta(\omega - \omega_0)(M_{\text{S}} U_{\text{S}}, -L_{\text{S}} U_{\text{S}})\, e^{i(L_{\text{S}}x+M_{\text{S}}y)} \tag{6.112}$$

and

$$(\bar{u}_{\text{P}}, \bar{v}_{\text{P}}) = 2\pi\, \delta(\omega - \omega_0)(L_{\text{P}} U_{\text{P}}, M_{\text{P}} U_{\text{P}}) e^{i(L_{\text{P}}x+M_{\text{P}}y)}. \tag{6.113}$$

The total displacement vector involves the superposition of all three types of components, (6.111)–(6.113), as

$$\bar{u} = \bar{u}_{\text{Si}} + \bar{u}_{\text{S}} + \bar{u}_{\text{P}} \qquad \text{and} \qquad \bar{v} = \bar{v}_{\text{Si}} + \bar{v}_{\text{S}} + \bar{v}_{\text{P}}.$$

The equations of motion are satisfied by (6.98) which now require

$$L_{\text{Si}}^2 + M_{\text{Si}}^2 = L_{\text{S}}^2 + M_{\text{S}}^2 = \rho\omega_0^2/\mu^* \quad \text{and} \quad L_{\text{P}}^2 + M_{\text{P}}^2 = \rho\omega_0^2/(\lambda^* + 2\mu^*). \tag{6.114}$$

The traction free boundary conditions on $x = 0$ are stated by

$$\sigma_{xx}(x, y, t) = 0 \qquad \text{at} \quad x = 0 \tag{6.115}$$

and

$$\sigma_{xy}(x, y, t) = 0 \qquad \text{at} \quad x = 0. \tag{6.116}$$

Condition (6.116), in terms of the transformed displacements, is written as

$$\frac{\partial \bar{u}(x, y, \omega)}{\partial y} + \frac{\partial \bar{v}(x, y, \omega)}{\partial x} = 0 \qquad \text{at} \quad x = 0. \tag{6.117}$$

We use (6.111)–(6.113) in (6.117), which gives an equation involving exponential dependence upon y. For this to be satisfied for all values of y we must require that

$$M_{Si} = M_S = M_P = M \tag{6.118}$$

where M now is still to be determined. Boundary condition (6.115) also requires that (6.118) apply. Using this result in (6.114) we get

$$L_{Si}^2 = L_S^2 = (\rho\omega_0^2/\mu^*) - M^2$$

and

$$L_P^2 = [\rho\omega_0^2/(\lambda^* + 2\mu^*)] - M^2. \tag{6.119}$$

Since L_{Si} refers to the incident shear wave and L_S refers to the reflected shear wave, they must be of opposite sign, and using the first part of (6.119), we get $L_S = -L_{Si}$. Using this and (6.118), see we that

$$L_S/M_S = -L_{Si}/M_{Si}\,. \tag{6.120}$$

But the ratio L_{Si}/M_{Si} is real for the incident shear wave being simple, according to the definition of this term given in connection with (6.103). From (6.120) the reflected shear wave is also simple, and this reflected shear wave experiences attenuation in the direction of propagation. From the ratio of the two equations in (6.99) and from (6.120), we find that

$$\tan \alpha_S = M_S/L_S = -M_{Si}/L_{Si} = -\tan \alpha_i \tag{6.121}$$

and the angle of reflection is equal to the angle of incidence for the two shear waves.

The characteristics of the reflected P wave are now obtained. Since the incident wave is simple, (6.99) and (6.118) give

$$M_P = (\rho^{1/2}\omega_0 \sin \alpha_i)/(\mu^*)^{1/2}. \tag{6.122}$$

This combines with the last of (6.119) as

$$\frac{L_P^2}{M_P^2} = \frac{\mu^*}{(\lambda^* + 2\mu^*) \sin^2 \alpha_i} - 1. \tag{6.123}$$

The reflected P wave will be simple if L_P/M_P is real which, along with (6.123), requires

$$\mu^*/(\lambda^* + 2\mu^*) \quad \text{must be real,} \qquad \text{and} \qquad \mu^*/(\lambda^* + 2\mu^*) \geqslant \sin^2 \alpha_i\,. \tag{6.124}$$

If (6.124) is not satisfied the reflected P wave will not be simple. This case will be investigated later. Writing μ^* and λ^* in real and imaginary parts, the requirement that the ratio in (6.124) be real implies

$$\lambda''(\omega_0)/\lambda'(\omega_0) = \mu''(\omega_0)/\mu'(\omega_0) \tag{6.125}$$

where μ' and λ' are the real parts, and μ'' and λ'' are the imaginary parts of μ^* and λ^*, respectively. Relation (6.124) combined with (6.125) leads to the requirement that

$$\sin^2 \alpha_i \leqslant \mu'(\omega_0)/[\lambda'(\omega_0) + 2\mu'(\omega_0)]. \tag{6.126}$$

The inequality in (6.126) defines a critical angle of incidence. If α_i is less than this critical angle and if (6.125) is satisfied, the reflected P wave will exhibit an attenuation in the direction of propagation. And if either of these two conditions are not met, then the reflected P wave will not be simple. This case is now investigated.

When (6.125) is not satisfied a more general approach must be followed. By combining the last of (6.119) with $M = M_i$ from (6.99), we get

$$L_P{}^2 = [\rho\omega_0{}^2/(\lambda^* + 2\mu^*)] - (\rho\omega_0{}^2/\mu^*) \sin^2 \alpha_i\,. \tag{6.127}$$

The angle of reflection of the P wave is given by

$$\tan \alpha_P = M_P{}'/L_P{}' \tag{6.128}$$

where $M_P{}'$ and $L_P{}'$ are the real parts of M_P and L_P, respectively. The ratio of the imaginary parts, M_P''/L_P'', determines the direction of attenuation. To calculate (6.128), L_P is taken from (6.127) and M_P is taken from (6.122).

More general information must be obtained from the complete solution of the boundary condition equations (6.115) and (6.116). These two equations, utilizing the results already obtained, assume the form

$$(M^2 - L_{Si}^2)\, U_S + 2ML_P U_P = -(M^2 - L_{Si}^2)\, U_{Si} \tag{6.129}$$

and

$$2L_{Si}MU_S + (M^2 - L_{Si}^2)\, U_P = 2L_{Si}MU_{Si}\,. \tag{6.130}$$

With L_P from (6.127), $M = M_{Si} = M_P$ from (6.122), and L_{Si} the same as (6.122), but with $\sin \alpha_i$ replaced by $\cos \alpha_i$, U_S and U_P can be determined from (6.129) and (6.130). These will be the complex constants which give the amplitudes, and phase angles of the reflected waves.

Similar types of results can be obtained in the case when the incident wave is a P wave rather than an S wave as was considered here. It can be shown that the reflected P wave is simple and has an angle of reflection equal to the angle of incidence. For the reflected S wave to be simple, it is necessary that $\lambda''/\lambda' = \mu''/\mu'$ and that the angle of incidence be less than some critical value, which can be determined. When $\lambda''/\lambda' \neq \mu''/\mu'$ the reflected S wave will not be simple; that is, it will experience an attenuation in a direction different from the direction of propagation.

The more general problem of the reflection and refraction of harmonic waves at the interface between two different viscoelastic media has been considered by Lockett [6.22]. These types of problems have also been studied by Cooper and Reiss [6.23] and Cooper [6.24]. The problem of surface wave generation was also considered in these references. With regard to Rayleigh type surface waves in viscoelastic media, Tsai and Kolsky [6.25] have presented experimental results. A theoretical analysis is given in Section 6.6.

6.5. MOVING LOADS ON A VISCOELASTIC HALF SPACE

A wide class of dynamic response problems involves moving loads on viscoelastic bodies. One such representative example is given here. This problem involves a discontinuous pressure moving at a constant velocity over a viscoelastic half space. The pressure is taken to depend upon a single spatial coordinate. Consequently there are just two independent spatial coordinates in the response problem. The velocity of the moving loads is taken to be sufficiently great that no disturbance can propagate ahead of the pressure discontinuity. Since the loads are moving at a steady rate, the problem, when viewed from a coordinate system moving with the loads, is of a steady nature with no dependence upon time. These simplifications allow many of the interesting features of the problem to be determined, although a complete analytical solution is not available. The analysis to be given here is from the work of Chu [6.26].

We let the boundary plane be specified by $x_2 = 0$, with reference to rectangular Cartesian coordinates x_i. The loading and, consequently, the response of the problem are independent of x_3. The boundary conditions are then

$$\sigma_{22}(x_1, x_2, t) = -ph(x_1 + vt) \quad \text{at} \quad x_2 = 0$$

and

$$\sigma_{12}(x_1, x_2, t) = 0 \quad \text{at} \quad x_2 = 0. \tag{6.131}$$

Thus, there is a step function pressure progressing over the surface of the half space at velocity v moving in the negative x_1 direction.

We let v_i designate the velocity components. The viscoelastic stress strain relations (2.2), combined with the equations of motion (2.6), give

$$\int_{-\infty}^{t} \mu(t-\tau)\, v_{i,jj}(x_i, \tau)\, d\tau + \int_{-\infty}^{t} [\lambda(t-\tau) + \mu(t-\tau)]\, v_{j,ij}(x_i \tau)\, d\tau = \rho \frac{\partial v_i(x_i, t)}{\partial t} \tag{6.132}$$

where $i = 1, 2$.

For the steady state conditions imposed by the load moving at constant velocity, it is required that v_i involve x_1 and t only in the combination $x_1 + vt$, thus

$$v_i = f_i(x_1 + vt, x_2). \tag{6.133}$$

Since v_i has the form of (6.133), it is logical to take a coordinate change such that

$$\begin{aligned} x &= x_1 + vt \\ y &= x_2 . \end{aligned} \tag{6.134}$$

The coordinate system x, y is moving with the load with $x = 0$ giving the location of the pressure discontinuity. With the change of coordinates (6.134), the equations of motion (6.132) become

$$\begin{aligned} &\int_{-\infty}^{x} \left[\mu\left(\frac{x-\xi}{v}\right)\right]\left(\frac{\partial^2}{\partial \xi^2} + \frac{\partial^2}{\partial y^2}\right) v_x(\xi, y)\, d\xi \\ &\quad + \int_{-\infty}^{x} \left[\lambda\left(\frac{x-\xi}{v}\right) + \mu\left(\frac{x-\xi}{v}\right)\right] \frac{\partial}{\partial \xi}\left[\frac{\partial v_x(\xi, y)}{\partial \xi} + \frac{\partial v_y(\xi, y)}{\partial y}\right] d\xi \\ &= \rho v^2 \frac{\partial v_x(x, y)}{\partial x} \end{aligned} \tag{6.135}$$

and

$$\begin{aligned} &\int_{-\infty}^{x} \left[\mu\left(\frac{x-\xi}{v}\right)\right]\left(\frac{\partial^2}{\partial \xi^2} + \frac{\partial^2}{\partial y^2}\right) v_y(\xi, y)\, d\xi \\ &\quad + \int_{-\infty}^{x} \left[\lambda\left(\frac{x-\xi}{v}\right) + \mu\left(\frac{x-\xi}{v}\right)\right] \frac{\partial}{\partial y}\left[\frac{\partial v_x(\xi, y)}{\partial \xi} + \frac{\partial v_y(\xi, y)}{\partial y}\right] d\xi \\ &= \rho v^2 \frac{\partial v_y(x, y)}{\partial x} \end{aligned} \tag{6.136}$$

where v_x and v_y are the components of velocity referred to the x, y coordinate system.

The appropriate boundary conditions are now

$$\begin{aligned} \sigma_{yy} &= -ph(x) \qquad \text{at} \quad y = 0 \\ \text{and} \qquad \sigma_{xy} &= 0 \qquad \text{at} \quad y = 0. \end{aligned} \tag{6.137}$$

The equations of motion are uncoupled following the classical elasticity approach. Using the Helmholtz representation in two dimensions,

$$\begin{aligned} v_x &= \frac{\partial \varphi(x, y)}{\partial x} + \frac{\partial \psi(x, y)}{\partial y} \\ \text{and} \qquad v_y &= \frac{\partial \varphi(x, y)}{\partial y} - \frac{\partial \psi(x, y)}{\partial x} \end{aligned} \tag{6.138}$$

we substitute (6.138) into the equations of motion (6.135) and (6.136) to get

$$\rho v^2 \frac{\partial}{\partial x}\left(\frac{\partial \varphi}{\partial x} + \frac{\partial \psi}{\partial y}\right) = \frac{\partial}{\partial x}\int_0^x \left[\lambda\left(\frac{x-\xi}{v}\right) + 2\mu\left(\frac{x-\xi}{v}\right)\right] \nabla^2\varphi(\xi, y)\, d\xi$$

$$+ \frac{\partial}{\partial y}\int_0^x \mu\left(\frac{x-\xi}{v}\right) \nabla^2\psi(\xi, y)\, d\xi \qquad (6.139)$$

and

$$\rho v^2 \frac{\partial}{\partial x}\left(\frac{\partial \varphi}{\partial y} - \frac{\partial \psi}{\partial x}\right) = \frac{\partial}{\partial y}\int_0^x \left[\lambda\left(\frac{x-\xi}{v}\right) + 2\mu\left(\frac{x-\xi}{v}\right)\right] \nabla^2\varphi(\xi, y)\, d\xi$$

$$- \frac{\partial}{\partial x}\int_0^x \mu\left(\frac{x-\xi}{v}\right) \nabla^2\psi(\xi, y)\, d\xi \qquad (6.140)$$

where $\nabla^2 = (\partial^2/\partial\xi^2) + (\partial^2/\partial y^2)$. The lower limit of $-\infty$ in (6.135) and (6.136) has been changed to 0, since it is assumed that the pressure is moving sufficiently fast that $v_x = v_y = 0$ for $x \leqslant 0$. Also, in arriving at the forms in (6.139) and (6.140), relationships of the following type have been used:

$$\int_0^x \mu\left(\frac{x-\xi}{v}\right) \frac{\partial}{\partial \xi}\psi(\xi, y)\, d\xi = \frac{\partial}{\partial x}\int_0^x \mu\left(\frac{x-\xi}{v}\right) \psi(\xi, y)\, d\xi$$

which also uses $\psi(0, y) = 0$.

By adding the derivative with respect to x of (6.139) to the derivative with respect to y of (6.140), we find that

$$\rho v^2 \frac{\partial \varphi}{\partial x} = \int_0^x \left[\lambda\left(\frac{x-\xi}{v}\right) + 2\mu\left(\frac{x-\xi}{v}\right)\right] \nabla^2\varphi(\xi, y)\, d\xi. \qquad (6.141)$$

Similarly, we find from (6.139) and (6.140) that

$$\rho v^2 \frac{\partial \psi}{\partial x} = \int_0^x \mu\left(\frac{x-\xi}{v}\right) \nabla^2\psi(\xi, y)\, d\xi. \qquad (6.142)$$

Equations (6.141) and (6.142) are conveniently solved through the use of the s multiplied Laplace transform, the transform of which, $\bar{f}(s)$ of $f(x)$, is defined as

$$\bar{f}(s) = s\int_0^\infty f(x)\, e^{-sx}\, dx. \qquad (6.143)$$

The s multiplied Laplace transform with respect to x of (6.141) and (6.142) is given through

$$(\bar{\lambda} + 2\bar{\mu})\frac{\partial^2\bar{\varphi}}{\partial y^2} = (\rho v^2 - \bar{\lambda} - 2\bar{\mu})\, s^2\bar{\varphi} \qquad (6.144)$$

and

$$\tilde{\mu}\,\frac{\partial^2 \bar{\psi}}{\partial y^2} = (\rho v^2 - \tilde{\mu})\, s^2 \bar{\psi}. \tag{6.145}$$

In these relations the s multiplied Laplace transform of $\mu(x/v)$ is given by

$$\mathscr{L}[\mu(x/v)] = \bar{\mu}(sv) = \tilde{\mu} \tag{6.146}$$

with $\bar{\mu}(s)$ being the s multiplied Laplace transform with respect to time of $\mu(t)$. A relation similar to (6.146) exists for λ as

$$\mathscr{L}[\lambda(x/v)] = \bar{\lambda}(sv) = \tilde{\lambda}. \tag{6.147}$$

In obtaining (6.144) and (6.145), $\varphi = \psi = 0$ for $x \leqslant 0$ is used, consistent with $v_i = 0$ for $x \leqslant 0$.

The boundary conditions (6.137) when Laplace transformed are given by

$$\begin{aligned} \bar{\sigma}_{yy}(s, y) &= -p \qquad \text{at} \quad y = 0 \\ \text{and} \qquad & \\ \bar{\sigma}_{xy}(s, y) &= 0 \qquad \text{at} \quad y = 0. \end{aligned} \tag{6.148}$$

The stresses, for use in the boundary conditions, are found from (2.2), (6.138), and the strain displacement (or strain rate–velocity) relations. The transform of these stresses are found as

$$\begin{aligned} \bar{\sigma}_{yy} &= \frac{2\tilde{\mu}}{sv}\left(\frac{\partial^2 \bar{\varphi}}{\partial y^2} - s\,\frac{\partial \bar{\psi}}{\partial y}\right) + \frac{\tilde{\lambda}}{sv}\left(s^2 \bar{\varphi} + \frac{\partial^2 \bar{\varphi}}{\partial y^2}\right) \\ \text{and} \qquad & \\ \bar{\sigma}_{xy} &= \frac{\tilde{\mu}}{sv}\left(2s\,\frac{\partial \bar{\varphi}}{\partial y} + \frac{\partial^2 \bar{\psi}}{\partial y^2} - s^2 \bar{\psi}\right). \end{aligned} \tag{6.149}$$

The solution of (6.144) and (6.145), satisfying the boundary conditions (6.148), is given through

$$\bar{\varphi} = -\frac{pv}{\tilde{\mu}s}\left[\frac{\gamma_s^2 - 1}{(\gamma_s^2 - 1)^2 + 4\gamma_f \gamma_s}\right] e^{-s\gamma_f y} \tag{6.150}$$

and

$$\bar{\psi} = -\frac{pv}{\tilde{\mu}s}\left[\frac{2\gamma_f}{(\gamma_s^2 - 1)^2 + 4\gamma_f \gamma_s}\right] e^{-s\gamma_s y} \tag{6.151}$$

where

$$\gamma_f = \left(\frac{\rho v^2}{\tilde{\lambda} + 2\tilde{\mu}} - 1\right)^{1/2} \tag{6.152}$$

and

$$\gamma_s = \left(\frac{\rho v^2}{\tilde{\mu}} - 1\right)^{1/2} \tag{6.153}$$

and where the other two constants involved in the solution of (6.144) and (6.145) are taken equal to zero for a bounded solution as $y \to \infty$. The transformed velocity solution corresponding to (6.150) and (6.151) is, from (6.138), found to be

$$\frac{\bar{v}_x}{v} = -\frac{p}{\bar{\mu}}\left[\frac{(\gamma_s^2 - 1)e^{-s\gamma_f y} - 2\gamma_f\gamma_s e^{-s\gamma_s y}}{(\gamma_s^2 - 1)^2 + 4\gamma_f\gamma_s}\right] \tag{6.154}$$

and

$$\frac{\bar{v}_y}{v} = \frac{p}{\bar{\mu}}\left[\frac{\gamma_f(\gamma_s^2 - 1)e^{-s\gamma_f y} + 2\gamma_f e^{-s\gamma_s y}}{(\gamma_s^2 - 1)^2 + 4\gamma_f\gamma_s}\right]. \tag{6.155}$$

The transformed stress solution may also readily be found.

We use the initial value theorem, appropriate to the s multiplied Laplace transform, to write

$$\lim_{s\to\infty} \bar{\mu}(s) = \mu(0) = \mu_0$$

and (6.156)

$$\lim_{s\to\infty} \bar{\lambda}(s) = \lambda(0) = \lambda_0 \,.$$

We adopt the following notation:

$$c_{s0} = (\mu_0/\rho)^{1/2}, \qquad c_{f0} = [\lambda_0 + 2\mu_0)/\rho]^{1/2} \tag{6.157}$$

and

$$\gamma_{s0} = [(v^2/c_{s0}^2) - 1]^{1/2}, \qquad \gamma_{f0} = [(v^2/c_{f0}^2) - 1]^{1/2}. \tag{6.158}$$

Now, for sufficiently large s the solution parameters have a simple form, which is shown below; for example, in the case of $\bar{v}_x$

$$\bar{v}_x \cong C_1 e^{-s\gamma_{f0} y} + C_2 e^{-s\gamma_{s0} y}, \qquad s \gg 1 \tag{6.159}$$

where C_i and γ_{f0} and γ_{s0} are independent of s. By using the shifting theorem and the initial value theorem, we find that the inversion of the two terms in (6.159) is respectively given by

$$C_1 h(x - \gamma_{f0}\, y)|_{\text{at } x-\gamma_{f0} y=0}$$

and

$$C_2 h(x - \gamma_{s0}\, y)|_{\text{at } x-\gamma_{s0} y=0} \,.$$

This, and the other comparable forms, shows that the velocity and stress fields experience discontinuous changes across the lines

$$x = \gamma_{f0} y \qquad \text{and} \qquad x = \gamma_{s0} y.$$

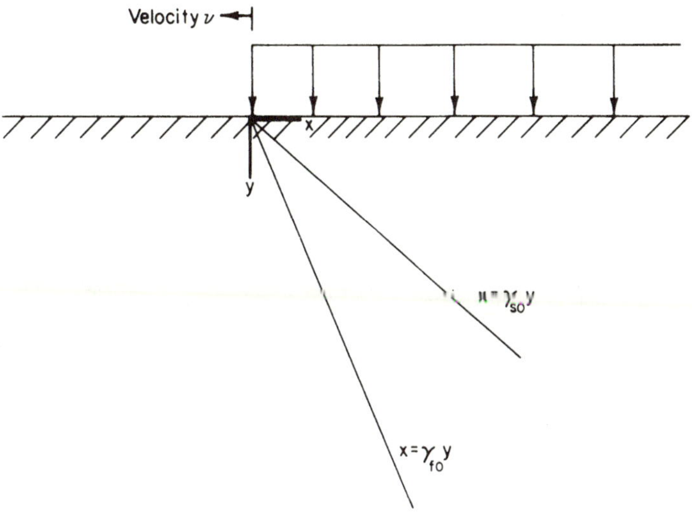

Fig. 6.3. Moving loads on half space.

The situation is as depicted in Fig. 6.3. Remembering that the x, y coordinate system is moving with the load, we recognize that the two lines, $x = \gamma_{f0} y$ and $x = \gamma_{s0} y$, are the wave fronts, the first due to the irrotational disturbance, and the second due to the shear disturbance. We now see that if the velocity were not sufficiently great such that the radicals in (6.158) are real, waves could propagate ahead of the moving pressure discontinuity. This type of problem would be much more difficult to analyze than the one considered here of the fast moving (superseismic) load.

To get more explicit results, we expand the relaxation functions in a Taylor's series about $t = 0$, as

$$\mu(t) = \mu_0 - \mu_1 t + (\mu_2 t^2/2) - \cdots$$

and (6.160)

$$\lambda(t) = \lambda_0 - \lambda_1 t + (\lambda_2 t^2/2) - \cdots .$$

The s multiplied Laplace transform, with respect to time of these, is given by

$$\bar{\mu}(s) = \mu_0 - (\mu_1/s) + (\mu_2/s^2) - \cdots$$

and (6.161)

$$\bar{\lambda}(s) = \lambda_0 - (\lambda_1/s) + (\lambda_2/s^2) - \cdots .$$

The $\tilde{\mu}$ and $\tilde{\lambda}$ forms from (6.146) and (6.147) are now

$$\tilde{\mu} = \mu_0 - (\mu_1/sv) + (\mu_2/s^2v^2) - \cdots$$

and (6.162)

$$\tilde{\lambda} = \lambda_0 - (\lambda_1/sv) + (\lambda^2/s^2v^2) - \cdots .$$

We substitute (6.162) into (6.153) and use the binomial expansion to obtain

$$s\gamma_s = s\gamma_{s0} + \frac{(\gamma_{s0}^2 + 1)(\mu_1/\mu_0)}{2\gamma_{s0}} + \cdots \tag{6.163}$$

where γ_{s0} is from (6.158), and the terms not written in (6.163) are of higher order in $1/s$. By using (6.163), we see that

$$\lim_{s\to\infty} e^{-s\gamma_s y} = \lim_{s\to\infty} e^{-s\gamma_{s0} y} \exp\left[-\frac{(\gamma_{s0}^2 + 1)\,\mu_1}{2\gamma_{s0}\mu_0} y\right]. \tag{6.164}$$

We let $[\![v_x]\!]_s$ designate the discontinuity in v_x across the shear front $x = \gamma_{s0} y$. From (6.154), (6.163), (6.164), and the initial value theorem and the shifting theorem, it follows that

$$\frac{[\![v_x]\!]_s}{v} = \frac{2p\gamma_{f0}\gamma_{s0}}{\mu_0[(\gamma_{s0}^2 - 1)^2 + 4\gamma_{f0}\gamma_{s0}]} \exp\left[-\frac{(\gamma_{s0}^2 + 1)\,\mu_1}{2\gamma_{s0}\mu_0} y\right]. \tag{6.165}$$

The discontinuity of v_x across $x = \gamma_{f0} y$ and v_y across $x = \gamma_{s0} y$ and $x = \gamma_{f0} y$ follow in a similar manner. We find that the strengths of these discontinuities are given by

$$\frac{[\![v_x]\!]_f}{v} = \frac{-p(\gamma_{s0}^2 - 1)}{\mu_0[(\gamma_{s0}^2 - 1)^2 + 4\gamma_{f0}\gamma_{s0}]} \exp\left[-\frac{(\gamma_{f0}^2 + 1)(\lambda_1 + 2\mu_1)}{2\gamma_{f0}(\lambda_0 + 2\mu_0)} y\right], \tag{6.166}$$

$$\frac{[\![v_y]\!]_s}{v} = \frac{2p\gamma_{f0}}{\mu_0[(\gamma_{s0}^2 - 1)^2 + 4\gamma_{f0}\gamma_{s0}]} \exp\left[-\frac{(\gamma_{s0}^2 + 1)\,\mu_1}{2\gamma_{s0}\mu_0} y\right], \tag{6.167}$$

and

$$\frac{[\![v_y]\!]_f}{v} = \frac{p\gamma_{f0}(\gamma_{s0}^2 - 1)}{\mu_0[(\gamma_{s0}^2 - 1)^2 + 4\gamma_{f0}\gamma_{s0}]} \exp\left[-\frac{(\gamma_{f0}^2 + 1)(\lambda_1 + 2\mu_1)}{2\gamma_{f0}(\lambda_0 + 2\mu_0)} y\right]. \tag{6.168}$$

Similar expressions can be found for the discontinuities in the stress components. Relations (6.165)–(6.168) are similar to the results for the analogous elasticity problem. In fact, if the exponentials in these relations are deleted, the results would directly apply to the comparable elasticity problem. We see that for the viscoelastic constitutive relations the strength of the discontinuities in the field variables diminishes exponentially with depth, in contrast to the situation for elastic constitutive relations where these discontinuities do not diminish with depth. It is remarkable that only the initial values and initial slopes of the relaxation functions are needed to calculate these discontinuities, in the viscoelastic case.

The similar problem of a load moving over a finite thickness viscoelastic slab, with the resulting wave reflection considerations, was also considered by Chu [6.26].

6.6. VISCOELASTIC RAYLEIGH WAVES

To complete this treatment of viscoelastic wave propagation, we consider the motion of Rayleigh waves near the surface of a viscoelastic half space. As we shall see, there are significant effects that do not even exist in the comparable elastic problem. The approach is from that of Currie *et al.* [6.27].

Take the general form for a traveling wave as

$$u_i(t) = U_i e^{i\omega(-s_p x_p + t)} \tag{6.169}$$

where s_p is the complex slowness vector in the direction of wave travel, and its magnitude is the reciprocal of wave velocity. Let the half space have a Cartesian coordinate system at the surface with x_3 normal to the surface. The plane wave motion will be described in terms of x_1 and x_3. Superimpose the two waves

$$u_i = \sum_{n=1}^{2} U_i^{(n)} e^{i\omega(-s_p^{(n)} x_p + t)} \tag{6.170}$$

where for transverse waves

$$s_p^{(1)} s_p^{(1)} = \rho/\mu^* = 1/c_s^2 \tag{6.171}$$

with

$$U_p s_p = 0 \tag{6.172}$$

and for longitudinal waves

$$s_p^{(2)} s_p^{(2)} = \rho/(\lambda^* + 2\mu^*) = 1/c_l^2 \tag{6.173}$$

with

$$U_i = U s_i\,. \tag{6.174}$$

On the free surface

$$\sigma_{13} = \sigma_{33} = 0 \qquad \text{on} \quad x_3 = 0. \tag{6.175}$$

To satisfy the boundary conditions (6.175) both types of waves must travel at the same velocity in the x_1 direction, and thus

$$s_1^{(1)} = s_1^{(2)} = s_1\,. \tag{6.176}$$

Now, using (6.176) in (6.171)–(6.174) there results

$$[s_3^{(1)}]^2 = \frac{\rho}{\mu^*} - s_1^2 \qquad \text{with} \quad U_3^{(1)} = -\frac{s_1}{s_3^{(1)}} U_1^{(1)} \tag{6.177}$$

and

$$[s_3^{(2)}]^2 = \frac{\rho}{\lambda^* + 2\mu^*} - s_1^2 \quad \text{with} \quad U_3^{(2)} = \frac{s_3^{(2)}}{s_1} U_1^{(2)}. \tag{6.178}$$

Using the steady state harmonic stress strain relations (2.49), (2.50), the strain–displacement relations (2.1), and (6.170) in the boundary conditions (6.175), we get

$$\frac{U_1^{(1)}}{U_1^{(2)}} = \frac{2s_3^{(2)}s_3^{(1)}}{2s_1^2 - \rho/\mu^*} \tag{6.179}$$

and

$$-4s_3^{(1)}s_3^{(2)} = s_1^2(2 - \rho/\mu^* s_1^2)^2. \tag{6.180}$$

Let

$$c = \rho/\mu^* s_1^2. \tag{6.181}$$

Then with (6.177) and (6.178), (6.180) becomes after squaring

$$c^3 - 8c^2 + \left(24 - \frac{16\mu^*}{\lambda^* + 2\mu^*}\right) c - 16\left(1 - \frac{\mu^*}{\lambda^* + 2\mu^*}\right) = 0. \tag{6.182}$$

The wave motion then follows from the solution of this equation. On writing the slowness vector in terms of real and imaginary parts according to $s = s' + is''$, it follows that the solutions must satisfy

$$\begin{aligned} s_1'' &\leqslant 0 \\ s_3^{(1)''} &< 0 \\ s_3^{(2)''} &< 0. \end{aligned} \tag{6.183}$$

The first of these is necessary for the wave not to grow as it propagates along the surface, and the last two conditions are required for the wave to decay with depth. Also, relation (6.180) must be satisfied explicitly to eliminate any extraneous roots of (6.182) that were introduced by squaring.

Consider first the particle paths. The real part of the displacement, from (6.177)–(6.180), becomes

$$u_i' = [A_i'(x_3) \cos \omega(-s_1'x_1 + t) - A_i''(x_3) \sin \omega(-s_1'x_1 + t)]\, e^{\omega s_1'' x_1} \tag{6.184}$$

where with no loss in generality $U^{(1)} = 1$, and

$$\begin{aligned} A_1 &= e^{-i\omega s_3^{(1)} x_3} \quad \left(\frac{2}{c-2}\right) e^{-i\omega s_3^{(2)} x_3} \\ A_3 &= -\frac{s_1}{s_3^{(1)}} \left[e^{-i\omega s_3^{(1)} x_3} + \left(\frac{c-2}{2}\right) e^{-i\omega s_3^{(2)} x_3} \right]. \end{aligned} \tag{6.185}$$

A_i' and A_i'' in (6.184) are the real and imaginary parts of A_i in (6.185). Eliminate time between u_1' and u_3' in (6.184) to get the equation of the particle path orbit,

$$(A_3'' u_1' - A_1'' u_3')^2 + (A_3' u_1' - A_1' u_3')^2 = (A_1' A_3'' - A_3' A_1'')^2 \, e^{2\omega s_1 x_1}. \tag{6.186}$$

This is the path of an ellipse. If

$$A_1' A_3'' - A_2' A_1'' = 0 \tag{6.187}$$

then the ellipse degenerates into a straight line. Furthermore, the direction of tracing the ellipse changes as

$$A_1' A_3'' - A_3' A_1'' \lessgtr 0. \tag{6.188}$$

The path is said to be retrograde when the sense is counterclockwise; otherwise it is direct. The angle made by one of the axes of the ellipse with the coordinate axis is given by θ, where

$$\tan 2\theta = \frac{2(A_3'' A_1'' + A_3' A_1')}{(A_1')^2 + (A_1'')^2 - (A_3')^2 - (A_3'')^2} \, .$$

When the ellipse degenerates into a straight line, θ is given by

$$\tan \theta = A_3''/A_1'' = A_3'/A_1' \, .$$

Now using A_1 and A_3 from (6.185) in the degenerate condition (6.187) gives

$$\begin{aligned} &[s_1(c-2)/s_3^{(1)}]'' \{(c'-2) \cosh \omega(s_3^{(2)} - s_3^{(1)})'' \, x_3 \\ &\quad + [\tfrac{1}{4}(\bar{c}-2)(c-2)+1] \cos \omega(s_3^{(2)} - s_3^{(1)})' \, x_3\} \\ &\quad - [s_1(c-s)/s_3^{(1)}]' \{c'' \sinh \omega(s_3^{(2)} - s_3^{(1)})'' \, x_3 \\ &\quad + [\tfrac{1}{4}(\bar{c}-2)(c-2)-1] \sin \omega(s_3^{(2)} - s_3^{(1)})' \, x_3\} = 0. \end{aligned} \tag{6.189}$$

where the primes denote real and imaginary parts, and $\bar{c}$ is the complex conjugate of c. We see that this equation has the possibility of many roots for x_3. However, there can only be a finite number, since for x_3 sufficiently large the hyperbolic terms in (6.189) will dominate. In an elastic material there is only one change from retrograde motion near the surface to direct motion at some depth. This is a fundamental distinction in behavior between the elastic and viscoelastic cases.

Small Damping Example

Next we consider an example to illustrate the detailed effects of the solution. Take

$$\lambda^* = \mu'(1 + i\epsilon a); \qquad \mu^* = \mu'(1 + i\epsilon b) \tag{6.190}$$

where for small damping

$$\epsilon a \ll 1, \qquad \epsilon b \ll 1.$$

We see that we have also taken $\lambda' = \mu'$ in this example, which implies ν' will be near the value 1/4 for ν'' small. For positive imaginary parts in μ^* and k^* it follows that

$$b \geqslant 0 \qquad \text{and} \qquad 3a + 2b \geqslant 0 \tag{6.191}$$

where ϵ is a positive constant. Expand (6.182) in powers of ϵ and discard terms of order ϵ^2 and higher. The roots of (6.182), with small damping, are then given by

$$\begin{aligned} c_1 &= 4 + 2i\epsilon(b - a) \\ c_2 &= 2 + \frac{2}{\sqrt{3}} + i\epsilon(a - b)\left(\frac{3\sqrt{3} + 5}{3\sqrt{3}}\right) \\ c_3 &= 2 - \frac{2}{\sqrt{3}} + i\epsilon(a - b)\left(\frac{3\sqrt{3} - 5}{3\sqrt{3}}\right). \end{aligned} \tag{6.192}$$

The roots in (6.192) will now be examined separately. For root c_1, using (6.181) and (6.192) we get

$$s_1 = \sqrt{\frac{\rho}{4\mu'}}\left(1 + i\epsilon\frac{a - 3b}{4}\right). \tag{6.193}$$

The admissibility condition (6.183) then gives

$$3b \geqslant a. \tag{6.194}$$

From (6.177) and (6.178) we find

$$\begin{aligned} s_3^{(1)} &= \pm\sqrt{\frac{3\rho}{4\mu'}}\left(1 - i\epsilon\frac{a + 5b}{12}\right) \\ s_3^{(2)} &= \pm\sqrt{\frac{\rho}{12\mu'}}\left(1 + i\epsilon\frac{11b - 17a}{12}\right). \end{aligned} \tag{6.195}$$

Substitute the later two admissibility criteria from (6.183) into (6.180), giving

$$(17a - 11b)(a + 5b) < 0. \tag{6.196}$$

With (6.196) satisfied, the proper signs in (6.195) are taken to satisfy the last two of (6.183). Combining the requirements (6.191), (6.194), and (6.196) gives the inclusive requirement that

$$-\tfrac{2}{3} \leqslant a/b < \tfrac{11}{17}. \tag{6.197}$$

One proceeds similarly for the other two roots in (6.192). It is found for c_2 that

$$\frac{5 + 13\sqrt{3}}{11 + 7\sqrt{3}} < \frac{a}{b} < \frac{39 + 23\sqrt{3}}{9 + 5\sqrt{3}} \tag{6.198}$$

while for the root c_3 all values of a and b are permissible.

It is important to compare these results with those in the elastic case, $a = b = 0$, then roots c_1 and c_2 do not give waves that decay with depth; thus these waves are not admissible. Root c_3 in the elastic case gives

$$s_1 = \sqrt{\frac{\rho}{\mu(2 - 2/\sqrt{3})}}$$

$$s_3^{(1)}/s_1 = -i\sqrt{2/\sqrt{3} - 1}$$

$$s_3^{(2)}/s_1 = -i\sqrt{2/\sqrt{3} + 1}.$$

The latter two values imply disturbances that decay exponentially with depth.

Returning to the viscoelastic example, the root c_3 mainly decays exponentially with depth, and thus it corresponds to elasticlike behavior. The roots c_1 and c_2, when they exist according to (6.197) and (6.198), provide an oscillatory behavior with respect to depth, and they also decay slowly with depth. From (6.197) and (6.198), c_1 will exist when

$$-0.67 \leqslant a/b < 0.65$$

and c_2 when

$$1.19 \leqslant a/b < 4.46.$$

Thus we see that these two roots cannot exist together. It can be shown that the wave speeds corresponding to the roots (6.192) have rank order according to

$$\tilde{c}_3 < c_T < \tilde{c}_2 < c_L < \tilde{c}_1$$

where c_T and c_L are the transverse and longitudinal wave speeds

$$c_T = \sqrt{\mu'/\rho}, \qquad c_L = \sqrt{3\mu'/\rho}.$$

For a simple numerical example, take $\epsilon a = 0.02$ and $\epsilon b = 0.04$, implying larger damping in shear than in longitudinal motion. It is found that for root c_1

$$s_1 = 0.499 - 0.012i$$

$$s_3^{(1)} = 0.866 - 0.016i$$

$$s_3^{(2)} = -0.289 - 0.002i$$

while for root c_3,

$$s_1 = 1.087 - 0.021i$$

$$s_3^{(1)} = -0.007 - 0.428i$$

$$s_3^{(2)} = -0.019 - 0.921i.$$

The important thing to note is that root c_1 corresponds to a wave that decays more slowly with horizontal distance traveled than does the elasticlike c_3 mode (compare 0.012 with 0.021). This could be very important in practical seismological applications. Other features that sharply distinguish the viscoelastic results from the elastic results are given by Currie *et al.* [6.27].

Incompressible Example

In the incompressible case, let $\lambda \to \infty$, and allow a reactive hydrostatic pressure. There is no restriction to small damping. The governing equation for c is found to be

$$c^3 - 8c^2 + 24c - 16 = 0. \tag{6.199}$$

The roots of (6.199) are given by

$$c_1 = 3.54 + 2.23i$$

$$c_2 = 3.54 - 2.23i$$

$$c_3 = 0.912.$$

Thus the root c_1 is not admissible since it does not decay with depth. The root c_2 is permissible if

$$\mu''/\mu' > 0.159$$

and the root c_3 is always permissible.

Rayleigh waves in materials governed by simple mechanical models have been considered; see, for example, Currie [6.28].

PROBLEMS

6.1. Obtain the response of a semi-infinite rod, idealized as a Maxwell model material, when subjected to a suddenly applied and thereafter maintained constant end stress. Use the direct application of the integral transform method.

6.2. Relative to the uniaxial wave propagation problem, obtain the asymptotic expansion for the stress in the region of the wave front by retaining one more term than is done in the derivation of Section 6.1.

6.3. Relative to the shear flow response problem of Section 6.2, obtain the complete derivation of the stress solution, for a general viscoelastic material. Specialize these results to those of the Maxwell model representation of the relaxation function and compare the result with (6.61). Finally, obtain the complete solution for the corresponding elastic problem having a shear modulus equal to the initial value of the relaxation function in shear.

6.4. Using the method of time dependent boundary conditions, obtain the response of a simply supported uniform viscoelastic beam (classical beam theory assumptions) to a suddenly applied uniform load which thereafter remains constant with time.

6.5. Extend the method of Section 6.4 to account for the reflection and refraction of incident harmonic waves at the interface between two viscoelastic materials. See Ref. [6.22].

6.6. By neglecting inertia terms, obtain the quasi static solution for the problem of Section 6.5, involving a moving load on the viscoelastic half space (plane strain).

6.7. Adapt the Rayleigh wave solution of Section 6.6 to model the case of small damping in μ^* but none in the bulk complex modulus k^*, which is thus a real constant k. Complete a numerical example, using $\epsilon b = 0.04$ and compare the results with those of Section 6.6.

REFERENCES

6.1. Thomas, T. Y., *Plastic Flow and Fracture in Solids.* Academic Press, New York, 1961.
6.2. Lee, E. H., and J. A. Morrison, "A Comparison of the Propagation of Longitudinal Waves in Rods of Viscoelastic Materials," *J. Polym. Sci.* **19**, 93 (1956).
6.3. Achenbach, J. D., and D. P. Reddy, "Note on Wave Propagation in Linearly Viscoelastic Media," *Z. Angew. Math. Phys.* **18**, 141 (1967).
6.4. Berry, D. S., "A Note on Stress Pulses in Viscoelastic Rods, *Phil. Mag.* [8] **3**, 100 (1958).
6.5. Chu, B. T., "Stress Waves in Isotropic Linear Viscoelastic Materials," *J. Mec.* **1**, 439 (1962).
6.6. Sackman, J. L., and I. Kaya, "On the Propagation of Transient Pulses in Linearly Viscoelastic Media," *J. Mech. Phys. Solids* **16**, 349 (1968).

6.7. Valanis, K. C., and S. Chang, "Stress Wave Propagation in a Finite Viscoelastic Thin Rod with a Constitutive Law of the Hereditary Type," *in Developments in Theoretical and Applied Mechanics* (T. C. Huang and M. W. Johnson, Jr., eds.). Wiley, New York, 1965.

6.8. Knauss, W. G., "Uniaxial Wave Propagation in a Viscoelastic Material Using Measured Material Properties," *J. Appl. Mech.* **35**, 449 (1968).

6.9. Fisher, G. M. C., and M. E. Gurtin, "Wave Propagation in the Linear Theory of Viscoelasticity," *Quart. Appl. Math.* **23**, 257 (1965).

6.10. Valanis, K. C., "Propagation and Attenuation of Waves in Linear Viscoelastic Solids," *J. Math. Phys.* **44**, 227 (1965).

6.11. Lee, E. H., and I. Kanter, "Wave Propagation in Finite Rods of Viscoelastic Material," *J. of Appl. Phys.* **24**, 1115 (1953).

6.12. Mindlin, R. D., and L. E. Goodman, "Beam Vibrations with Time-Dependent Boundary Conditions," *J. Appl. Mech.* **17**, 377 (1950).

6.13. Herrmann, G., "Forced Motion of Elastic Rods," *J. Appl. Mech.* **21**, 221 (1954).

6.14. Berry, J. G., and P. M. Naghdi, "On the Vibration of Elastic Bodies Having Time-Dependent Boundary Conditions," *Quart. Appl. Math.* **14**, 43 (1956).

6.15. Achenbach, J. D., "Vibrations of a Viscoelastic Body," *AIAA J.* **5**, 1213 (1967).

6.16. Christensen, R. M., "Application of the Method of Time-Dependent Boundary Conditions in Linear Viscoelasticity," *J. Appl. Mech.* **34**, 503 (1967).

6.17. Valanis, K. C., and C. T. Sun, "Axisymmetric Wave Propagation in a Solid Viscoelastic Sphere," *Int. J. Eng. Sci.* **5**, 939 (1967).

6.18. Hunter, S. C., "Tentative Equations for the Propagation of Stress, Strain, and Temperature Fields in Viscoelastic Solids," *J. Mech. Phys. Solids* **9**, 39 (1961).

6.19. Petrof, R. C., and S. Gratch, "Wave Propagation in a Viscoelastic Material with Temperature-Dependent Properties and Thermomechanical Coupling," *J. Appl. Mech.* **31**, 423 (1964).

6.20. Wolosewik, R. M., and S. Gratch, "Transient Response in a Viscoelastic Material with Temperature-Dependent Properties and Thermomechanical Coupling," *J. Appl. Mech.* **32**, 620 (1965).

6.21. Hunter, S. C., "The Transient Temperature Distribution in a Semi-Infinite Viscoelastic Rod Subject to Longitudinal Oscillations," *Int. J. Eng. Sci.* **5**, 119 (1967).

6.22. Lockett, F. J., "The Reflection and Refraction of Waves at an Interface between Viscoelastic Materials," *J. Mech. Phys. Solids* **10**, 53 (1962).

6.23. Cooper, H. F., Jr., and E. L. Reiss, "Reflection of Plane Viscoelastic Waves from Plane Boundaries," *J. Acoust. Soc. Amer.* **39**, 1133 (1966).

6.24. Cooper, H. F., Jr., "Reflection and Transmission of Oblique Plane Waves at a Plane Interface between Viscoelastic Media," *J. Acoust. Soc. Amer.* **42**, 1064 (1967).

6.25. Tsai, Y. M., and H. Kolsky, "Surface Wave Propagation for Linear Viscoelastic Solids," *J. Mech. Phys. Solids* **16**, 99 (1968).

6.26. Chu, B. T., "Response of Various Material Media to High-Velocity Loadings; I. Linear Elastic and Viscoelastic Materials," *J. Mech. Phys. Solids* **13**, 165 (1965).

6.27. Currie, P. K., M. A. Hayes, and P. M. O'Leary, "Viscoelastic Rayleigh Waves," *Quart. Appl. Math.* **35**, 35 (1977).

6.28. Currie, P. K., "Viscoelastic Surface Waves on a Standard Linear Solid," *Quart. Appl. Math.* **37**, 332 (1979).

Chapter VII General Theorems and Formulations

As the title of this chapter indicates, it is comprised of a collection of results which have general applicability and usefulness. For the most part these results are extensions to viscoelasticity of some of the well-known theorems of elasticity. For example the classical variational theorems and minimum theorems of elasticity will be shown to possess several different types of generalizations to viscoelasticity. The first item of study here is the establishment of the uniqueness of solution of the coupled thermoviscoelastic boundary value problem.

7.1. UNIQUENESS OF SOLUTION OF COUPLED THERMOVISCOELASTIC BOUNDARY VALUE PROBLEM

Now we give a general proof which establishes the uniqueness of solution for linear thermoviscoelastic problems. This proof of the uniqueness of solution is a generalization to the nonisothermal case of the isothermal uniqueness theorem given by Onat and Breuer [7.1].

The relevant linear equations from Chapter 3, which govern the dynamic, nonisothermal, anisotropic, coupled viscoelasticity theory, are given by (3.45)–(3.52), where the equations of motion (3.45), with inertia terms included, are taken. The Laplace transform of these relations, with zero initial conditions for $t < 0$ are assumed to exist. These relations are given by (3.53)–(3.57), and they are repeated here for convenience. The equations of motion, the strain displacement relations, and the stress constitutive relations are respectively given by

$$\bar{\sigma}_{ij,j} + \bar{F}_i = s^2 \rho \bar{u}_i \tag{7.1}$$

$$\bar{\epsilon}_{ij} = \tfrac{1}{2}(\bar{u}_{i,j} + \bar{u}_{j,i}) \tag{7.2}$$

and

$$\bar{\sigma}_{ij} = s\bar{G}_{ijkl}\bar{\epsilon}_{kl} - s\bar{\varphi}_{ij}\bar{\theta} \tag{7.3}$$

where s is the transform variable and $\theta(x_i, t)$ is the infinitesimal temperature deviation from the base temperature T_0. The heat conduction equation is

$$(k_{ij}/T_0)\,\bar{\theta}_{,ij} = s^2\bar{m}\bar{\theta} + s^2\bar{\varphi}_{ij}\bar{\epsilon}_{ij} \tag{7.4}$$

where k_{ij} and the relaxation functions $m(t)$ and $\varphi_{ij}(t)$ are mechanical properties of the material. The boundary conditions are given by

$$\bar{\sigma}_{ij}n_j = \bar{S}_i \quad \text{on} \quad B_\sigma, \tag{7.5}$$

$$\bar{u}_i = \bar{\Delta}_i \quad \text{on} \quad B_u, \tag{7.6}$$

$$\bar{\theta} = \bar{\hat{\theta}} \quad \text{on} \quad B_1 \tag{7.7}$$

and

$$k_{ij}\bar{\theta}_{,i}n_j = 0 \qquad \text{on} \quad B_2 \tag{7.8}$$

where B_σ, B_u, B_1, and B_2 are the appropriate subregions of the boundary, assumed to be constant with time.

The uniqueness theorem appropriate to thermoviscoelastic boundary value problems is now stated.

The isotropic coupled thermoviscoelastic boundary value problem, governed by field equations and boundary conditions (3.45)–(3.52), *possesses a unique solution, provided the Laplace transform of all field variables is assumed to exist with transformed field equations and boundary conditions* (7.1)–(7.8), *and provided the initial values of the relaxation functions are positive definite and the thermal conductivity is positive semidefinite, as*

$$G_{ijkl}(0)\,\gamma_{ij}\gamma_{kl} > 0, \qquad m(0) > 0, \qquad k_{ij}\gamma_i\gamma_j \geqslant 0$$

where γ_{ij} is any symmetric second order tensor.

To prove this theorem, we assume that two separate solutions of the field equations, boundary conditions, and initial conditions exist, with the field variables given by

$$\begin{matrix} u_i^{(1)} & & u_i^{(2)} \\ \epsilon_{ij}^{(1)} & & \epsilon_{ij}^{(2)} \\ \sigma_{ij}^{(1)} & \text{and} & \sigma_{ij}^{(2)} \\ \theta^{(1)} & & \theta^{(2)} \end{matrix}$$

where these are assumed to be continuous functions of time and of the spatial coordinates. We designate the difference solution, u_i^d, ϵ_{ij}^d, σ_{ij}^d, and θ^d, by

$$\begin{aligned} u_i^d(x_i, t) &= u_i^{(1)}(x_i, t) - u_i^{(2)}(x_i, t), \\ \epsilon_{ij}^d(x_i, t) &= \epsilon_{ij}^{(1)}(x_i, t) - \epsilon_{ij}^{(2)}(x_i, t), \\ \sigma_{ij}^d(x_i, t) &= \sigma_{ij}^{(1)}(x_i, t) - \sigma_{ij}^{(2)}(x_i, t) \end{aligned} \tag{7.9}$$

and

$$\theta^d(x_i, t) = \theta^{(1)}(x_i, t) - \theta^{(2)}(x_i, t).$$

We note that the difference solution, $u_i^d(x_i, t)$, $\epsilon_{ij}^d(x_i, t)$, $\sigma_{ij}^d(x_i, t)$, and $\theta^d(x_i, t)$, satisfies homogeneous boundary conditions. Because of this, we can write the following relation:

$$\int_B \bar{\sigma}_{ij}^d n_j \bar{u}_i^d \, da = 0 \tag{7.10}$$

relative to the Laplace transformed variables. We apply the divergence theorem to (7.10) and use the form of (7.1) and (7.2) appropriate to the difference solution to get

$$\int_V [\rho s^2 \bar{u}_i{}^d \bar{u}_i{}^d + \bar{\sigma}_{ij}^d \bar{\epsilon}_{ij}^d]\, dv = 0. \tag{7.11}$$

We now use the difference form of (7.3) in (7.11) to obtain

$$\int_V [\rho s^2 \bar{u}_i{}^d \bar{u}_i{}^d + s\bar{G}_{ijkl} \bar{\epsilon}_{ij}^d \bar{\epsilon}_{kl}^d - s\bar{\varphi}_{ij} \bar{\epsilon}_{ij}^d \bar{\theta}^d]\, dv = 0. \tag{7.12}$$

The last term in (7.12) will be eliminated by using the heat conduction equation (7.4) in difference form and multiplied by $\bar{\theta}^d$:

$$(k_{ij}/T_0)\, \bar{\theta}^d \bar{\theta}^d_{,ij} = s^2 \bar{m} (\bar{\theta}^d)^2 + s^2 \bar{\varphi}_{ij} \bar{\epsilon}_{ij}^d \bar{\theta}^d. \tag{7.13}$$

It is necessary to express the first term in (7.13) in an alternate form. To do this we note the following identity:

$$\int_B k_{ij} \bar{\theta}^d \bar{\theta}^d_{,i} n_j \, da = 0 \tag{7.14}$$

since $\bar{\theta}^d = 0$ on B_1 and $k_{ij} \bar{\theta}^d_{,i} n_j = 0$ on B_2, where B_1 and B_2 are complementary parts of the boundary B. By applying the divergence theorem to (7.14), we find that

$$k_{ij} \bar{\theta}^d \bar{\theta}^d_{,ij} = -k_{ij} \bar{\theta}^d_{,i} \bar{\theta}^d_{,j}\,. \tag{7.15}$$

Now we solve for $\bar{\epsilon}_{ij}^d \bar{\theta}^d$ from (7.13), using (7.15), and substitute it into (7.12). There results

$$\int_V \left[\rho s^2 \bar{u}_i{}^d \bar{u}_i{}^d + s\bar{G}_{ijkl} \bar{\epsilon}_{ij}^d \bar{\epsilon}_{kl}^d + \frac{k_{ij}}{sT_0} \bar{\theta}^d_{,i} \bar{\theta}^d_{,j} + s\bar{m} (\bar{\theta}^d)^2\right] dv = 0. \tag{7.16}$$

We consider real values of s such that $s > s_0$, where s_0 is the solution's furthest singularity to the right on the real axis. We recall that

$$\lim_{s\to\infty} s\bar{G}_{ijkl}(s) = G_{ijkl}(0)$$

and (7.17)

$$\lim_{s\to\infty} s\bar{m}(s) = m(0).$$

Now it is required that the initial values of the relaxation functions be positive definite; thus,

$$G_{ijkl}(0)\, \gamma_{ij} \gamma_{kl} > 0$$

$$m(0) > 0. \tag{7.18}$$

It then follows that for s sufficiently large, $s\bar{G}_{ijkl}(s)\,\gamma_{ij}\gamma_{kl} > 0$ and $s\bar{m}(s) > 0$. If it is also required that $k_{ij}\gamma_i\gamma_j \geqslant 0$, and if s is sufficiently large, then Eq. (7.16) is composed of all nonnegative terms, which to be satisfied requires

$$\bar{u}_i^{\,d} = \bar{\theta}^d = 0 \qquad \text{for} \quad s > s_0 \quad \text{and sufficiently large.} \tag{7.19}$$

A theorem from Doetsch [7.2] states that if the Laplace transform of a continuous function is zero on an infinite set of points on the real axis, which constitute an arithmetic progression, the function itself is identically zero. Applying this to the present situation reveals that the difference functions $u_i^{\,d}$ and θ^d must be identically zero, as

$$u_i^{\,d} = \theta^d = 0. \tag{7.20}$$

Since the difference solution is identically zero, the solution of the coupled thermoviscoelastic boundary value problem must be unique. This sufficiency proof follows from the assumptions $k_{ij}\gamma_i\gamma_j \geqslant 0$, $m(0) > 0$, and $G_{ijkl}(0)\gamma_{ij}\gamma_{kl} > 0$. It is remarkable that only the initial values of the relaxation functions are involved in the proof.

The use of the Laplace transformation to establish the uniqueness theorem here is, in a sense, artificial. Although integral transform techniques are a highly valuable tool in solving boundary value problems, their use in establishing general theorems is less appealing because of the implied restrictions which must accompany their use. The Laplace transform is used in this section only because of the simplicity it affords in this particular proof; it is not an essential part of a uniqueness proof. The isothermal uniqueness theorem given in Section 2.2 did not use an integral transform method.

References were given in Section 2.2 to several different isothermal uniqueness theorems. Sternberg and Gurtin [7.3] have given a nonisothermal uniqueness theorem, applicable to different conditions from those considered here.

7.2. REPRESENTATION IN TERMS OF DISPLACEMENT FUNCTIONS

In the theory of elasticity, two particularly useful procedures for uncoupling the equations of equilibrium are those of the Galerkin vector and the Papkovitch–Neuber stress function. We turn now to the extension to viscoelasticity of these elasticity uncoupling procedures. This extension depends upon the use of the Stieltjes convolution notation introduced in Section 1.2.

We recall that the elasticity formulation with no body forces has the displacement form of the equations of equilibrium as

$$\nabla^2\mathbf{u} + \frac{1}{1-2\nu}\nabla\nabla\cdot\mathbf{u} = 0 \tag{7.21}$$

where $\mathbf{u}$ is the displacement vector, ∇ is the Laplacian operator, and $\nabla \cdot \mathbf{u}$ is the scalar product. The Galerkin vector solution of these equations is given by

$$\mathbf{u} = (1/2\mu)[2(1-\nu)\nabla^2\mathbf{g} - \nabla\nabla \cdot \mathbf{g}] \tag{7.22}$$

where

$$\nabla^4\mathbf{g} = 0. \tag{7.23}$$

Thus, the solution of the elastic equations of equilibrium is reduced to the problem of finding a biharmonic vector function $\mathbf{g}$. An alternative procedure is that of the Papkovitch–Neuber stress function, whereby the displacement vector is taken as

$$\mathbf{u} = (1/2\mu)[\nabla(\varphi + \mathbf{r} \cdot \boldsymbol{\psi}) - 4(1-\nu)\boldsymbol{\psi}] \tag{7.24}$$

where $\mathbf{r}$ is the position vector from the origin of the coordinate system and

$$\nabla^2\varphi = \nabla^2\boldsymbol{\psi} = 0. \tag{7.25}$$

In this case, the integration of the equations of equilibrium is reduced to the problem of finding four harmonic functions.

For viscoelasticity, the equations of (quasi-static) equilibrium, in a form comparable to (7.21), are given by

$$\nabla^2\mathbf{u} * dG_1 + \tfrac{1}{3}\nabla\nabla \cdot \mathbf{u} * d(G_1 + 2G_2) = 0 \tag{7.26}$$

where the Stieltjes convolution notation of Section 1.2 is used. This form is easily verified, using $\sigma_{ij,j} = 0$ along with (2.3), $G_1(t) = 2\mu(t)$, $G_2(t) = 3k(t)$, and the strain displacement relations.

The generalization of the elastic Galerkin's vector to viscoelasticity is given by

$$\mathbf{u} = 2\nabla^2\mathbf{g} * d(2G_1 + G_2) - \nabla\nabla \cdot \mathbf{g} * d(G_1 + 2G_2) \tag{7.27}$$

where

$$\nabla^4\mathbf{g} = 0. \tag{7.28}$$

To prove that this representation satisfies the equations of equilibrium, (7.27) is directly substituted into (7.26). In order to do this, we use (7.27) to obtain the following two forms:

$$\nabla^2\mathbf{u} = 2(\nabla^4\mathbf{g}) * d(2G_1 + G_2) - \nabla^2[\nabla\nabla \cdot \mathbf{g} * d(G_1 + 2G_2)] \tag{7.29}$$

and

$$\nabla\nabla \cdot \mathbf{u} = \nabla\nabla \cdot [2\nabla^2\mathbf{g} * d(2G_1 + G_2) - \nabla\nabla \cdot \mathbf{g} * d(G_1 + 2G_2)]. \tag{7.30}$$

Now we note that the first term on the right-hand side of (7.29) vanishes, owing to (7.28). Then by substituting (7.29) and (7.30) into (7.26) we get the form

$$\begin{aligned} &-\nabla^2[\nabla\nabla \cdot \mathbf{g} * d(G_1 + 2G_2)] * dG_1 \\ &\quad + \tfrac{1}{3}\nabla\nabla \cdot [2\nabla^2\mathbf{g} * d(2G_1 + G_2)] * d(G_1 + 2G_2) \\ &\quad - \tfrac{1}{3}\nabla\nabla \cdot [\nabla\nabla \cdot \mathbf{g} * d(G_1 + 2G_2)] * d(G_1 + 2G_2) = 0. \end{aligned} \tag{7.31}$$

We use the vector analysis identity, $\nabla^2\nabla$ div $= \nabla$ div $\nabla^2 = \nabla$ div ∇ div, and the associativity and distributivity properties of the Stieltjes convolutions (see Section 1.2) in (7.31). This reduces (7.31) to the simple form

$$\nabla^2\nabla \operatorname{div} \mathbf{g} * [-d(G_1 + 2G_2) * dG_1 + d(G_1) * d(G_1 + 2G_2)] = 0. \quad (7.32)$$

Finally, we use the commutivity property of Stieltjes convolutions to show that (7.32) is satisfied, and the proof is complete.

The generalization of the elastic Papkovich–Neuber stress function to viscoelasticity is given by

$$\mathbf{u} = \nabla(\varphi + \mathbf{r} \cdot \boldsymbol{\psi}) * d(G_1 + 2G_2) - 4\boldsymbol{\psi} * d(2G_1 + G_2) \quad (7.33)$$

where

$$\nabla^2\varphi = \nabla^2\boldsymbol{\psi} = 0. \quad (7.34)$$

The proof of this representation follows, as with the Galerkin vector, by substituting (7.33) into (7.26), using the properties of the Stieltjes convolutions, and using (7.34).

The comparable representation forms, when the body forces are nonzero, are given in Gurtin and Sternberg [7.4].

These two Stieltjes convolutions displacement representations in viscoelasticity reduce the integration problem to simple forms. An alternative procedure to these methods would be to take an integral transform of the governing viscoelastic field equations and boundary conditions and apply the elasticity-type Galerkin vector representation, or the Papkovich–Neuber representation, directly in the transform plane. The uncoupling procedure would be formally the same as in elasticity; and, indeed, an elasticity solution obtained by this means could directly be converted to the transformed viscoelastic solution. This latter procedure would, in general, be easier to apply than the Stieltjes convolutions forms of (7.27) and (7.33). But there is an important distinction which must be drawn between these two alternative methods of uncoupling the equations of equilibrium. The Stieltjes convolution forms could be applied in problems in which integral transform methods are not applicable. For example, the contact problems discussed in Section 5.1 are problems for which integral transform methods do not apply. The problem of the indentation of a viscoelastic half space by a curved indentor was discussed in Section 5.1. For this problem the complete and general analysis was given by Graham [7.5], using the Stieltjes convolution form of the Papkovich–Neuber stress function just discussed.

7.3. RECIPROCAL THEOREM

A viscoelactic reciprocal theorem will be proved which is analogous to the one in elasticity theory. In doing this we find it convenient to follow the treatment of Gurtin and Sternberg [7.4] and utilize the formalized notation for Stieltjes convolutions, as given in Section 2.2.

The viscoelastic reciprocal theorem is now stated, and the usual notation is used whereby F_i, σ_i, and u_i refer to the components of body forces, boundary stress vector components, and displacements respectively.

An isotropic viscoelastic body, when subjected to two different states of loading with corresponding body forces, surface stresses, and displacements, F_i, σ_i, u_i *and* F_i', σ_i', u_i', *respectively, has a field variable solution which satisfies the relationship*

$$\int_B [\sigma_i * du_i']\, da + \int_V [F_i * du_i']\, dv = \int_B [\sigma_i' * du_i]\, da + \int_V [F_1' * du_i]\, dv. \tag{7.35}$$

The proof of this theorem begins by applying the divergence theorem to (7.35) and using the equations of equilibrium to write the left-hand side of (7.35), symbolized by L, as

$$L = \int_B [\sigma_i * du_i']\, da + \int_V [F_i * du_i']\, dv = \int_V [\sigma_{ij} * d\epsilon_{ij}']\, dv. \tag{7.36}$$

We write (7.36) in deviatoric and dilatational component form to get

$$L = \int_V [s_{ij} * de_{ij}' + \tfrac{1}{3}\sigma_{kk} * d\epsilon_{jj}']\, dv. \tag{7.37}$$

Now we use the stress strain relations (1.22) in (7.37) to obtain

$$L = \int_V [G_1 * de_{ij} * de_{ij}' + \tfrac{1}{3} G_2 * d\epsilon_{kk} * d\epsilon_{jj}']\, dv. \tag{7.38}$$

We use the associativity and commutivity properties of Stieltjes convolutions (Section 1.2) to obtain the forms

$$\begin{aligned} G_1 * de_{ij} * de_{ij}' &= G_1 * d(e_{ij} * de_{ij}') \\ &= G_1 * d(e_{ij}' * de_{ij}) \\ &= G_1 * de_{ij}' * de_{ij}\,. \end{aligned} \tag{7.39}$$

Relation (7.39) and a similar one for the second term in (7.38) then gives (7.38) as

$$L = \int_V [G_1 * de_{ij}' * de_{ij} + \tfrac{1}{3} G_2 * d\epsilon_{kk}' * d\epsilon_{jj}]\, dv$$

or

$$L = \int_V [\sigma_{ij}' * d\epsilon_{ij}]\, dv. \tag{7.40}$$

The comparison of (7.36) and (7.40) completes the proof of (7.35).

An interesting special case of this theorem results when the two loading systems have a form involving a separation of the spatial and time portions of the variable. In this case, the two systems of loading terms have the forms:

Unprimed System

$$\begin{aligned} F_i(x_i, t) &= \tilde{F}_i(x_i)\, g(t) \\ u_i(x_i, t) &= \tilde{\Delta}_i(x_i)\, g(t) \qquad \text{on} \quad B_u \\ \sigma_{ij}(x_i, t) n_j &= \tilde{s}_i(x_i)\, g(t) \qquad \text{on} \quad B_\sigma \end{aligned} \tag{7.41}$$

and

Primed System

$$\begin{aligned} F_i(x_i, t) &= \tilde{F}_i'(x_i)\, g(t) \\ u_i(x_i, t) &= \tilde{\Delta}_i'(x_i)\, g(t) \qquad \text{on} \quad B_u \\ \sigma_{ij}(x_i, t) n_j &= \tilde{s}_i'(x_i)\, g(t) \qquad \text{on} \quad B_\sigma . \end{aligned} \tag{7.42}$$

It must be noted that, even though these forms involve a separation of variables, the resulting solutions of the viscoelastic boundary value problems would not in general have a separation of variables form. As we argued in Section 2.3, for a separation of variables solution to exist, the displacement and stress boundary conditions must have a related but different time variation, which is in contrast to the situation here.

The viscoelastic reciprocal theorem (7.35) appropriate to these two loading states asserts that

$$\begin{aligned} &\int_{B_u} [\sigma_i * dg]\, \tilde{\Delta}_i'\, da + \int_{B_\sigma} \tilde{S}_i[u_i' * dg]\, da + \int_V \tilde{F}_i[u_i' * dg]\, dv \\ &\quad = \int_{B_u} [\sigma_i' * dg]\, \tilde{\Delta}_i\, da + \int_{B_\sigma} \tilde{S}_i'[u_i * dg]\, da + \int_V \tilde{F}_i'[u_i * dg]\, dv \end{aligned} \tag{7.43}$$

where the commutivity of the Stieltjes convolutions has been used.

We invert the order of space and time integration in (7.43) to obtain the form

$$\begin{aligned} &\Big[\int_{B_u} \sigma_i \tilde{\Delta}_i'\, da + \int_{B_\sigma} \tilde{S}_i u_i'\, da + \int_V \tilde{F}_i u_i'\, dv \\ &\quad - \int_{B_u} \sigma_i' \tilde{\Delta}_i\, da - \int_{B_\sigma} \tilde{S}_i' u_i\, da - \int_V \tilde{F}_i' u_i\, dv\Big] * dg = 0. \end{aligned} \tag{7.44}$$

For (7.44) to be satisfied for all values of time, it is necessary that the term in the brackets vanish. When that form is multiplied by $g(t)$, it gives

$$\int_B \sigma_i u_i'\, da + \int_V F_i u_i'\, dv = \int_B \sigma_i' u_i\, da + \int_V F_i' u_i\, dv. \tag{7.45}$$

This special case of the viscoelastic reciprocal theorem, applicable only when (7.41) and (7.42) are appropriate, has exactly the same form as the elastic reciprocal theorem.

A particularly interesting application of the viscoelastic reciprocal theorem is found in deriving a formula for the volume change of a body, analogous to the comparable elasticity formula, given for example, by Sokolnikoff [7.6]. This is given as a problem at the end of the chapter.

Sternberg [7.7] has shown how the viscoelastic reciprocal theorem and several other general theorems can be adapted to nonisothermal conditions through the use of the body force analogy.

7.4. VARIATIONAL THEOREMS

There have been several derivations of variational theorems in quasi-static viscoelasticity; for example, Onat [7.8], Schapery [7.9], Biot [7.10], Christensen [7.11], and Gurtin [7.12]. Probably the derivation by Gurtin is the most general treatment. In his approach, the Stieltjes convolution notation is used in a manner similar to that just employed in the previous two sections. The variational theorems derived by Christensen, however, are the only ones for which associated minimum theorems have been established. Accordingly, both of these types of theorems are discussed here, for quasi-static conditions. Also, undisturbed conditions are assumed for $t < 0$.

The relations of Section 2.1 which define the quasi-static viscoelastic boundary value problem are rewritten here for convenience as

$$\epsilon_{ij} = \tfrac{1}{2}(u_{i,j} + u_{j,i}), \tag{7.46}$$

$$\sigma_{ij} = \int_0^t G_{ijkl}(t-\tau)\,\frac{\partial \epsilon_{kl}(\tau)}{\partial \tau}\,d\tau, \qquad \sigma_{ij} = G_{ijkl} * d\epsilon_{kl}\,, \tag{7.47}$$

or

$$\epsilon_{ij} = \int_0^t J_{ijkl}(t-\tau)\,\frac{\partial \sigma_{kl}(\tau)}{\partial \tau}\,d\tau, \qquad \epsilon_{ij} = J_{ijkl} * d\sigma_{kl}\,, \tag{7.48}$$

$$\sigma_{ij,j} + F_i = 0, \tag{7.49}$$

$$\sigma_{ij} n_j = S_i \quad \text{on} \quad B_\sigma\,, \tag{7.50}$$

and

$$u_i = \Delta_i \quad \text{on} \quad B_u \tag{7.51}$$

where S_i and Δ_i are the prescribed stresses and displacements on the boundary, and the general anisotropic case is being considered. In the variational theorems to be formulated here we will assume that the stresses, strains, and displacements are continuous and continuously differentiable. Under certain conditions this restriction can be relaxed somewhat, see for example Ref. [7.12].

The procedure now is to establish a functional F, for which Eqs. (7.46)–(7.51) imply that its variation δF vanishes; that is, $\delta F = 0$. Exactly what is meant by the variation of a functional is now defined. We take, for example, a functional F of a single history parameter $\epsilon(t-s)$, as

$$F = \mathop{\psi}_{s=0}^{\infty} \left(\epsilon(t-s), \epsilon(t)\right) \tag{7.52}$$

with the usual dependence upon the current value $\epsilon(t)$. We assume that this functional, for another history $\epsilon(t-s) + [\delta\epsilon(t-s)]$, can be expanded about that for the history $\epsilon(t-s)$ to give

$$\begin{aligned} \mathop{\psi}_{s=0}^{\infty} \left(\epsilon(t-s) + \delta\epsilon(t-s), \epsilon(t)\right) &= \mathop{\psi}_{s=0}^{\infty} \left(\epsilon(t-s), \epsilon(t)\right) + \delta \mathop{\psi}_{s=0}^{\infty} \left(\epsilon(t-s), \epsilon(t) \mid \delta\epsilon(t-s)\right) \\ &\quad + \tfrac{1}{2}\delta^2 \mathop{\psi}_{s=0}^{\infty}\left(\epsilon(t-s), \epsilon(t) \mid \delta\epsilon(t-s), \delta\epsilon(t-s)\right) \\ &\quad + \cdots . \end{aligned} \tag{7.53}$$

Such an expansion will be assumed to exist for the functionals of interest here.[1] It may be noted that (7.53) is an expansion for functionals which is completely analogous to the Taylor series expansion for functions. In (7.53), $\delta \mathop{\psi}_{s=0}^{\infty}(\ )$ symbolizes a functional of histories $\epsilon(t-s)$ and $\delta\epsilon(t-s)$ which is linear in $\delta\epsilon(t-s)$ and is called the first order Fréchet differential. Similarly, $\delta^2 \mathop{\psi}_{s=0}^{\infty}(\ )$ is the second order Fréchet differential, and is quadratic in the history $\delta\epsilon(t-s)$. The higher order functionals are correspondingly defined.

We define the first variation of the functional F, or simply δF, as

$$\delta F = \frac{d}{d\alpha}\left[\mathop{\psi}_{s=0}^{\infty} \left(\epsilon(t-s) + \delta\epsilon(t-s)\alpha, \epsilon(t)\right) + \mathop{\psi}_{s=0}^{\infty} \left(\epsilon(t-s), \epsilon(t) + \delta\epsilon(t)\alpha\right)\right]_{\alpha=0} \tag{7.54}$$

where α is a real constant. Now we perform the operation indicated in (7.54) upon the functional in (7.52), using (7.53), to show that

$$\delta F = \delta \mathop{\psi}_{s=0}^{\infty} \left(\epsilon(t-s), \epsilon(t) \mid \delta\epsilon(t-s)\right) + \frac{\partial}{\partial \epsilon(t)} \mathop{\psi}_{s=0}^{\infty} \left(\epsilon(t-s), \epsilon(t)\right) \delta\epsilon(t). \tag{7.55}$$

[1] Sufficient conditions for the existence of the expansion (7.53) are discussed in Chapter 8.

Thus, the first variation of the functional is nothing more than the first order Fréchet differential in the expansion of the functional of the history $\epsilon(t-s)+\delta\epsilon(t-s)$ about that for the history $\epsilon(t-s)$, plus the partial derivative term. Functionals having more than one tensor valued history argument, as will be considered here, have variations defined similar to that in (7.54) and (7.55).

We define the functional F_1 by

$$F_1 = \int_V [\tfrac{1}{2}G_{ijkl} * d\epsilon_{ij} * d\epsilon_{kl} - \sigma_{ij} * d\epsilon_{ij} - (\sigma_{ij,j} + F_i) * du_i]\, dv$$

$$+ \int_{B_u} [\sigma_i * d\Delta_i]\, da + \int_{B_\sigma} [(\sigma_i - S_i) * du_i]\, da \tag{7.56}$$

where F_i, Δ_i, and S_i are given quantities, $\sigma_i = \sigma_{ij}n_j$ on the boundary, and the coordinate dependence of all variables is understood. We are now in a position to state the first variational theorem as:

The first variation δF_1 *of the functional* F_1, *defined by* (7.56), *vanishes if and only if the field equations and boundary conditions* (7.46)–(7.51) *are satisfied.*

To prove this theorem we begin by taking the variations in the histories $u_i(\tau)$, $\epsilon_{ij}(\tau)$, and $\sigma_{ij}(\tau)$ as being given by

$$u_i(\tau) + \delta u_i(\tau)\alpha,$$

$$\epsilon_{ij}(\tau) + \delta\epsilon_{ij}(\tau)\alpha, \tag{7.57}$$

and

$$\sigma_{ij}(\tau) + \delta\sigma_{ij}(\tau)\alpha$$

where α is a real number, and $\delta u_i(\delta)$, $\delta\epsilon_{ij}(\tau)$, and $\delta\sigma_{ij}(\tau)$ are the arbitrary, but sufficiently smooth, changes in the histories.

The resulting first variation of the functional F_1, from (7.56), is then given by

$$\delta F_1 = \int_V [G_{ijkl} * d\epsilon_{kl} * d\delta\epsilon_{ij} - \sigma_{ij} * d\delta\epsilon_{ij} - \delta\sigma_{ij} * d\epsilon_{ij}$$

$$- (\sigma_{ij,j} + F_i) * d\delta u_i - \delta\sigma_{ij,j} * du_i]\, dv$$

$$+ \int_{B_u} [\delta\sigma_i * d\Delta_i]\, da + \int_{B_\sigma} [(\sigma_i - S_i) * d\delta u_i + \delta\sigma_i * du_i]\, da \tag{7.58}$$

where, in the first term, the commutativity property $G_{ijkl} * d\delta\epsilon_{ij} * d\epsilon_{kl} = G_{ijkl} * d\epsilon_{kl} * d\delta\epsilon_{ij}$ and the symmetry relation $G_{ijkl} = G_{klij}$ have both been used.

The term $\int_{B_u} [\delta\sigma_i * du_i]\, da$ is subtracted from the integral over B_u and added to the last term in the B_σ integral. By using the divergence theorem, (7.58) can be rearranged into the form

$$\delta F_1 = \int_V [(G_{ijkl} * d\epsilon_{kl} - \sigma_{ij}) * d\delta\epsilon_{ij} - (\sigma_{ij,j} + F_i) * d\delta u_i$$
$$- (\epsilon_{ij} - \tfrac{1}{2}u_{i,j} - \tfrac{1}{2}u_{j,i}) * d\delta\sigma_{ij}]\, dv$$
$$+ \int_{B_\sigma} [(\sigma_i - S_i) * d\delta u_i]\, da + \int_{B_u} [(\Delta_i - u_i) * d\delta\sigma_i]\, da. \quad (7.59)$$

The satisfaction of the field equations and boundary conditions (7.46)–(7.51) then requires that the first variation of F_1 vanish, giving

$$\delta F_1 = 0. \quad (7.60)$$

It may also be argued that, for the variations δu_i, $\delta\epsilon_{ij}$, and $\delta\sigma_{ij}$ taken as arbitrarily specified, the variational equation $\delta F_1 = 0$ gives as Euler equations the field equations and boundary conditions (7.46)–(7.51). This completes the proof of the first variational theorem.

The variational theorem just given is a generalization to viscoelasticity of the Hu–Washizu variational theorem in elasticity. It is consequently expected that the Hellinger–Reissner elasticity variational theorem could be generalized to account for viscoelastic effects. This is now done. We taken the functional F_2 as defined by

$$F_2 = \int_V [\tfrac{1}{2}J_{ijkl} * d\sigma_{ij} * d\sigma_{kl}$$
$$- \tfrac{1}{2}\sigma_{ij} * d(u_{i,j} + u_{j,i}) + F_i * du_i]\, dv$$
$$+ \int_{B_u} [\sigma_i * d(u_i - \Delta_i)]\, da + \int_{B_\sigma} [S_i * du_i]\, da \quad (7.61)$$

where now the variations in u_i and σ_{ij} are taken to be arbitrary.

The second variational theorem is stated as:

The first variation δF_2 of the functional F_2, defined by (7.61), *vanishes if and only if the field equations and boundary conditions* (7.48)–(7.51) *are satisfied, with $\epsilon_{ij} = \frac{1}{2}(u_{i,j} + u_{j,i})$ given.*

To effect the proof we take variations in the histories $u_i(\tau)$ and $\sigma_{ij}(\tau)$ as being given by

$$u_i(\tau) + \delta u_i(\tau)\alpha$$

and

$$\sigma_{ij}(\tau) + \delta\sigma_{ij}(\tau)\alpha$$

where α is a real number and $\delta u_i(\tau)$ and $\delta\sigma_{ij}(\tau)$ are the arbitrary changes in the histories.

The functional F_2 has the resulting first variation given by

$$\delta F_2 = \int_V [J_{ijkl} * d\sigma_{kl} * d\delta\sigma_{ij} - \tfrac{1}{2}\sigma_{ij} * d(\delta u_{i,j} + \delta u_{j,i})$$
$$- \tfrac{1}{2}\delta\sigma_{ij} * d(u_{i,j} + u_{j,i}) + F_i * d\delta u_i]\, dv$$
$$+ \int_{B_u} [\delta\sigma_i * d(u_i - \Delta_i) + \sigma_i * d\delta u_i]\, da + \int_{B_\sigma} [S_i * d\delta u_i]\, da \qquad (7.62)$$

where, in the first term, the commutativity property of Stieltjes convolutions has been used, along with $J_{ijkl} = J_{klij}$.

The term $\int_{B_\sigma} \sigma_i * d\,\delta u_i$ is added and subtracted respectively from the last two terms in (7.62). By using the divergence theorem we can put (7.62) into the form

$$\delta F_2 = \int_V \{[-\tfrac{1}{2}(u_{i,j} + u_{j,i}) + J_{ijkl} * d\sigma_{kl}] * d\delta\sigma_{ij}$$
$$+ (\sigma_{ij,j} + F_i) * d\delta u_i\}\, dv$$
$$+ \int_{B_u} [(u_i - \Delta_i) * d\delta\sigma_i]\, da - \int_{B_\sigma} [(\sigma_i - S_i) * d\delta u_i]\, da. \qquad (7.63)$$

We see that the satisfaction of the field equations (7.48) and (7.49) with $\epsilon_{ij} = \tfrac{1}{2}(u_{i,j} + u_{j,i})$ along with the boundary conditions (7.50) and (7.51), gives the first variation of F_2 as vanishing; thus,

$$\delta F_2 = 0. \qquad (7.64)$$

It may also be shown that $\delta F_2 = 0$ gives (7.48)–(7.51) as Euler equations, presuming $\epsilon_{ij} = \tfrac{1}{2}(u_{i,j} + u_{j,i})$. This completes the proof of the second variational theorem.

A variational theorem can be stated which is related to the first theorem given here but which has the simplified functional, given by

$$\int_V [\tfrac{1}{2}G_{ijkl} * d\epsilon_{ij} * d\epsilon_{kl} - F_i * du_i]\, dv - \int_{B_\sigma} (S_i * du_i)\, da$$

where $\epsilon_{ij} = \tfrac{1}{2}(u_{i,j} + u_{j,i})$. In this case only the displacements are varied, subject to the requirement that the displacement boundary conditions (7.51) are satisfied. Such a variational theorem is an extension to viscoelasticity of the theorem of stationary potential energy in elasticity. Similarly, a variational theorem can be stated which is related to the second theorem given here but which has a simplified functional given by

$$+\tfrac{1}{2}\int_V (J_{ijkl} * d\sigma_{ij} * d\sigma_{kl})\, dv - \int_{B_u} (\sigma_i * d\Delta_i)\, da.$$

In this case, only the stresses are varied, subject only to the requirement that the equations of equilibrium (7.49) and the stress boundary conditions (7.50) are satisfied. This type of variational theorem is an extension to viscoelasticity, of the theorem of stationary complementary energy in elasticity. These restricted variational theorems are proved in Ref. [7.12].

The first and second variational theorems just proved provide the rigorous means of extending to viscoelasticity the approximate elasticity theories which are derived by elasticity variational theorems. For example, the Reissner elastic plate theory would have a viscoelastic counterpart which would be derived in the same manner as in the elastic case, but would use the second variational theorem given here. Actually, this type of formal use of these variational theorems is seldom made, since it is usually much easier to appeal to integral transform methods to directly obtain an approximate viscoelastic theory from the corresponding approximate elasticity theory. That is, elasticity variational theorems can be reinterpreted as viscoelastic variational theorems involving integral transformed field variables; and, thereby, the Euler equations obtained from a particular elasticity variational theorem can be reinterpreted as integral transformed viscoelastic Euler equations. It is important to note, however, that the direct use of the viscoelastic variational theorems is a rigorous and general means of establishing theories of mechanical behavior, while proceeding through the use of integral transform methods combined with the corresponding elasticity theories is necessarily a much less rigorous procedure.

The variational theorems just given are from Gurtin [7.12]. Now we derive two different types of variational theorems, not using Stieltjes convolutions.

These second two variational theorems are restricted to applications where separation of the time and spatial variables are allowed. Sufficient conditions to allow a separation of variables are outlined in Section 2.3. The corresponding restrictions on the mechanical properties are that the material be isotropic and that viscoelastic Poisson's ratio be a real constant. In the relations (7.46)–(7.51), governing the viscoelastic boundary value problems, the anisotropic constitutive relations (7.47) and (7.48) must be replaced by the isotropic forms as

$$\sigma_{ii}(t) = \frac{2(1+\nu)}{(1-2\nu)}\int_0^t \mu(t-\tau)\frac{\partial \epsilon_{ii}(\tau)}{\partial \tau}\, d\tau \tag{7.65}$$

and

$$s_{ij}(t) = 2\int_0^t \mu(t-\tau)\frac{\partial e_{ij}(\tau)}{d\tau}\, d\tau \tag{7.66}$$

or

$$\epsilon_{ii}(t) = \frac{(1 - 2\nu)}{2(1 + \nu)} \int_0^t J(t - \tau) \frac{\partial \sigma_{ii}(\tau)}{\partial \tau} d\tau \tag{7.67}$$

and

$$e_{ij}(t) = \tfrac{1}{2} \int_0^t J(t - \tau) \frac{\partial s_{ij}(\tau)}{\partial \tau} d\tau. \tag{7.68}$$

We define the functional F_3 by

$$\begin{aligned} F_3 = 2 \int_V \int_0^t \Big\{ &-\sigma_{ij}(\tau) \frac{\partial \epsilon_{ij}(\tau)}{\partial \tau} + \frac{\sigma_{ij}(\tau)}{2} \left[\frac{\partial u_{i,j}(\tau)}{\partial \tau} + \frac{\partial u_{j,i}(\tau)}{\partial \tau} \right] \\ &- F_i(\tau) \frac{\partial u_i(\tau)}{\partial \tau} \Big\} d\tau\, dv + \int_V \int_0^t \left[\rho \frac{\partial A(\tau)}{\partial \tau} + \Lambda(\tau) \right] d\tau\, dv \\ &- 2 \int_{B_\sigma} \int_0^t S_i(\tau) \frac{\partial u_i(\tau)}{\partial \tau} d\tau\, da \\ &- 2 \int_{B_u} \int_0^t \sigma_i(\tau) \left[\frac{\partial u_i(\tau)}{\partial \tau} - \frac{\partial \Delta_i(\tau)}{\partial \tau} \right] d\tau\, da \end{aligned} \tag{7.69}$$

where $\rho A(t)$ is the isothermal form of the stored energy, and $\Lambda(t)$ is the corresponding rate of dissipation of energy.

At this point, it is worth noting that if the medium under consideration were elastic, the time integrations in (7.69) could be performed, and the resulting functional would be that of the Hu–Washizu variational theorem in elasticity, as was the case with F_1.

The forms of $\rho A(t)$ and $\Lambda(t)$ are taken from the isothermal specialization of (3.22) and (3.25), from the thermodynamical derivation of Section 3.1. These are

$$\begin{aligned} \rho A(t) = &\frac{(1 + \nu)}{3(1 - 2\nu)} \int_0^t \int_0^t \mu(2t - \tau - \eta) \frac{\partial \epsilon_{ii}(\tau)}{\partial \tau} \frac{\partial \epsilon_{jj}(\eta)}{\partial \eta} d\tau\, d\eta \\ &+ \int_0^t \int_0^t \mu(2t - \tau - \eta) \frac{\partial e_{ij}(\tau)}{\partial \tau} \frac{\partial e_{ij}(\eta)}{\partial \eta} d\tau\, d\eta \end{aligned} \tag{7.70}$$

and

$$\begin{aligned} \Lambda(t) = &-\frac{(1 + \nu)}{3(1 - 2\nu)} \int_0^t \int_0^t \frac{\partial}{\partial t} \mu(2t - \tau - \eta) \frac{\partial \epsilon_{ii}(\tau)}{\partial \tau} \frac{\partial \epsilon_{jj}(\eta)}{\partial \eta} d\tau\, d\eta \\ &- \int_0^t \int_0^t \frac{\partial}{\partial t} \mu(2t - \tau - \eta) \frac{\partial e_{ij}(\tau)}{\partial \tau} \frac{\partial e_{ij}(\eta)}{\partial \eta} d\tau\, d\eta \end{aligned} \tag{7.71}$$

where the two possible independent arguments of the relaxation function are taken in the additive form, and the viscoelastic Poisson's ratio is taken as a real constant.

We substitute (7.70) and (7.71) into (7.69) to get the form of the functional F_3 as

$$
\begin{aligned}
F_3 = 2\int_V \int_0^t \Big\{ &-\sigma_{ij}(\tau)\frac{\partial \epsilon_{ij}(\tau)}{\partial \tau} + \frac{\sigma_{ij}(\tau)}{2}\left[\frac{\partial u_{i,j}(\tau)}{\partial \tau} + \frac{\partial u_{j,i}(\tau)}{\partial \tau}\right] \\
&- F_i(\tau)\frac{\partial u_i(\tau)}{\partial \tau} + \frac{(1+\nu)}{3(1-2\nu)}\frac{\partial \epsilon_{ii}(\tau)}{\partial \tau}\int_0^\tau \mu(\tau-\eta)\frac{\partial \epsilon_{jj}(\eta)}{\partial \eta}\,d\eta \\
&+ \frac{\partial e_{ij}(\tau)}{\partial \tau}\int_0^\tau \mu(\tau-\eta)\frac{\partial e_{ij}(\eta)}{\partial \eta}\,d\eta \Big\}\,d\tau\,dv \\
&- 2\int_{B_\sigma}\int_0^t S_i(\tau)\frac{\partial u_i(\tau)}{\partial \tau}\,d\tau\,da \\
&- 2\int_{B_u}\int_0^t \sigma_i(\tau)\left[\frac{\partial u_i(\tau)}{\partial \tau} - \frac{\partial \Delta_i(\tau)}{\partial \tau}\right]d\tau\,da. \qquad (7.72)
\end{aligned}
$$

We take the first variation of F_3 in (7.72) and use the divergence theorem to put it into the form

$$
\begin{aligned}
\delta F_3 = 2\int_V \int_0^t \{ &[-\dot{\epsilon}_{ij} + \tfrac{1}{2}(\dot{u}_{i,j} + \dot{u}_{j,i})]\,\delta\sigma_{ij} \\
&- \sigma_{ij}\,\delta\dot{\epsilon}_{ij} + \frac{(1+\nu)}{3(1-2\nu)}\,\delta\dot{\epsilon}_{ii}\int_0^\tau \mu(\tau-\eta)\,\dot{\epsilon}_{jj}(\eta)\,d\eta \\
&+ \frac{(1+\nu)}{3(1-2\nu)}\,\dot{\epsilon}_{ii}\int_0^\tau \mu(\tau-\eta)\,\delta\dot{\epsilon}_{jj}(\eta)\,d\eta \\
&+ \delta\dot{e}_{ij}\int_0^\tau \mu(\tau-\eta)\dot{e}_{ij}(\eta)\,d\eta + \dot{e}_{ij}\int_0^\tau \mu(\tau-\eta)\,\delta\dot{e}_{ij}(\eta)\,d\eta \\
&+ (-\sigma_{ij,j} - F_i)\,\delta\dot{u}_i\}\,d\tau\,dv \\
&- 2\int_{B_\sigma}\int_0^t (S_i - \sigma_i)\,\delta\dot{u}_i\,d\tau\,da - 2\int_{B_u}\int_0^t \delta\sigma_i(\dot{u}_i - \Delta_i)\,d\tau\,da \qquad (7.73)
\end{aligned}
$$

where the time arguments, in addition to the coordinate dependence, are now understood, and the superimposed dot is differentiation with respect to the time variable.

The conditions under which viscoelastic boundary value problems admit a separation of variables solution are discussed in Section 2.3. Such conditions will be assumed here, so that the field variables can be taken in the form

$$
\begin{aligned}
u_i(x_i, t) &= \tilde{u}_i(x_i)\,u(t), \\
\epsilon_{ij}(x_i, t) &= \tilde{\epsilon}_{ij}(x_i)\,u(t), \qquad (7.74)
\end{aligned}
$$

and

$$\sigma_{ij}(x_i, t) = \tilde{\sigma}_{ij}(x_i)\,\sigma(t).$$

Using (7.74) in the terms in (7.73) involving integrations with respect to τ, and taking variations only with respect to the spatial part of the field variables, we get

$$
\begin{aligned}
\delta F_3 = 2 \int_V \int_0^t \Big\{ & [-\dot{\epsilon}_{ij} + \tfrac{1}{2}(\dot{u}_{i,j} + \dot{u}_{j,i})]\, \delta\sigma_{ij} \\
& + \delta\tilde{e}_{ij}(x_i)\, \dot{u}(\tau) \left[-s_{ij}(x_i, \tau) + 2 \int_0^\tau \mu(\tau - \eta)\, \dot{e}_{ij}(x_i, \eta)\, d\eta \right] \\
& + \delta\tilde{\epsilon}_{kk}(x_i)\, \dot{u}(\tau) \left[-\frac{\sigma_{kk}(x_i, \tau)}{3} + \frac{2(1+\nu)}{3(1-2\nu)} \int_0^\tau \mu(\tau - \eta)\, \dot{\epsilon}_{kk}(x_i, \eta)\, d\eta \right] \\
& - (\sigma_{ij,j} + F_i)\, \delta\dot{u}_i \Big\}\, d\tau\, dv \\
& - 2 \int_{B_\sigma} \int_0^t (S_i - \sigma_i)\, \delta\dot{u}_i\, d\tau\, da - 2 \int_{B_u} \int_0^t \delta\sigma_i(\dot{u}_i - \dot{\Delta}_i)\, d\tau\, da. \qquad (7.75)
\end{aligned}
$$

We see that the separation of variables condition provides the key which allows the integrals in (7.73) to be combined into the simplified form in (7.75).

Equations (7.46), (7.49)–(7.51), (7.65), and (7.66) show that the first variation of F_3 vanishes. That is, when these relations are satisfied,

$$\delta F_3 = 0. \qquad (7.76)$$

It follows from (7.69) that the variational requirement, $\delta F_3 = 0$, gives (7.46), (7.49)–(7.51), (7.65), and (7.66) as Euler equations.

These results are now stated as the third variational theorem:

For viscoelastic boundary value problems which admit a separation of variables form of solution, the first variation δF_3 of the functional F_3, defined by (7.69), (7.70), *and* (7.71), *vanishes if and only if the field equations and boundary conditions* (7.46), (7.49)–(7.51), (7.65), *and* (7.66) *are satisfied. Variations are allowed only in the spatial parts of the field variables.*

A fourth type of variational theorem is derived, which, as with the second theorem, corresponds to the Hellinger–Reissner variational theorem of elasticity. In this case, we define the functional F_4 by

$$
\begin{aligned}
F_4 = \int_V \int_0^t \Big\{ & -\frac{\partial \sigma_{ij}(\tau)}{\partial \tau} [u_{i,j}(\tau) + u_{j,i}(\tau)] \\
& + 2 \frac{\partial F_i(\tau)}{\partial \tau} u_i(\tau) + \rho \frac{\partial W(\tau)}{\partial \tau} - \Gamma(\tau) \Big\}\, d\tau\, dv \\
& + 2 \int_{B_\sigma} \int_0^t \frac{\partial S_i(\tau)}{\partial \tau} u_i(\tau)\, d\tau\, da \\
& + 2 \int_{B_u} \int_0^t \frac{\partial \sigma_i(\tau)}{\partial \tau} [u_i(\tau) - \Delta_i(\tau)]\, d\tau\, da \qquad (7.77)
\end{aligned}
$$

where

$$\rho W(t) = \frac{(1-2\nu)}{12(1+\nu)} \int_0^t \int_0^t J(2t-\tau-\eta) \frac{\partial\sigma_{ii}(\tau)}{\partial\tau} \frac{\partial\sigma_{jj}(\eta)}{\partial\eta} \, d\tau \, d\eta + \tfrac{1}{4} \int_0^t \int_0^t J(2t-\tau-\eta) \frac{\partial s_{ij}(\tau)}{\partial\tau} \frac{\partial s_{ij}(\eta)}{\partial\eta} \, d\tau \, d\eta \tag{7.78}$$

and

$$\Gamma(\tau) = \frac{(1-2\nu)}{12(1+\nu)} \int_0^t \int_0^t \frac{\partial}{\partial t} J(2t-\tau-\eta) \frac{\partial\sigma_{ii}(\tau)}{\partial\tau} \frac{\partial\sigma_{jj}(\eta)}{\partial\eta} \, d\tau \, d\eta + \tfrac{1}{4} \int_0^t \int_0^t \frac{\partial}{\partial t} J(2t-\tau-\eta) \frac{\partial s_{ij}(\tau)}{\partial\tau} \frac{\partial s_{ij}(\eta)}{\partial\eta} \, d\tau \, d\eta. \tag{7.79}$$

The thermodynamical significance of $\rho W(t)$ and $\Gamma(t)$ has been established in Section 3.7. We substitute (7.78) and (7.79) into (7.77) to get

$$\begin{aligned} F_4 = 2 \int_V \int_0^t \Big\{ & -\tfrac{1}{2} \frac{\partial\sigma_{ij}(\tau)}{\partial\tau} [u_{i,j}(\tau) + u_{j,i}(\tau)] + \frac{\partial F_i(\tau)}{\partial\tau} u_i(\tau) \\ & + \frac{1}{4} \frac{\partial s_{ij}(\tau)}{\partial\tau} \int_0^\tau J(\tau-\eta) \frac{\partial s_{ij}(\eta)}{\partial\eta} \, d\eta \\ & + \frac{(1-2\nu)}{12(1+\nu)} \frac{\partial\sigma_{ii}(\tau)}{\partial\tau} \int_0^\tau J(\tau-\eta) \frac{\partial\sigma_{jj}(\eta)}{\partial\eta} \, d\eta \Big\} \, d\tau \, dv \\ & + 2 \int_{B_\sigma} \int_0^t \frac{\partial S_i(\tau)}{\partial\tau} u_i(\tau) \, d\tau \, da \\ & + 2 \int_{B_u} \int_0^t \frac{\partial\sigma_i(\tau)}{\partial\tau} [u_i(\tau) - \Delta_i(\tau)] \, d\tau \, da. \end{aligned} \tag{7.80}$$

By taking the first variation of F_4 from (7.80), and using the divergence theorem, we get

$$\begin{aligned} \delta F_4 = 2 \int_V \int_0^t \Big\{ & -\tfrac{1}{2} \delta\dot{\sigma}_{ij}(u_{i,j} + u_{j,i}) + \dot{F}_i \, \delta u_i \\ & + \tfrac{1}{4} \delta\dot{s}_{ij} \int_0^\tau J(\tau-\eta) \, \dot{s}_{ij} \, d\eta + \tfrac{1}{4} \dot{s}_{ij} \int_0^\tau J(\tau-\eta) \, \delta\dot{s}_{ij} \, d\eta \\ & + \frac{(1-2\nu)}{12(1+\nu)} \delta\dot{\sigma}_{ii} \int_0^\tau J(\tau-\eta) \, \dot{\sigma}_{jj} \, d\eta \\ & + \frac{(1-2\nu)}{12(1+\nu)} \dot{\sigma}_{ii} \int_0^\tau J(\tau-\eta) \, \delta\dot{\sigma}_{jj} \, d\eta + \dot{\sigma}_{ij,j} \, \delta u_i \Big\} \, d\tau \, dv \\ & + 2 \int_{B_\sigma} \int_0^t (\dot{S}_i - \dot{\sigma}_i) \, \delta u_i \, d\tau \, da + 2 \int_{B_u} \int_0^t \delta\dot{\sigma}_i (u_i - \Delta_i) \, d\tau \, da \end{aligned} \tag{7.81}$$

where, as before, the time arguments are understood and the superimposed dot is differentiation with respect to the time variable.

As with the variational theorem for F_3, conditions are assumed such that a separation of variables solution is admitted in the form of (7.74). By using this in the terms in (7.81), which involve integrations with respect to τ, and admitting variations only in the spatial part of the field variables, we obtain

$$\delta F_4 = 2\int_V \int_0^t \Big\{(\dot{\sigma}_{ij,j} + \dot{F}_i)\,\delta u_i$$
$$+ \delta s_{ij}(x_i)\,\dot{\sigma}(\tau)\Big[-\frac{(u_{i,j}+u_{j,i})}{2} + \frac{\delta_{ij}}{3}u_{k,k} + \tfrac{1}{2}\int_0^\tau J(\tau-\eta)\,\dot{s}_{ij}(x_i,\eta)\,d\eta\Big]$$
$$+ \delta\tilde{\sigma}_{ii}(x_i)\,\dot{\sigma}(\tau)\Big[-\frac{u_{k,k}}{3} + \frac{(1-2\nu)}{6(1+\nu)}\int_0^\tau J(\tau-\eta)\,\dot{\sigma}_{jj}(x_i,\eta)\,d\eta\Big]\Big\}\,d\tau\,dv$$
$$+ 2\int_{B_\sigma}\int_0^t (\dot{S}_i - \dot{\sigma}_i)\,\delta u_i\,d\tau\,da + 2\int_{B_u}\int_0^t \delta\dot{\sigma}_i(u_i - \Delta_i)\,d\tau\,da. \tag{7.82}$$

Equations (7.46), (7.49)–(7.51), (7.67), and (7.68) show that the first variation of F_4 must vanish. Thus,

$$\delta F_4 = 0. \tag{7.83}$$

It follows from (7.77) that the variational equation $\delta F_4 = 0$ gives as Euler equations (7.49)–(7.51) and

$$\tfrac{1}{2}(u_{i,j} + u_{j,i}) - \frac{\delta_{ij}}{3}u_{k,k} = \tfrac{1}{2}\int_0^t J(t-\tau)\,\frac{\partial s_{ij}(\tau)}{\partial\tau}\,d\tau \tag{7.84}$$

and

$$u_{k,k} = \frac{(1-2\nu)}{2(1+\nu)}\int_0^t J(t-\tau)\,\frac{\partial\sigma_{kk}(\tau)}{\partial\tau}\,d\tau. \tag{7.85}$$

These results are now combined in the statement of the fourth variational theorem:

For viscoelastic boundary value problems which admit a separation of variables form of solution, the first variation δF_4 of the functional F_4, defined by (7.77), (7.78), *and* (7.79), *vanishes if and only if the field equations and boundary conditions* (7.49)–(7.51), (7.84), *and* (7.85) *are satisfied. Variations are allowed only in the spatial parts of the field variables.*

Although these last two variational theorems are more restrictive than the first two, the conditions for their applicability are revealing. That is, through the employment of the separation of variables technique these variational theorems are found as straightforward extensions of the corresponding elasticity theorems, but the revealing aspect of this is that it can only occur under the conditions which permit a separation of variables solution. Furthermore, this close relationship between elastic and viscoelastic variational theorems under separation of variables conditions suggests that under the same conditions elastic and viscoelastic minimum theorems might be similarly related. This possibility is next investigated, and it is indeed found that viscoelastic minimum theorems can be established. However, as will be seen, these viscoelastic minimum theorems are not the simple generalizations of the corresponding elastic theorems as occurred in the case of variational theorems.

7.5. MINIMUM THEOREMS

The variational theorems just considered establish the stationary value of certain functionals. Under certain restrictive conditions, the much stronger result will be obtained whereby particular functionals can be shown to not only have a stationary character, but also a minimum character. The minimum theorems to be proved here correspond to the last two variational theorems of the preceding section. The viscoelastic boundary value problem is again posed by the relations of the preceding section, namely (7.46), (7.49)–(7.51), and (7.65)–(7.68). As in the previous variational theorems, it is assumed that the stresses, strains, and displacements are continuous and continuously differentiable.

The first minimum theorem is associated with the theorem of minimum potential energy in elasticity and in fact reduces to it in the case of elasticity. We define the functional M_1 as

$$M_1 = \int_V \int_0^t \left[\rho \frac{\partial A(\tau)}{\partial \tau} + \Lambda(\tau) - 2F_i(\tau) \frac{\partial u_i(\tau)}{\partial \tau}\right] d\tau\, dv - 2 \int_{B_\sigma} \int_0^t S_i(\tau) \frac{\partial u_i(\tau)}{\partial \tau}\, d\tau\, da \tag{7.86}$$

where $\rho A(t)$ and $\Lambda(t)$ are defined by (7.70) and (7.71), respectively. Using these, as in the third variational theorem derivation, M_1 is written as

$$M_1 = \int_V \int_0^t \left[\dot{\epsilon}_{ii}(\tau) \frac{2(1+\nu)}{3(1-2\nu)} \int_0^\tau \mu(\tau - \eta)\, \dot{\epsilon}_{jj}(\eta)\, d\eta + 2\dot{e}_{ij}(\tau) \int_0^\tau \mu(\tau-\eta)\, \dot{e}_{ij}(\eta)\, d\eta - 2F_i(\tau)\, \dot{u}_i(\tau)\right] d\tau\, dv - 2 \int_{B_\sigma} \int_0^t S_i(\tau)\, \dot{u}_i(\tau)\, d\tau\, da. \tag{7.87}$$

Conditions are assumed, such that a separation of variables solution applies, in the form of (7.74). See Section 2.3 for the statement of such conditions.

Furthermore, we take variations from the solution of the boundary value problem by varying only the spatial parts of the field variables in (7.74). Under these conditions the first minimum theorem can be stated as:

For a viscoelastic boundary value problem, which admits a separation of variables form of solution, define as admissible states only those states for which (a) *the displacements satisfy the displacement boundary conditions*, (7.51), *and* (b) *the displacements differ from the state of the solution only in the spatial rather than in the time part of the solution. Of these admissible states, the one which is the solution of the boundary value problem posed by* (7.46), (7.49) (7.51), (7.65), *and* (7.66) *renders the functional M_1, defined by* (7.86), (7.70), *and* (7.71), *a minimum, compared with M_1 evaluated for any other admissible state, if*

$$\int_0^t \sigma_{ij}(\tau) \frac{\partial \epsilon_{ij}(\tau)}{\partial \tau} \, d\tau \geqslant 0$$

for the process under consideration.

The proof begins by letting the admissible displacement field $\mathring{u}_i(x_i, t)$, in accordance with (7.74), have the form

$$\mathring{u}_i(x_i, t) = \tilde{u}_i(x_i) \, u(t) + \delta\tilde{u}_i(x_i) \, u(t) \tag{7.88}$$

where $\tilde{u}_i(x_i) \, u(t)$ is the exact solution displacement field, and the varied displacement is seen to involve variations only in the spatial portions of the displacement field.

Under these conditions, the minimum theorem to be proved is the statement that

$$\mathring{M}_1 - M_1 \geqslant 0 \tag{7.89}$$

where M_1 is defined by (7.87) for the field variables of the solution of the viscoelastic boundary value problem, while $\mathring{M}_1$ represents the value of (7.87) for any admissible displacement field.

We expand (7.89), using (7.87) and (7.88). Then,

$$\begin{aligned} \mathring{M}_1 - M_1 = \int_V \int_0^t \Big\{ & \frac{2(1+\nu)}{3(1-2\nu)} [2\tilde{\epsilon}_{ii}\delta\tilde{\epsilon}_{jj} + \delta\tilde{\epsilon}_{ii}\delta\tilde{\epsilon}_{jj}] \, \dot{u}(\tau) \int_0^\tau \mu(\tau - \eta) \, \dot{u}(\eta) \, d\eta \\ & + 2[2\tilde{e}_{ij}\delta\tilde{e}_{ij} + \delta\tilde{e}_{ij}\delta\tilde{e}_{ij}] \, \dot{u}(\tau) \int_0^\tau \mu(\tau - \eta) \, \dot{u}(\eta) \, d\eta \\ & - 2F_i(\tau) \, \delta\tilde{u}_i \dot{u}(\tau) \Big\} \, d\tau \, dv - 2 \int_{B_\sigma} \int_0^t S_i(\tau) \, \delta\tilde{u}_i \dot{u}(\tau) \, d\tau \, da. \end{aligned} \tag{7.90}$$

The last integral is written as an integral over the entire boundary, since $\delta\tilde{u}_i = 0$

over B_u, and then it is converted to a volume integral through the divergence theorem. This results in

$$\begin{aligned}\mathring{M}_1 - M_1 = \int_V \int_0^t \Big\{ & \Big[\frac{4(1+\nu)}{3(1-2\nu)} \int_0^\tau \mu(\tau-\eta)\, \dot{\epsilon}_{jj}(\eta)\, d\eta - \tfrac{2}{3}\sigma_{jj}(\tau)\Big] \delta\tilde{\epsilon}_{ii}\dot{u}(\tau) \\ & + \Big[4 \int_0^\tau \mu(\tau-\eta)\, \dot{e}_{ij}(\eta)\, d\eta - 2s_{ij}(\tau)\Big] \delta\tilde{e}_{ij}\dot{u}(\tau) \\ & - 2[\sigma_{ij,j} + F_i]\, \delta\tilde{u}_i\dot{u}(\tau) \\ & + \frac{2(1+\nu)}{3(1-2\nu)} \delta\tilde{\epsilon}_{ij}\dot{u}(\tau) \int_0^\tau \mu(\tau-\eta)\, \delta\tilde{\epsilon}_{jj}\dot{u}(\eta)\, d\eta \\ & + 2\delta\tilde{e}_{ij}\dot{u}(\tau) \int_0^\tau \mu(\tau-\eta)\, \delta\tilde{e}_{ij}\dot{u}(\eta)\, d\eta \Big\}\, d\tau\, dv. \end{aligned} \tag{7.91}$$

The first three bracketed terms vanish since the solution must satisfy the stress strain relations (7.65) and (7.66), and the equations of equilibrium, (7.49). Equation (7.91) is left as

$$\mathring{M}_1 - M_1 = \int_V \int_0^t [\delta\dot{\epsilon}_{ij}(\tau)\, \delta\sigma_{ij}(\tau)]\, d\tau\, dv \tag{7.92}$$

where

$$\delta\sigma_{ij}(t) = \delta_{ij} \frac{2(1+\nu)}{3(1-2\nu)} \int_0^t \mu(t-\tau) \frac{\partial \delta\epsilon_{kk}(\tau)}{\partial\tau}\, d\tau + 2 \int_0^t \mu(t-\tau) \frac{\partial \delta e_{ij}(\tau)}{\partial\tau}\, d\tau. \tag{7.93}$$

Replacing the symbol $\delta\epsilon_{ij}$ by ϵ_{ij} and $\delta\sigma_{ij}$ by σ_{ij} in (7.93) it has the form of the viscoelastic stress strain relation. It then follows from (7.92) and (7.93) that (7.89) is satisfied if

$$\int_0^t \sigma_{ij}(\tau) \frac{\partial \epsilon_{ij}(\tau)}{\partial\tau}\, d\tau \geqslant 0 \tag{7.94}$$

for the process or processes examined. Thus, the minimum theorem is true for all materials for which the work done is nonnegative. The implications of a nonnegative work requirement have been discussed in Section 3.3. It was found there that if the stored energy and the rate of dissipation of energy are taken as nonnegative then the nonnegative work requirement (7.94) is always satisfied. Furthermore, for relaxation functions represented by a positive constant plus a series of positive decaying exponentials, as in (2.68), these conditions are always satisfied and this minimum theorem is applicable to all processes.

It is reasonable to expect the existence of a second minimum theorem which corresponds to the theorem of minimum complementary energy in elasticity and reduces to it under the proper conditions. To establish such a theorem we define the functional M_2 as

$$M_2 = \int_V \int_0^t \left[\rho \frac{\partial W(\tau)}{\partial \tau} - \Gamma(\tau)\right] d\tau\, dv - 2 \int_{B_u} \int_0^t \frac{\partial \sigma_i(\tau)}{\partial \tau} \Delta_i(\tau)\, d\tau\, da \quad (7.95)$$

where $\rho W(\tau)$ and $\Gamma(\tau)$ are defined by (7.78) and (7.79) respectively.

As with the first minimum theorem, conditions are assumed such that a separation of variables solution applies in the form of (7.74). The variations are permitted only in the spatial parts of the field variables. With these agreements the second minimum theorem can be stated as:

For a viscoelastic boundary value problem, which admits a separation of variables form of solution, define as admissible states only those states for which (a) *the stresses satisfy the equilibrium equations* (7.49), (b) *the stresses satisfy the stress boundary conditions* (7.50), *and* (c) *the stresses differ from the state of the solution only in the spatial rather than in the time part of the solution. Of these admissible states, the one which is the solution of the boundary value problem posed by* (7.46), (7.49)–(7.51), (7.67), *and* (7.68), *renders the functional* M_2, *defined by* (7.95), (7.78), *and*(7.79), *a minimum, compared with* M_2 *evaluated for any admissible state, if*

$$\int_0^t \frac{\partial \sigma_{ij}(\tau)}{\partial \tau} \epsilon_{ij}(\tau)\, d\tau \geqslant 0$$

for the process under consideration.

Under these conditions the minimum theorem to be proved is the statement that

$$\mathring{M}_2 - M_2 \geqslant 0 \quad (7.96)$$

where M_2 is defined by (7.95) for the field variables of the solution of the viscoelastic boundary value problem, while $\mathring{M}_2$ represents the value of (7.95) for any admissible stress field.

We let the admissible stress field $\mathring{\sigma}_{ij}(x_i, t)$ have the form

$$\mathring{\sigma}_{ij}(x_i, t) = \tilde{\sigma}_{ij}(x_i)\, \sigma(t) + \delta\tilde{\sigma}_{ij}(x_i)\, \sigma(t) \quad (7.97)$$

where $\tilde{\sigma}_{ij}(x_i)\, \sigma(t)$ is the exact solution result and the varied stress field is seen to involve only a spatial variation from that of the solution.

To obtain the conditions under which this minimum theorem is true, we start by expanding (7.96), using (7.95). This gives

$$\begin{aligned} \mathring{M}_2 - M_2 = \int_V \int_0^t \Big\{ &\frac{(1-2\nu)}{6(1+\nu)} [2\tilde{\sigma}_{ii}\delta\tilde{\sigma}_{jj} + \delta\tilde{\sigma}_{ii}\delta\tilde{\sigma}_{jj}]\, \dot{\sigma}(\tau) \int_0^\tau J(\tau-\eta)\, \dot{\sigma}(\eta)\, d\eta \\ &+ \tfrac{1}{2}[2\tilde{s}_{ij}\delta\tilde{s}_{ij} + \delta\tilde{s}_{ij}\delta\tilde{s}_{ij}]\, \dot{\sigma}(\tau) \int_0^\tau J(\tau-\eta)\, \dot{\sigma}(\eta)\, d\eta \Big\}\, d\tau\, dv \\ &- 2\int_{B_u} \int_0^t \delta\tilde{\sigma}_{ij} n_j \dot{\sigma}(\tau)\, \Delta_i(\tau)\, d\tau\, da. \end{aligned} \quad (7.98)$$

The last integral is written as an integral over the entire boundary, since $\delta\tilde{\sigma}_{ij} = 0$ over B_σ. Using the divergence theorem and $\delta\tilde{\sigma}_{ij,j} = 0$, (7.98) is written as

$$\begin{aligned} \mathring{M}_2 - M_2 = \int_V \int_0^t \Big\{ \Big[\frac{(1-2\nu)}{3(1+\nu)} \int_0^\tau J(\tau-\eta)\, \dot{\sigma}_{jj}(\eta)\, d\eta - \tfrac{2}{3}\epsilon_{jj}(\tau) \Big] \delta\tilde{\sigma}_{ii}\dot{\sigma}(\tau) \\ + \Big[\int_0^\tau J(\tau-\eta)\, \dot{s}_{ij}(\eta)\, d\eta - 2e_{ij}(\tau) \Big] \delta\tilde{s}_{ij}\dot{\sigma}(\tau) \\ + \frac{(1-2\nu)}{6(1+\nu)} \delta\tilde{\sigma}_{ij}\dot{\sigma}(\tau) \int_0^\tau J(\tau-\eta)\, \delta\tilde{\sigma}_{ij}\dot{\sigma}(\eta)\, d\eta \\ + \tfrac{1}{2}\delta\tilde{s}_{ij}\dot{\sigma}(\tau) \int_0^\tau J(\tau-\eta)\, \delta\tilde{s}_{ij}\dot{\sigma}(\eta)\, d\eta \Big\}\, d\tau\, dv. \end{aligned} \tag{7.99}$$

The first two bracketed terms in (7.99) vanish since the solution satisfies the stress strain relations (7.67) and (7.68). Relation (7.99) becomes

$$\mathring{M}_2 - M_2 = \int_V \int_0^t [\delta\dot{\sigma}_{ij}(\tau)\, \delta\epsilon_{ij}(\tau)]\, d\tau\, dv \tag{7.100}$$

where

$$\delta\epsilon_{ij}(t) = \frac{1}{2} \int_0^t J(t-\tau) \frac{\partial \delta s_{ij}(\tau)}{\partial \tau}\, d\tau + \delta_{ij} \frac{(1-2\nu)}{6(1+\nu)} \int_0^t J(t-\tau) \frac{\partial \delta\sigma_{kk}(\tau)}{\partial \tau}\, d\tau. \tag{7.101}$$

We replace the symbols $\delta\epsilon_{ij}$ by ϵ_{ij} and $\delta\sigma_{ij}$ by σ_{ij} in (7.101) to give (7.101) the form of the viscoelastic stress strain relation. It follows from (7.100) and (7.101) that (7.96) is satisfied if

$$\int_0^t \frac{\partial \sigma_{ij}(\tau)}{\partial \tau} \epsilon_{ij}(\tau)\, d\tau \geqslant 0 \tag{7.102}$$

for the process being studied.

The requirement (7.102) is not as easily interpreted as the nonnegative work requirement (7.94) of the first minimum theorem. Certainly, in the limiting case of an elastic material, (7.102) implies $J \geqslant 0$ where J is the elastic shear compliance. But, in the case of viscoelastic materials, it is easy to show that (7.102) can be violated. For example, using nothing more complicated than a one-dimensional Maxwell-model stress strain relation, for two consecutive step function applications of stress, it can be shown that (7.102) is violated. On the other hand, (7.102) is readily seen to be satisfied under the conditions of a creep test, if $J(0) \geqslant 0$, and Poisson's ratio is suitably restricted.

Although the first minimum theorem is applicable to typical viscoelastic materials, the conditions under which the second minimum theorem is applicable for realistic mechanical properties characterization imposes restrictions upon the processes for which (7.102) can be satisfied. Thus, the usefulness of this second minimum theorem is somewhat limited. Nevertheless, there are interesting special conditions under which it is applicable. The conditions of the creep test have been mentioned as one such case. Another is that of the condition of steady state harmonic oscillation. Although it is, in principle, possible to show this latter application by directly restricting the processes in the two theorems just given, in practice it is easier to derive two new minimum theorems, directly appropriate to steady state harmonic oscillation conditions. We now do this to arrive at the third and fourth minimum theorems.

The proof of the next theorem is only briefly outlined, and that of the last theorem is left as an exercise, since they have many elements in common with the two previous theorems. The viscoelastic boundary value problem now posed is given by the adaptation to steady state harmonic conditions of the relations (7.46), (7.49)–(7.51), and (7.65)–(7.68).

As with the first minimum theorem, the one to be given now is associated with the theorem of minimum potential energy in elasticity. We define the functional M_3 as

$$M_3 = \frac{1}{e^{2i\omega t}} \left\{ \mu^*(i\omega) \int_V \left[\frac{(1+\nu)}{3(1-2\nu)} \epsilon_{ii}\epsilon_{jj} + e_{ij}e_{ij} - F_i u_i \right] dv - \int_{B_\sigma} S_i u_i \, da \right\} \tag{7.103}$$

where ω is the frequency of oscillation, $\mu^*(i\omega)$ is the complex modulus in shear, and viscoelastic Poisson's ratio is again a real constant. The field variables are all harmonic functions of time, as must be the specified boundary displacements and tractions and the body forces. Variations from the solution of the boundary value problem are taken only in the spatial parts of the field variables and not in the harmonic function of time part.

The third minimum theorem is now stated as:

For a harmonic oscillation viscoelastic boundary value problem, define as admissible states only those states for which (a) *the displacements satisfy the displacement boundary conditions and* (b) *the displacements differ from the state of the solution only in the spatial rather than the time part of the solution. With*

$$\operatorname{Re} \mu^*(i\omega) \geqslant 0, \qquad \operatorname{Im} \mu^*(i\omega) \geqslant 0, \qquad \text{and} \qquad -1 \leqslant \nu \leqslant 1/2$$

then

$$\operatorname{Re}(\mathring{M}_3 - M_3) \geqslant 0 \qquad \textit{and} \qquad \operatorname{Im}(\mathring{M}_3 - M_3) \geqslant 0$$

where M_3 is the functional (7.103) *evaluated for the solution of the boundary value problem and $\mathring{M}_3$ is* (7.103) *evaluated for any admissible displacement state.*

We begin the proof of this theorem by defining an admissible displacement field $\mathring{u}_i(x_i, t)$ in the form

$$\mathring{u}_i(x_i, t) = [\tilde{u}_i(x_i) + \delta\tilde{u}_i(x_i)]\, e^{i\omega t} \tag{7.104}$$

where $\tilde{u}_i(x_i)\, e^{i\omega t}$ designates the solution of the boundary value problem and necessarily $\delta\tilde{u}_i(x_i) = 0$ on B_u. By following a lengthy procedure similar to that of the first minimum theorem, (7.103) and (7.104) are combined to give

$$\mathring{M}_3 - M_3 = \mu^*(i\omega) \int_V \left[\frac{(1+\nu)}{3(1-2\nu)}\, \delta\tilde{\epsilon}_{ii}\delta\tilde{\epsilon}_{jj} + \delta\tilde{e}_{ij}\delta\tilde{e}_{ij}\right] dv \qquad (7.105)$$

where $\delta\tilde{\epsilon}_{ij}(x_i)$ has the obvious identification $\delta\tilde{\epsilon}_{ij} = \frac{1}{2}(\delta\tilde{u}_{i,j} + \delta\tilde{u}_{j,i})$ with the spatial parts of the displacement expression in (7.104), and $\delta\tilde{\epsilon}_{ij}$ has been decomposed into dilatational and deviatoric parts.

The proof of the theorem follows directly from (7.105) for $\delta\tilde{\epsilon}_{ij}$ being real. Since $\delta\tilde{u}_i(x_i)$ represents deviations from the spatial part of the solution $\tilde{u}_i(x_i)$, for $\delta\tilde{u}_i(x_i)$ and the spatial part of the admissable displacement field to be real $\tilde{u}_i(x_i)$ must be real. This is true under the conditions of the theorem. It can be reasoned that, if viscoelastic Poisson's ratio were not a real constant or if the medium were not homogeneous, $\tilde{u}_i(x_i)$ would be complex, and the minimum theorem would be invalid under these conditions. It has been implicitly assumed that in boundary value problems having both prescribed stresses and displacements on the surface and having prescribed body forces, the phase angle between the prescribed surface tractions and body forces on the one hand, and the prescribed surface displacements on the other hand, must have the particular value which allows the spatial part of the displacement field to be real. This phase angle is given by $\tan^{-1}(\text{Im}\,\mu^*/\text{Re}\,\mu^*)$. The restriction here to homogeneous materials is in sharp contrast to the corresponding minimum theorem in elasticity, where the application to nonhomogeneous conditions is allowed.

The last minimum theorem is the steady state harmonic counterpart of the second minimum theorem and the theorem of minimum complementary energy in elasticity. We define the functional M_4 through

$$M_4 = \frac{1}{e^{2i\omega t}} \left\{ \frac{J^*(i\omega)}{4} \int_V \left[\frac{(1-2\nu)}{3(1+\nu)}\, \sigma_{ii}\sigma_{jj} + s_{ij}s_{ij}\right] dv - \int_{B_u} \sigma_i \Delta_i \, da \right\} \qquad (7.106)$$

where ω is the frequency of oscillation, $J^*(i\omega)$ is the complex compliance in shear, and again viscoelastic Poisson's ratio is taken to be a real constant. The field variables are all harmonic functions of time, and variations from the solution of the boundary value problem are taken only in the spatial parts of the field variables and not in the harmonic function of time part.

The fourth minimum theorem is now stated as:

For a harmonic oscillation viscoelastic boundary value problem, define as admissible states only those states for which (a) *the stresses satisfy the stress boundary conditions,* (b) *the stresses satisfy the equations of equilibrium* $\sigma_{ij,j} + F_i = 0$, *and* (c) *the stresses differ from the state of the solution only in the spatial rather than the time part of the solution. With*

$$\mathrm{Re}\, J^*(i\omega) \geqslant 0, \qquad \mathrm{Im}\, J^*(i\omega) \leqslant 0, \qquad \textit{and} \qquad -1 \leqslant \nu \leqslant \tfrac{1}{2}$$

then

$$\mathrm{Re}(\mathring{M}_4 - M_4) \geqslant 0 \qquad \textit{and} \qquad \mathrm{Im}(\mathring{M}_4 - M_4) \leqslant 0$$

where M_4 *is the functional* (7.106) *evaluated for the solution of the boundary value problem and* $\mathring{M}_4$ *is* (7.106) *evaluated for any admissible stress state.*

The proof of this theorem follows the lines of reasoning given in proving the previous minimum theorems.

As an application of these minimum theorems, Christensen [7.13] has used them to derive bounds upon the effective complex moduli for two special types of composite viscoelastic materials. That is, in some cases where it is not possible to obtain exact solutions for the effective complex moduli of composite viscoelastic materials, these minimum theorems may be used to obtain bounds upon the real and imaginary parts of these moduli through the use of admissible displacement and stress fields. The two special types of composite materials in this application are those of a homogeneous material containing voids or perfectly rigid inclusions, both cases of which qualify under the dual categories of homogeneous materials and composite materials. This resolves the seeming inconsistency of the application of homogeneous material minimum theorems to composite media.

There have been attempts to formulate minimum theorems using integral transformed field variables. Typically, these procedures err in failing to take into account the complex variable nature of the formulation.

7.6. OPTIMAL STRAIN HISTORY

There are many types of problems in viscoelasticity that can be formulated as an optimal path problem. Such optimality problems have a close resemblance to the standard formulations of the calculus of variations. The minimum theorems of Section 7.5 were of such a type; another type will be given here. The complication over and above that of the calculus of variations is in accounting for the effect of the strain history of viscoelastic behavior.

The specific problem to be studied here is that of finding the optimal strain history path in going from one strain state to another while minimizing the work done during a given time interval. This problem was solved by Breuer [7.14]. A different approach was given later to the same problem by Gurtin *et al.* [7.15]. We follow the latter procedure. We pose the problem as follows. In a particular viscoelastic material, the strain state is taken to be zero up until time $t = 0$.

At time T the strain state has a value ϵ_0 that is constant thereafter. Find the strain history $\epsilon(t)$ with

$$\epsilon(0) = 0, \qquad \epsilon(T) = \epsilon_0$$

in order to minimize the work done in deforming the material. By definition the work is

$$W = \int_0^T \sigma(t) \frac{d\epsilon(t)}{dt}\, dt. \tag{7.107}$$

Write the one dimensional stress strain relation as

$$\sigma(t) = \int_{-\infty}^t G(t-\tau) \frac{d\epsilon(\tau)}{d\tau}\, d\tau. \tag{7.108}$$

Combining (7.107) and (7.108) gives

$$W(\epsilon) = \int_0^T \int_0^t G(t-\tau)\, \dot{\epsilon}(\tau)\, \dot{\epsilon}(t)\, d\tau\, dt. \tag{7.109}$$

In seeking to minimize (7.109) we do not know if $\epsilon(t)$ is continuous on the closed interval $[0, T]$. Certainly we cannot assume continuity at the outset of the derivation.

Begin by integrating (7.108) by parts to obtain the alternative form

$$\sigma(t) = G(0)\, \epsilon(t) + \int_0^t \dot{G}(t-\tau)\, \epsilon(\tau)\, d\tau \tag{7.110}$$

where $\dot{G}(t-\tau)$ denotes the time derivative

$$\dot{G}(t-\tau) = \frac{d}{dt} G(t-\tau).$$

Substitute (7.110) into (7.107), with the result

$$W(\epsilon) = \tfrac{1}{2} G(0)\, \epsilon_0^2 + \int_0^T \dot{\epsilon}(t) \int_0^t \dot{G}(t-\tau)\, \epsilon(\tau)\, d\tau\, dt. \tag{7.111}$$

Integrate (7.111) by parts using $\epsilon(0) = 0$ and Leibnitz's rule to obtain

$$\begin{aligned} W(\epsilon) = \tfrac{1}{2} G(0)\, \epsilon_0^2 &+ \epsilon_0 \int_0^T \dot{G}(T-t)\, \epsilon(t)\, dt \\ &- \dot{G}(0) \int_0^T \epsilon^2(t)\, dt - \int_0^T \epsilon(t) \left[\int_0^t \ddot{G}(t-\tau)\, \epsilon(\tau)\, d\tau\right] dt \end{aligned} \tag{7.112}$$

where

$$\ddot{G}(t-\tau) = \frac{d^2}{dt^2} G(t-\tau).$$

The last term in (7.112) can be put into alternative form. To do this, begin with the following form and break the second integration into the intervals shown:

$$\Theta = -\tfrac{1}{2}\int_0^T \int_0^T \ddot{G}(|t-\tau|)\,\epsilon(\tau)\,\epsilon(t)\,d\tau\,dt$$

$$= -\tfrac{1}{2}\int_0^T \int_0^t \ddot{G}(t-\tau)\,\epsilon(\tau)\,\epsilon(t)\,d\tau\,dt$$

$$-\tfrac{1}{2}\int_0^T \int_t^T \ddot{G}(\tau-t)\,\epsilon(\tau)\,\epsilon(t)\,d\tau\,dt. \tag{7.113}$$

Invert the orders of the limits in the last term and change the variables in it to obtain

$$\Theta = -\tfrac{1}{2}\int_0^T \epsilon(t)\left[\int_0^t \ddot{G}(t-\tau)\,\epsilon(\tau)\,d\tau\right]dt$$

$$-\tfrac{1}{2}\int_0^T \epsilon(t')\left[\int_0^{t'} \ddot{G}(t'-\tau')\,\epsilon(\tau')\,d\tau'\right]dt' \tag{7.114}$$

where

$$\tau = T - \tau', \qquad t = T - t'.$$

With a further change of dummy variables in the last term in (7.114) it is seen that the two terms are identical and can be combined to eliminate the factor of $\tfrac{1}{2}$. Furthermore, the ensuing term is identical to the last term in (7.112) and thus can be replaced by the term in (7.113). There results

$$W(\epsilon) = \tfrac{1}{2}G(0)\,\epsilon_0^2 + \epsilon_0\int_0^T \dot{G}(T-t)\,\epsilon(t)\,dt - \dot{G}(0)\int_0^T \epsilon^2(t)\,dt$$

$$-\tfrac{1}{2}\int_0^T \int_0^T \ddot{G}(|t-\tau|)\,\epsilon(\tau)\,\epsilon(t)\,d\tau\,dt. \tag{7.115}$$

Write (7.115) as

$$W(\epsilon) = A + L(\epsilon) + Q(\epsilon, \epsilon) \tag{7.116}$$

where

$$A = \tfrac{1}{2}G(0)\,\epsilon_0^2$$

$$L(\epsilon) = \epsilon_0\int_0^T \dot{G}(T-t)\,\epsilon(t)\,dt \tag{7.117}$$

$$Q(\epsilon, \beta) = -\dot{G}(0)\int_0^T \epsilon(t)\,\beta(t)\,dt - \tfrac{1}{2}\int_0^T \int_0^T \ddot{G}(|t-\tau|)\,\epsilon(\tau)\,\beta(t)\,d\tau\,dt.$$

The second form in (7.117) is linear, the last form bilinear and symmetric in the strain variables. Now form

$$W(\epsilon + \beta) - W(\epsilon) = Q(\beta, \beta) + 2Q(\epsilon, \beta) + L(\beta) \tag{7.118}$$

Since $Q(\beta, \beta) \geqslant 0$, for $W(\epsilon)$ to be a minimum we must have

$$2Q(\epsilon, \beta) + L(\beta) = 0 \tag{7.119}$$

because these terms can depend upon the sign of β.

Writing out (7.119), using (7.117), gives

$$\int_0^T \beta(t) \left[-2\dot{G}(0)\, \epsilon(t) - \int_0^T \ddot{G}(| t - \tau |)\, \epsilon(\tau)\, d\tau + \epsilon_0 \dot{G}(T - t) \right] dt = 0.$$

The Euler equation for this problem is thus given by setting the term in brackets equal to zero, yielding

$$2\dot{G}(0)\, \epsilon(t) + \int_0^T \ddot{G}(| t - \tau |)\, \epsilon(\tau)\, d\tau = \epsilon_0 \dot{G}(T - t). \tag{7.120}$$

We have made considerable progress. The minimum problem has now been reduced to the solution of an Euler equation in the strain history $\epsilon(t)$.

To solve the Euler equation, begin by examining the general characteristics of the solution. Let

$$f(t) = \epsilon_0 - \epsilon(T - t). \tag{7.121}$$

It can be verified that $f(t)$ satisfies the same Euler equation, (7.120), as $\epsilon(t)$; thus,

$$\epsilon(t) = \epsilon_0 - \epsilon(T - t)$$

and the strain path is antisymmetric about $T/2$. To proceed further requires specialization to a particular model of viscoelastic stress strain behavior.

As an example, take the standard linear solid described in Problem 1.1. The relaxation function is given by

$$G(t) = (G_0 - G_\infty)\, e^{-t/\tau_1} + G_\infty\,. \tag{7.122}$$

The other terms in (7.120), $\dot{G}(0)$, $\dot{G}(T - t)$, $\ddot{G}(t - \tau)$, and $\ddot{G}(\tau - t)$, can readily be formed from (7.122). With (7.122), the Euler equation (7.120) becomes

$$-\frac{2}{\tau_1}\, \epsilon(t) + \frac{e^{-t/\tau_1}}{\tau_1^2} \int_0^t e^{\tau/\tau_1} \epsilon(\tau)\, d\tau + \frac{e^{t/\tau_1}}{\tau_1^2} \int_t^T e^{-\tau/\tau_1} \epsilon(\tau)\, d\tau = -\frac{\epsilon_0}{\tau_1}\, e^{-T/\tau_1} e^{t/\tau_1}. \tag{7.123}$$

As a trial solution, try the linear function

$$\epsilon(t) = c_0 + c_1 t. \tag{7.124}$$

Substituting (7.124) in (7.123) gives terms that are independent of time, expressed as a coefficient of $(t)^0$, and coefficients of t, e^{-t/τ_1}, and e^{t/τ_1}. Setting these coefficients equal to zero gives four equations, only the last two of which are not identically satisfied; i.e.,

$$-\frac{c_0}{\tau_1} + c_1 = 0$$
$$-\frac{c_0}{\tau_1} - \frac{T}{\tau_1} c_1 - c_1 = -\frac{\epsilon_0}{\tau_1}. \tag{7.125}$$

The solution of (7.125) is

$$c_0 = \frac{\epsilon_0}{T/\tau_1 + 2}$$

and

$$c_1 = \frac{\epsilon_0}{\tau_1(T/\tau_1 + 2)}. \tag{7.126}$$

The complete solution for the optimal strain history is then

$$\epsilon(t) = \epsilon_0 \frac{1 + t/\tau_1}{2 + T/\tau_1} \tag{7.127}$$

and the work done is given by

$$W(\epsilon) = \frac{\epsilon_0^2}{2}\left(G_0 + \frac{G_\infty - G_0}{1 + 2\tau_1/T}\right). \tag{7.128}$$

The important thing to observe from the solution (7.127) is that the optimal strain path involves jump discontinuities at $t = 0$ and $t = T$. The discontinuous nature of the optimal strain path appears to be inherent in typical problems of viscoelastic optimization. For $T \to 0$, $W = \frac{1}{2}G_0\epsilon_0^2$, whereas for $T \to \infty$, $W = \frac{1}{2}G_\infty\epsilon_0^2$, in accordance with physical expectations.

Having succeeded in finding the optimal strain history for the standard linear solid, we proceed to the generalized Maxwell model. Let

$$G(t) = \sum_{i=1}^{N} G_i e^{-t/\tau_i} + G_\infty. \tag{7.129}$$

The Euler equation (7.120) with the form (7.129) gives

$$2\epsilon(t) \sum_{i=1}^{N} \frac{1}{\tau_i} + \sum_{i=1}^{N} \frac{e^{-t/\tau_i}}{\tau_i^2} \int_0^t e^{t/\tau_i}\epsilon(\tau)\, d\tau + \sum_{i=1}^{N} \frac{e^{t/\tau_i}}{\tau_i^2} \int_t^T e^{-\tau/\tau_i}\epsilon(\tau)\, d\tau$$
$$= -\epsilon_0 \sum_{i=1}^{N} \frac{e^{-T/\tau_i} e^{t/\tau_i}}{\tau_i}. \tag{7.130}$$

If one tries the linear form of solution (7.124) in (7.130), and then sets the coefficients of the time terms equal to zero, there results more equations than there are unknowns. The linear form (7.124) is not correct, and it can be seen that exponential terms must occur in the solution. The general solution is given by Breuer [7.14]. The example given here shows the practicality of formulating optimal work and energy problems in viscoelasticity. Bounds on the work done in viscoelastic deformation have been given by Breuer [7.16], Day [7.17], and Spector [7.18]. A solution of maximum recoverable energy has been given by Breuer and Onat [7.19]. Other references to optimization problems, in relationship to residual stresses, are given in Section 3.8.

PROBLEMS

7.1. A uniqueness theorem for isothermal, anisotropic, dynamic, linear viscoelasticity has been established which does not require the existence of a Laplace transform, as was required in Section 7.1. This uniqueness theorem, using Volterra's method in the proof, is given in Ref. [7.20]. Formulate completely the proof of this theorem, following that given in Ref. [7.20].

7.2. Using the reciprocal theorem given in Section 7.3, find the total volume change of a viscoelastic body subjected to surface tractions and body forces. The resulting formula involves only these surface tractions, the body forces, and the volumetric creep function.

7.3. Derive viscoelastic variational theorems, using the Stieltjes convolution notation, which correspond to the theorems of stationary potential and complementary energy in elasticity.

7.4. Obtain the complete proofs for the third and fourth minimum theorems of Section 7.5.

7.5. Verify relation (7.122) in the optimal strain history problem of Section 7.6. Discuss the possible means of minimizing the maximum stored energy in going from one strain state to another in a given time interval.

REFERENCES

7.1. Onat, E. T., and S. Breuer, "On Uniqueness in Linear Viscoelasticity," *in Progress in Applied Mechanics* (D. C. Drucker, ed.) (The Prager Anniversary Volume), p. 349. Macmillan, New York, 1963.

7.2. Doetsch, G., *Handbuch der Laplace-Transformation*, Vol. 1, p. 74. Birkhäuser, Basel, 1950.

7.3. Sternberg, E., and M. E. Gurtin, "Uniqueness in the Theory of Thermorheologically Simple Ablating Viscoelastic Solids," *in Progress in Applied Mechanics* (D. C. Drucker, ed.) (The Prager Anniversary Volume), p. 373. Macmillan, New York, 1963.

7.4. Gurtin, M. E., and E. Sternberg, "On the Linear Theory of Viscoelasticity," *Arch. Ration. Mech. Anal.* **11**, 343 (1962).

7.5. Graham, G. A. C., "The Contact Problem in the Linear Theory of Viscoelasticity," *Int. J. Eng. Sci.* **3**, 27 (1965).

7.6. Sokolnikoff, I. S., *Mathematical Theory of Elasticity*, 2nd ed. McGraw-Hill, New York, 1956.

7.7. Sternberg, E., "On the Analysis of Thermal Stresses in Viscoelastic Solids," *Proc. 3rd Symp. Nav. Structural Mech.* 348. Macmillan, New York, 1964.

7.8. Onat, E. T., "On a Variational Principle in Linear Viscoelasticity," *J. Mech.* **1**, 135 (1962).

7.9. Schapery, R. A., "On the Time Dependence of Viscoelastic Variational Solutions," *Quart. Appl. Math.* **22**, 207 (1964).

7.10. Biot, M. A., "Linear Thermodynamics and the Mechanics of Solids," *Proc. 3rd U.S. Nat. Cong. Appl. Mech.* 1 (1958).

7.11. Christensen, R. M., "Variational and Minimum Theorems for the Linear Theory of Viscoelasticity," *Z. Angew. Math. Phys.* **19**, 233 (1968).

7.12. Gurtin, M. E., "Variational Principles in the Linear Theory of Viscoelasticity," *Arch. Ration. Mech. Anal.* **13**, 179 (1963).

7.13. Christensen, R. M., "Viscoelastic Properties of Heterogeneous Media," *J. Mech. Phys. Solids* **17**, 23 (1969).

7.14. Breuer, S., "The Minimizing Strain-Rate History and the Resulting Greatest Lower Bound on Work in Linear Viscoalasticity," *Z. Angew. Math. Mech.* **49**, 209 (1969).

7.15. Gurtin, M. E., R. C. MacCamy, and L. F. Murphy, "On Optimal Strain Paths in Linear Viscoelasticity," *Quart. Appl. Math.* **37**, 151 (1979).

7.16. Breuer, S., "Lower Bounds on Work in Linear Viscoelasticity," *Quart. Appl. Math.* **27**, 139 (1969).

7.17. Day, W. A., "Improved Estimates for Least Work in Linear Viscoelasticity," *Quart. J. Mech. Appl. Math.* **32**, 17 (1979).

7.18. Spector, S. J., "On Monotonicity of the Optimal Strain Path in Linear Viscoelasticity," *Quart. Appl. Math.* **38**, 369 (1980).

7.19. Breuer, S., and E. T. Onat, "On Recoverable Work in Linear Viscoelasticity," *Z. Angew. Math. Phys.* **15**, 12 (1964).

7.20. Edelstein, W. S., and M. E. Gurtin, "Uniqueness Theorems in the Linear Dynamic Theory of Anisotropic Viscoelastic Solids," *Arch. Ration. Mech. Anal.* **17**, 47 (1964).

Chapter VIII Nonlinear Viscoelasticity

Nonlinear viscoelasticity has many features in common with the linear theory and the most fundamental common aspect between the two theories is that of the memory hypothesis. In terms of a relation between stress and strain, this simply means that the current value of stress is determined not only by the current value of strain, but also by the complete past history of strain. This hypothesis, that the material has a memory for past deformation events, is the starting point in the development of the linear theory, and it also is the starting point here with the general nonlinear theory. It is to be expected that the means of derivation will be more involved here, since the theorems and techniques of dealing with linear functionals are not applicable. However, the end results justify these intermediate complications since we shall develop a complete nonlinear theory which is not restricted to infinitesimal deformation conditions but which can still be used to solve at least some simple boundary value problems. Two examples will be given which reveal the usefulness and generality of the nonlinear theory.

We are motivated to develop and examine a general nonlinear theory of viscoelasticity because, in the practical application of viscoelastic materials, such materials are often used under conditions which do not comply with the infinitesimal deformation assumptions of the linear theory. For some materials, the range of deformation beyond which superposition, and thereby linearity, does not remain valid is extremely limited and the need for a more general theory is apparent. The treatment of the nonlinear theory certainly could occupy much more space than two chapters in a book on viscoelasticity, but only a brief introductory outline of the subject is given here. Also only one particular type of nonlinear theory of viscoelasticity is considered, although several other types of theories will be mentioned in this chapter and in the next, which is on mechanical properties determination. However, no attempt is made to give a completely representative survey of such theories.

In dealing with a nonlinear continuum theory it is necessary to utilize a completely nonlinear treatment of the kinematical aspects of the problem. Correspondingly, the particular measures of deformation to be used here must have the proper coordinate invariance characteristics. More generally, it is necessary that all mathematical descriptions of physical quantities and events to be used here must be independent of the particular frame of reference to which they are referred. This requirement is sometimes called the principle of objectivity, or material frame indifference, and is discussed at length by

Truesdell and Noll [8.1]. All results presented here are consistent with that requirement.

The formulation to be given here is based upon the development by Coleman [8.2, 8.3] of a general thermodynamical theory of nonlinear viscoelasticity. Coleman's work specialized the theory of simple materials (Noll [8.4]) for specific application to materials with fading memory. Some details of the following work are taken from Christensen [8.5] and Laws [8.6] which in turn are based upon Coleman's work. Although Coleman's treatment is not restricted to isothermal conditions, for simplicity of presentation we restrict the following derivation to isothermal conditions. Also, we will construct separate theories for nonlinear viscoelastic solids and fluids. Finally, it should be mentioned that Truesdell and Noll [8.1] provide a general account of nonlinear continuum mechanics, with special treatments of viscoelastic solids and fluids.

8.1. DERIVATION OF CONSTITUTIVE RELATIONS FOR SOLIDS

Contrary to the situation in the infinitesimal theory, it is now necessary to distinguish between the coordinates of each particle of the material in some reference configuration and the coordinates of these particles in all possible deformed configurations. We let the coordinates of a typical particle in the reference configuration be designated by X_K, referred to a given fixed set of rectangular Cartesian axes. Now we let the deformed configuration coordinates of the same particle at any time be given by x_i, referred to the same set of fixed axes as were used to designate X_K. The complete deformation history is specified by the relationship

$$x_i(\tau) = x_i(X_K, \tau) \qquad (-\infty < \tau \leqslant t)$$

where t is the current time.

The deformation gradient is defined by

$$x_{i,L}(X_K, \tau) = \frac{\partial x_i(X_K, \tau)}{\partial X_L}.$$

These partial derivatives, with respect to the reference configuration coordinates, give a measure of the deformation of the material. The theory of simple materials provides an expression of the postulate that the current values of field variables, such as stress and stored energy, depend not only upon the current value of the deformation gradient, but also upon the past history of the deformation gradient. Thus, for example, the stress constitutive relation is expressed through the functional relationship

$$\sigma_{ij} = \underset{s=0}{\overset{\infty}{\chi_{ij}}} \left(x_{j,L}(t-s), x_{j,L}(t)\right) \tag{8.1}$$

where the dependence of the deformation gradient upon the X_K coordinates is implied. As it stands, this definition of simple materials has great generality. However, we shall further restrict the functional in (8.1) to be of a type that does not change with time, that is, which satisfies time translation invariance. For example, materials which exhibit the effects of a yield stress, or a yield stress criteria, are automatically excluded from the latter class of behavior. Such yielding type materials imply, among other things, that the actual nature or type of functional, involved in constitutive relations, changes in accordance with the yield criteria, and thereby changes with time. Another example is that of partially crystalline polymeric materials under sufficiently large deformation and rate of deformation conditions. In these circumstances, the amount and nature of the crystallinity in these materials change. Again, a single functional constitutive relation of the type of (8.1) is insufficient to characterize these materials under the conditions mentioned. Our interest here is confined to simple materials, of the special type mentioned, and, when we suitably restrict the simple material constitutive relations in accordance with a fading memory hypothesis, we arrive at a theory of viscoelastic materials.

Noll [8.4] developed simple material theory and obtained reduced forms of the stress constitutive relation (8.1) as necessitated by the objectivity requirements. We shall shortly present one such possible form, and, henceforth, in this and the next two sections, the application shall be made to solids, whereby the reference configuration is identified with the preferred undeformed configuration.

Since the deformation gradient is not objective, we begin by selecting a measure of deformation which is objective, that is, the strain measure

$$2E_{KL}(X_K, \tau) = x_{k,K}x_{k,L} - \delta_{KL} \tag{8.2}$$

where δ_{KL} is the usual Kronecker symbol. The deformation gradient $x_{k,K}$ used in (8.2) is obtained by direct differentiation when $x_k(X_K, \tau)$ is known. As seen from (8.2), lower case indices are used to designate tensor components referred to the deformed configuration coordinates while upper case indices designate tensor components referred to the reference configuration coordinates. The strain measure E_{KL} represents a measure of deformation with respect to the reference configuration.

The local balance of energy expression and the local entropy production inequality, both under isothermal conditions, are combined to give

$$-\rho\dot{A} + \sigma_{ij}d_{ij} \geqslant 0 \tag{8.3}$$

where ρ is the mass density at time t, A is the stored energy per unit mass, σ_{ij} are the components of the Cauchy stress tensor defined with respect to the deformed configuration, and d_{ij} is defined in terms of the velocity gradient as

$$2d_{ij} = v_{i,j} + v_{j,i} \tag{8.4}$$

where

$$v_i(\tau) = \dot{x}_i(\tau, X_K), \qquad v_i = v_i(t). \tag{8.5}$$

A superimposed dot is used to designate differentiation with respect to τ, holding X_K fixed; and, when the time argument is not specifically noted, it is understood to be current time t. The relation (8.3) is the isothermal generalization to the nonlinear case of the comparable form (3.5) used in the derivation of the linear theory of thermoviscoelasticity. In fact, the derivation here follows along the same lines as that given in Section 3.1 for the linear theory. Now, however, rather than taking a specific representation for the stored energy for use in (8.3), only certain characteristics of the stored energy need to be considered.

Consistent with the hypothesis of a memory effect, the stored energy is taken to be a functional of the past history of the deformation with one particular form being given by

$$A = \underset{s=0}{\overset{\infty}{\psi}} \left(E_{KL}(t-s), E_{KL}(t)\right) \tag{8.6}$$

where, as with the first consideration of a stress constitutive relation in Section 1.2, a specific dependence upon the current configuration is included. The functional (8.6) is assumed to be a continuous functional of the strain history; also $E_{KL}(t)$, as well as its first time derivative, is assumed to be continuous.

To use the stored energy from (8.6) in (8.3), it is necessary to obtain its time derivative. But this is not as easily accomplished as it was in the linear theory, Section 3.1, since (8.6) is not a specific representation but actually is just a formalism. In order to obtain the time derivative of A in (8.6), in a generalized sense, it is convenient to proceed in the following manner. We let the functions $E_{KL}(\tau)$, with the dependence upon X_K understood, constitute a history. We define the norm on the collection of histories by

$$\|\mathbf{E}\| = \left[\int_0^\infty E_{KL}(t-s)\, E_{KL}(t-s)\, h^2(s)\, ds\right]^{1/2} \tag{8.7}$$

where $h(s)$ is a monotonically decreasing function and is termed an influence function of order r such that

$$\lim_{s\to\infty} s^r h(s) = 0. \tag{8.8}$$

The collection of histories with a finite norm constitute a Hilbert space.

It is assumed that the stored energy functional $\underset{s=0}{\overset{\infty}{\psi}}(\ )$ is Fréchet differentiable in the Hilbert space corresponding to $h(s)$.

Thus,

$$\underset{s=0}{\overset{\infty}{\psi}} (E_{KL}(t-s) + \delta E_{KL}(t-s), E_{KL}(t)) = \underset{s=0}{\overset{\infty}{\psi}} (E_{KL}(t-s), E_{KL}(t))$$
$$+ \underset{s=0}{\overset{\infty}{\delta\psi}} (E_{KL}(t-s), E_{KL}(t) \mid \delta E_{KL}(t-s))$$
$$+ o \,\| \delta E_{KL}(t-s) \| \tag{8.9}$$

where $\underset{s=0}{\overset{\infty}{\delta\psi}}(\ )$, the first order Fréchet differential, is linear in $\delta E_{KL}(t-s)$, and is continuous in all variables. Relation (8.9) is simply an expansion of the functional of the history $E_{KL}(t-s) + \delta E_{KL}(t-s)$ about the functional of the history $E_{KL}(t-s)$. This expansion is the same type as was used in the variational theorems of Section 7.4, and, as mentioned there, it is analogous to a Taylor series expansion for functions. For the expansion (8.9) to exist it is sufficient that an influence function of the type of $h(s)$ exists. Thus, although $h(s)$ is not a mechanical property, the appropriateness of this type of characterization is an intrinsic property of the material. In fact, materials for which an influence function of this type exists such that the Fréchet differential $\underset{s=0}{\overset{\infty}{\delta\psi}}(\ )$ of $\underset{s=0}{\overset{\infty}{\psi}}(\ )$ exists, are said to obey the principle of fading memory [8.2], which we shall refer to as the fading memory hypothesis. The term fading memory is appropriate for use here since the effect of $h(s)$ is to make the norm of a history (8.7) more dependent upon the recent than the distant deformation events. The fading memory hypothesis just introduced should be compared with that given in Section 1.3 which was used in the linear theory. The comparison shows that the present hypothesis is, in a sense, less restrictive than that for the linear theory; however, the present hypothesis is sufficiently strong to suffice for the present purposes. An even less restrictive type of fading memory hypothesis is mentioned at the end of this derivation.

The availability of the Fréchet differential of the stored energy, in (8.9), now allows the derivative of the stored energy to be defined. The derivative of A is composed of the two partial derivatives shown below:

$$\dot{A} = \frac{\partial}{\partial E_{KL}(t)} \underset{s=0}{\overset{\infty}{\psi}} (E_{KL}(t-s), E_{KL}(t))\, \dot{E}_{KL}$$
$$+ \lim_{\substack{\delta E_{KL}(t-s)\to 0 \\ \delta t\to 0}} \delta t^{-1} \Big[\underset{s=0}{\overset{\infty}{\psi}} (E_{KL}(t-s) + \delta E_{KL}(t-s), E_{KL}(t))$$
$$- \underset{s=0}{\overset{\infty}{\psi}} (E_{KL}(t-s), E_{KL}(t)) \Big] \tag{8.10}$$

where $\delta E_{KL}(t-s) = E_{KL}(t+\delta t - s) - E_{KL}(t-s)$. The first term in (8.10) just involves the derivative of the functional $\overset{\infty}{\underset{s=0}{\psi}}(\ )$ with respect to $E_{KL}(t)$. This derivative is obtained from the rule of differentiation of functions since $\overset{\infty}{\underset{s=0}{\psi}}(\ )$ only has a function type dependence upon $E_{KL}(t)$. Using (8.9), the second term in (8.10) is rewritten as a Fréchet differential to obtain

$$\dot{A} = \frac{\partial}{\partial E_{KL}(t)} \overset{\infty}{\underset{s=0}{\psi}} \left(E_{KL}(t-s), E_{KL}(t)\right) \dot{E}_{KL}$$

$$+ \overset{\infty}{\underset{s=0}{\delta\psi}} \left(E_{KL}(t-s), E_{KL}(t) \,\middle|\, \frac{dE_{KL}(t-s)}{dt}\right). \tag{8.11}$$

The reason for requiring the continuity of $\dot{E}_{KL}(t)$, $\infty < t < \infty$, is now seen to follow because the Fréchet differential is defined only for continuous arguments. This is in contrast to the circumstances of the linear isothermal and nonisothermal derivations where only the strain histories were required to be continuous.

With the time derivative of the stored energy available from (8.11) it is now substituted into (8.3) giving

$$\sigma_{ij} d_{ij} - \rho \frac{\partial}{\partial E_{KL}(t)} \overset{\infty}{\underset{s=0}{\psi}} \left(E_{KL}(t-s), E_{KL}(t)\right) \dot{E}_{KL} + \rho \overset{\infty}{\underset{s=0}{\Lambda}} (\ ) \geqslant 0 \tag{8.12}$$

where

$$\overset{\infty}{\underset{s=0}{\Lambda}} (\ ) = \overset{\infty}{\underset{s=0}{\delta\psi}} \left(E_{KL}(t-s), E_{KL}(t) \,\middle|\, \frac{dE_{KL}(t-s)}{dt}\right). \tag{8.13}$$

The nonlinear kinematical relationship $\dot{E}_{KL} = d_{ij} x_{i,K} x_{j,L}$ is obtained by differentiating (8.2) and using (8.4). By using this in (8.12) it then becomes

$$\left[\sigma_{ij} - \rho \frac{\partial}{\partial E_{KL}(t)} \overset{\infty}{\underset{s=0}{\psi}} \left(E_{KL}(t-s), E_{KL}(t)\right) x_{i,K} x_{j,L}\right] d_{ij} + \rho \overset{\infty}{\underset{s=0}{\Lambda}} (\ ) \geqslant 0. \tag{8.14}$$

Coleman [8.2] shows, as a result of the fading memory hypothesis with an influence function $r > \frac{1}{2}$, that for a given history of deformation it is possible to take a (second) deformation history with the same current value as the given history and sufficiently close to the given history so that the norm of the difference between them is arbitrarily small even though the current rate of deformation of the second history remains arbitrary. Also, the norm of the difference between the rates of the two deformation histories can be made to be arbitrarily small.

It follows that the current rate of deformation tensor d_{ij} may be arbitrarily specified without changing the functionals in (8.14). The physical significance of this result is that the material possesses the "instantaneous elasticity" capability which is characteristic of the infinitesimal theory and is consistent with experimental observations. We now see that, with a given deformation history, for (8.14) to be satisfied it is necessary that the coefficient of d_{ij} vanish, giving

$$\sigma_{ij} = \rho \frac{\partial}{\partial E_{KL}(t)} \mathop{\psi}_{s=0}^{\infty} \left(E_{KL}(t-s), E_{KL}(t)\right) x_{i,K} x_{j,L} \tag{8.15}$$

which leaves (8.14) as

$$\rho \mathop{\Lambda}_{s=0}^{\infty} (\ ) \geqslant 0. \tag{8.16}$$

This last relationship states the result that the rate of dissipation of energy must be nonnegative. The similarly between the present nonlinear derivation and the linear theory derivation of Section 3.1 is to be noted.

Equation (8.15) provides the stress constitutive relation which we have been seeking. Relation (8.15) can be written in the more symbolic form as

$$\sigma_{ij} = \rho \frac{\partial A}{\partial E_{KL}} x_{i,K} x_{j,L} . \tag{8.17}$$

An alternative form of (8.15) can be obtained by noting from (8.11) that

$$\frac{\partial}{\partial E_{KL}(t)} \mathop{\psi}_{s=0}^{\infty} \left(E_{KL}(t-s), E_{KL}(t)\right) = \frac{\partial \dot{A}}{\partial \dot{E}_{KL}} . \tag{8.18}$$

Thus, we can write (8.15) as

$$\sigma_{ij} = \rho \frac{\partial \dot{A}}{\partial \dot{E}_{KL}} x_{i,K} x_{j,L} . \tag{8.19}$$

The two forms of the stress constitutive relation, (8.17) and (8.19), are entirely equivalent, and whichever one is to be used in a particular situation is merely a matter of convenience. Relation (8.19) involves a process of first obtaining $\dot{A}$ and then recognizing the coefficient of $\dot{E}_{KL}$ in this expression, as stated by the derivative $\partial \dot{A}/\partial \dot{E}_{KL}$. Relation (8.17) directly involves the derivatives $\partial A/\partial E_{KL}$. Both forms are useful; however, the form (8.19) has an additional appeal in as much as it has a direct counterpart form for viscoelastic fluids, as we shall see.

The preceeding derivation is based upon the definition of the norm (8.7) and the associated fading memory hypothesis. Results comparable to those just derived have been obtained by Coleman and Mizel [8.7], under more general conditions of adherence to certain properties of a Banach function space, without defining an explicit norm. We here prefer to retain the simpler but more restrictive derivation just given since the formula for the norm and the fading memory hypothesis help promote a better understanding of the physical nature of the behavior of materials for which this theory is applicable.

Some simplification is achieved for isotropic materials. Following Wineman and Pipkin [8.8], the stored energy for isotropic materials is taken to be expressed as a functional of the six invariants of the history $E_{KL}(\tau)$, given by

$$\begin{aligned} \mathrm{I}(\tau) &= \operatorname{tr} \mathbf{E}(\tau) \\ \mathrm{II}(\tau_1, \tau_2) &= \operatorname{tr} \mathbf{E}(\tau_1)\, \mathbf{E}(\tau_2) \\ &\;\vdots \\ \mathrm{VI}(\tau_1, \tau_2, \tau_3, \tau_4, \tau_5, \tau_6) &= \operatorname{tr} \mathbf{E}(\tau_1)\, \mathbf{E}(\tau_2)\, \mathbf{E}(\tau_3)\, \mathbf{E}(\tau_4)\, \mathbf{E}(\tau_5)\, \mathbf{E}(\tau_6) \end{aligned} \tag{8.20}$$

where tr denotes the trace of the matrix products of the matrices $\mathbf{E}(\tau)$ corresponding to $E_{KL}(\tau)$. The forms (8.20) are invariant with respect to any rotation of the X_K coordinate system, and thus a functional of the histories (8.20) satisfies the isotropy condition. The fact that the trace of matrix products higher than that of six matrices need not be considered was shown by Spencer and Rivlin [8.9]. The forms in (8.20) may be equivalently written in index notation by using the Kronecker symbol and the generalized Kronecker symbols.

The isotropic nonlinear functional for the stored energy thus involves the histories shown in (8.20), in all possible combinations. Also, the functional depends upon the current values of the invariants $\mathrm{I}(t)$, $\mathrm{II}(t, t)$,..., $\mathrm{VI}(t, t, t, t, t, t)$ as well as upon all possible combinations of current values and past histories, such as $\mathrm{II}(t, \tau_2)$, etc. A typical example of a functional representation for the stored energy is given in Section 8.3. The stress constitutive relation for an isotropic material still follows from either (8.17) or (8.19) where the stored energy is a functional of the histories in (8.20).

The general dependence of the isotropic material stored energy upon the invariants of the history is completely analogous to the isotropic infinitesimal theory situation. Indeed, the isothermal form of the stored energy in the infinitesimal isotropic theory can be written in a form such that one term is a quadratic functional of the history of the first invariant in (8.20), while the other term is a quadratic functional of the second history invariant in (8.20), but where the invariants are relative to the history of infinitesimal strain ϵ_{ij}. These terms have forms similar to the first two double integral terms in (3.22). To see this explicitly, we would have to integrate by parts the forms in (3.22) to get forms involving functionals of the history of the invariants rather than involving the rates of the history of the invariants. The dependence upon the current configuration could then be noted.

It is useful to formulate the nonlinear stress constitutive relation for incompressible materials. This is obtained from (8.19) by requiring that the mass per unit volume remains unchanged and by adding a reactive hydrostatic pressure p to give

$$\sigma_{ij} = -p\delta_{ij} + \rho \frac{\partial \dot{A}}{\partial \dot{E}_{KL}} x_{i,K} x_{j,L} \,. \tag{8.21}$$

We conclude this section with a discussion of a means of representing the stored energy A for use in the stress constitutive equations. The procedure to be stated here involves the use of an approximation theorem, and follows the outline given by Chacon and Rivlin [8.10]. Under the continuity assumptions already invoked, it follows, from the Stone–Weierstrass theorem, that a real continuous scalar valued functional of $E_{KL}(\tau)$ may be uniformly approximated by a polynomial in a set of real continuous linear scalar valued functionals of $E_{KL}(\tau)$. Accordingly, if we let $f_{(i)}$ designate the linear functionals

$$f_{(i)} = \underset{s=0}{\overset{\infty}{\zeta_{(i)}}} \left(E_{KL}(t-s), E_{KL}(t)\right), \qquad i = 1, 2, \ldots, N \tag{8.22}$$

the stored energy functional is approximated by the expansion

$$A = \sum_{i=1}^{N} f_{(i)} + \sum_{i=1}^{N} \sum_{j=1}^{N} f_{(i)} f_{(j)} + \cdots \tag{8.23}$$

where the polynomial expansion is truncated at the level involving N products. Now, in the same manner as was employed in Chapter 1, the Riesz representation theorem is used to represent the linear functionals used in the expansion as Stieltjes integrals. This gives the following representation for (8.22):

$$f_{(i)} = \int_0^\infty E_{KL}(t-s)\, dg_{KL}^{(i)}(s) \tag{8.24}$$

where

$$g_{KL}^{(i)}(\tau) = 0, \qquad \tau < 0$$

and where the integrating functions $g_{KL}^{(i)}(\tau)$ are taken to be continuous with continuous first derivatives for $\tau > 0$. Relation (8.24) is written in the alternative form

$$f_{(i)} = g_{KL}^{(i)}(0)\, E_{KL}(t) + \int_0^\infty E_{KL}(t-s) \frac{dg_{KL}^{(i)}(s)}{ds}\, ds \tag{8.25}$$

which gives an explicit display of a dependence upon the current state of deformation, through $E_{KL}(t)$. Alternatively, through integration by parts, (8.24) is written as

$$f_{(i)} = \int_{-\infty}^{t} g_{KL}^{(i)}(t-\tau) \frac{\partial E_{KL}(\tau)}{\partial \tau}\, d\tau. \tag{8.26}$$

Either form (8.25) or (8.26) can be used in the expansion (8.23). If (8.25) were used, the stress constitutive relation (8.17) would be appropriate for use with it. But, obviously, the form (8.26) is simpler and more compact than (8.25) for use in the expansion. Therefore, it will be used, and as might be expected the stress constitutive relation (8.19) will be more direct in application than (8.17) in this case.

We insert the representation (8.26) in the polynomial expansion (8.23) to get the explicit form

$$A = \int_{-\infty}^{t} g_{KL}(t-\tau)\,\frac{\partial E_{KL}(\tau)}{\partial \tau}\,d\tau + \int_{-\infty}^{t}\int_{-\infty}^{t} g_{KLMN}(t-\tau, t-\eta)\,\frac{\partial E_{KL}(\tau)}{\partial \tau}\,\frac{\partial E_{MN}(\eta)}{\partial \eta}\,d\tau\,d\eta + \cdots. \tag{8.27}$$

The integrating functions $g_{KL}(\tau)$, $g_{KLMN}(\tau, \eta)$, etc. in the present application are determined to model the behavior of the material and must have a form consistent with the fading memory hypothesis. For isotropic materials, the integrating functions in (8.27) must be restricted in accordance with the material symmetry requirements. This means that the functionals in (8.27) must be restricted such that they represent functionals of the invariants, (8.20), of the history $E_{KL}(\tau)$. This gives the terms shown explicitly in (8.27) as having the (isotropic) form

$$A = \int_{-\infty}^{t} g(t-\tau)\,\frac{\partial E_{KK}(\tau)}{\partial \tau}\,d\tau + \int_{-\infty}^{t}\int_{-\infty}^{t} \gamma(t-\tau, t-\eta)\,\frac{\partial E_{KK}(\tau)}{\partial \tau}\,\frac{\partial E_{LL}(\eta)}{\partial \eta}\,d\tau\,d\eta + \int_{-\infty}^{t}\int_{-\infty}^{t} \Delta(t-\tau, t-\eta)\,\frac{\partial E_{KL}(\tau)}{\partial \tau}\,\frac{\partial E_{KL}(\eta)}{\partial \eta}\,d\tau\,d\eta + \cdots. \tag{8.28}$$

The first terms in an expansion of this type are applied in the work of the next two sections.

8.2. REDUCTION TO LINEAR THEORY

The present nonlinear theory can directly be reduced to the infinitesimal theory forms used heretofore. The same formal argument, as was explained in some detail in Section 3.1, is employed here to approximate the stored energy by a polynomial expansion of linear functionals. Specifically, the form (8.28) for isotropic materials is used. The deformation is infinitesimal if $\epsilon \ll 1$, where

$$\epsilon = \sup_{\tau} \left| x_{i,K}(\tau) - \delta_{iK} \right| \tag{8.29}$$

and sup designates the least upper bound. The strain E_{KL} is then of $O(\epsilon)$. Coleman and Noll [8.11] showed that any function of $O(\epsilon^n)$ in the above sense is also a function of $O(\epsilon^n)$ with respect to the Hilbert space norm (8.7). Using this result, we now neglect terms of $O(\epsilon^3)$ in the expansion (8.28). Thus, the free energy is represented as

$$\begin{aligned}\rho_0 A = &\frac{1}{2}\int_{-\infty}^{t}\int_{-\infty}^{t} \lambda(2t-\tau-\eta)\frac{\partial E_{KK}(\tau)}{\partial\tau}\frac{\partial E_{LL}(\eta)}{\partial\eta}\,d\tau\,d\eta \\ &+\int_{-\infty}^{t}\int_{-\infty}^{t} \mu(2t-\tau-\eta)\frac{\partial E_{KL}(\tau)}{\partial\tau}\frac{\partial E_{KL}(\eta)}{\partial\eta}\,d\tau\,d\eta \end{aligned}\tag{8.30}$$

where this representation is for the free energy per unit mass in the deformed configuration multiplied by the mass density in the reference configuration. In obtaining (8.30) from (8.28) the first term in (8.28) has been dropped, since it can be shown that it does not provide any increase in generality. The arguments of the λ and μ relaxation functions in (8.30) are taken in the simplified form with arguments added; the simplification is not necessary, however.

We use Leibnitz's rule to take the time derivative of (8.30), thus,

$$\begin{aligned}\rho_0 \dot{A} = &\dot{E}_{KK}\int_{-\infty}^{t}\lambda(t-\tau)\frac{\partial E_{LL}(\tau)}{\partial\tau}\,d\tau + 2\dot{E}_{KL}\int_{-\infty}^{t}\mu(t-\tau)\frac{\partial E_{KL}(\tau)}{\partial\tau}\,d\tau \\ &+\frac{1}{2}\int_{-\infty}^{t}\int_{-\infty}^{t}\frac{\partial}{\partial t}\lambda(2t-\tau-\eta)\frac{\partial E_{KK}(\tau)}{\partial\tau}\frac{\partial E_{LL}(\eta)}{\partial\eta}\,d\tau\,d\eta \\ &+\int_{-\infty}^{t}\int_{-\infty}^{t}\frac{\partial}{\partial t}\mu(2t-\tau-\eta)\frac{\partial E_{KL}(\tau)}{\partial\tau}\frac{\partial E_{KL}(\eta)}{\partial\eta}\,d\tau\,d\eta. \end{aligned}\tag{8.31}$$

Relation (8.31) is substituted into the general nonlinear stress constitutive relation (8.19) to obtain

$$\sigma_{ij} = \frac{\rho}{\rho_0}\left[\delta_{KL}\int_{-\infty}^{t}\lambda(t-\tau)\frac{\partial E_{MM}(\tau)}{\partial\tau}\,d\tau + 2\int_{-\infty}^{t}\mu(t-\tau)\frac{\partial E_{KL}(\tau)}{\partial\tau}\right]x_{i,K}x_{j,L}\,. \tag{8.32}$$

Consistent with the assumption $\epsilon \ll 1$, where ϵ is given by (8.29), then

$$x_{i,K} = \delta_{iK} + O(\epsilon), \qquad x_{j,L} = \delta_{jL} + O(\epsilon),$$

and

$$\rho = \rho_0[1 + O(\epsilon)] \tag{8.33}$$

where ρ_0 is the mass density in the reference configuration.

Relations (8.33) when substituted into (8.32) give, neglecting $O(\epsilon)$ terms,

$$\sigma_{ij} = \delta_{ij} \int_{-\infty}^{t} \lambda(t-\tau) \frac{\partial E_{kk}(\tau)}{\partial \tau} d\tau + 2 \int_{-\infty}^{t} \mu(t-\tau) \frac{\partial E_{ij}(\tau)}{\partial \tau} d\tau. \quad (8.34)$$

We define the components of the displacement vector as $u_i = x_i - X_i$. From (8.2) and (8.29) we get

$$E_{ij} = \tfrac{1}{2}(u_{i,j} + u_{j,i}) + O(\epsilon^2). \quad (8.35)$$

In these last two steps, the lower case subscript notation is used on $E_{KL} - E_{kl}$ since the distinction between the coordinates in the deformed and reference configuration is now unimportant. Although the stress tensor σ_{ij} has been defined with respect to areas in the deformed configuration, it is herein interpreted as being defined with respect to the reference configuration, since now the deformed configuration is taken to be only an infinitesimal departure from the reference configuration.

Relations (8.34) and (8.35), with the $O(\epsilon^2)$ terms neglected, now define the reduced form of the general nonlinear viscoelastic stress constitutive relation, for infinitesimal deformation conditions. We recognize these forms as the linear relations which have been used throughout the previous chapters.

Coleman and Noll [8.11] gave a derivation of the linear theory of viscoelasticity, as deduced from the general nonlinear theory. In this reference they also derived a special nonlinear theory for solids known as finite linear viscoelasticity. This theory does not restrict the deformation to be small, but it does restrict the deformation to have been slowly changing in the recent past. In this manner the current value of stress is found to be determined by linear integrals of the history of deformation measured with respect to the current configuration; however, in contrast to the infinitesimal theory, the integrating functions in these integrals are nonlinear functions of the current state of deformation. This theory is stated in Section 9.6.

8.3. SIMPLE SHEAR DEFORMATION EXAMPLE

As an example of the general nonlinear theory, a particular problem is solved. The problem to be considered is that of the simple shear deformation of an incompressible isotropic material. To be sure, this problem, or any problem involving homogeneous deformations, is trivial for the infinitesimal theory. For reasons which will become apparent, the nonlinear counterpart problem is not trivial, and some interesting effects are revealed.

The deformation is specified by

$$\begin{aligned} x_1 &= X_1 + K(t)\, X_2, \\ x_2 &= X_2, \\ x_3 &= X_3, \end{aligned} \quad (8.36)$$

and

where $K(t)$ is the specified time dependent parameter which determines the amount of distortion, as shown in Fig. 8.1.

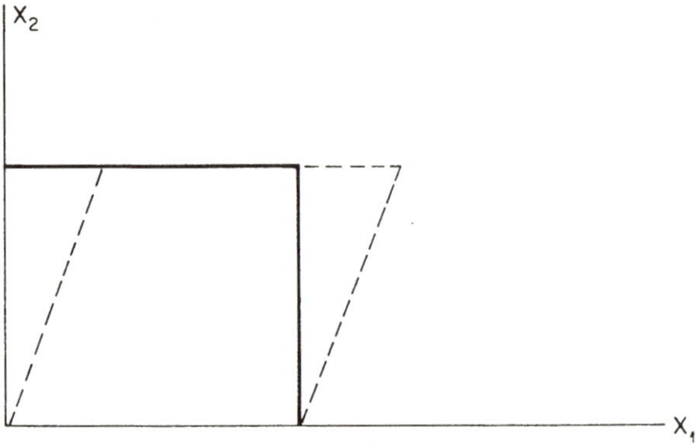

Fig. 8.1. Simple shear deformation.

The deformation gradient tensor, from the differentiation of (8.36), has the components given by

$$[x_{i,K}] = \begin{bmatrix} 1 & K(t) & 0 \\ 0 & 1 & 0 \\ 0 & 0 & 1 \end{bmatrix}. \tag{8.37}$$

Using this in (8.2), we find the strain components to be

$$[2E_{KL}] = \begin{bmatrix} 0 & K(t) & 0 \\ K(t) & K^2(t) & 0 \\ 0 & 0 & 0 \end{bmatrix}. \tag{8.38}$$

The stress constitutive relation for incompressible materials (8.21) when combined with (8.37) and (8.38) reduces the stress components to the forms

$$\begin{aligned} \sigma_{11}(t) &= -p + \rho \frac{\partial \dot{A}}{\partial \dot{E}_{11}} + K^2(t)\rho \frac{\partial \dot{A}}{\partial \dot{E}_{22}} + K(t)\rho \frac{\partial \dot{A}}{\partial \dot{E}_{12}} + K(t)\rho \frac{\partial \dot{A}}{\partial \dot{E}_{21}}, \\ \sigma_{22}(t) &= -p + \rho \frac{\partial \dot{A}}{\partial \dot{E}_{22}}, \\ \sigma_{33}(t) &= -p + \rho \frac{\partial \dot{A}}{\partial \dot{E}_{33}}, \\ \sigma_{12}(t) &= K(t)\rho \frac{\partial \dot{A}}{\partial \dot{E}_{22}} + \rho \frac{\partial \dot{A}}{\partial \dot{E}_{12}}, \end{aligned} \tag{8.39}$$

and

$$\sigma_{23}(t) = \sigma_{31}(t) = 0$$

where p is the indeterminate hydrostatic pressure. That $\sigma_{23} = \sigma_{31} = 0$ follows from

$$\frac{\partial \dot{A}}{\partial \dot{E}_{23}} = \frac{\partial \dot{A}}{\partial \dot{E}_{31}} = 0$$

which in turn follows from $E_{23} = E_{31} = 0$ and the fact that A is expressed as a functional of the histories in (8.20), as in (8.28).

The deformation in this problem of simple shear is isochoric; that is, there is no volume change, so the incompressibility condition is automatically satisfied. The last two terms in $\sigma_{11}(t)$ in (8.39) may be combined from symmetry considerations. As a practical boundary condition, the surfaces $x_3 =$ constant are considered to be free of tractions. This determines p from the condition that $\sigma_{33}(t) = 0$. Making these changes in relations (8.39) we find

$$\begin{aligned}
\sigma_{11}(t) &= \rho \frac{\partial \dot{A}}{\partial \dot{E}_{11}} + K^2(t)\rho \frac{\partial \dot{A}}{\partial \dot{E}_{22}} - \rho \frac{\partial \dot{A}}{\partial \dot{E}_{33}} + 2K(t)\rho \frac{\partial \dot{A}}{\partial \dot{E}_{12}}, \\
\sigma_{22}(t) &= \rho \frac{\partial \dot{A}}{\partial \dot{E}_{22}} - \rho \frac{\partial \dot{A}}{\partial \dot{E}_{33}}, \\
\sigma_{12}(t) &= K(t)\rho \frac{\partial \dot{A}}{\partial \dot{E}_{22}} + \rho \frac{\partial \dot{A}}{\partial \dot{E}_{12}},
\end{aligned} \tag{8.40}$$

and

$$\sigma_{23}(t) = \sigma_{31}(t) = \sigma_{33}(t) = 0.$$

These last forms for the stress components provide the exact solution for the simple shear deformation of a nonlinear incompressible isotropic viscoelastic solid. When the stored energy functional A is given a particular representation, the derivatives involved in (8.40) can be evaluated, and the solution becomes specialized for that particular material. The only nonzero stress component in the infinitesimal theory solution of this same problem is the shear stress σ_{12}. The fact that in the nonlinear theory the normal stresses σ_{11} and σ_{22} are nonzero should not be too surprising. One simple way of seeing this is through observing a line parallel to the X_2 direction inscribed on the material in the undeformed configuration. This line experiences a stretch in the deformed configuration and, consequently, associated normal stress effects are expected. If the angle of deformation $K(t)$ is infinitesimally small the amount of this stretch becomes a higher order infinitesimal than that of the distortional angle change, and the normal stress effects disappear. Such normal stress effects are commonplace in the solutions of nonlinear elasticity problems, and they must be expected to be commonplace in nonlinear viscoelasticity.

To complete the discussion we obtain the results for a particular type of material. We assume the following form for the stored energy:

$$\rho A(t) = \int_{-\infty}^{t} \int_{-\infty}^{t} \gamma(t - \tau, t - \eta) \frac{\partial E_{KK}(\tau)}{\partial \tau} \frac{\partial E_{LL}(\eta)}{\partial \eta} d\tau \, d\eta$$
$$+ \frac{1}{2} \int_{0}^{t} \int_{0}^{t} \Delta(t - \tau, t - \eta) \frac{\partial E_{KL}(\tau)}{\partial \tau} \frac{\partial E_{KL}(\eta)}{\partial \eta} d\tau \, d\eta \quad (8.41)$$

where $\gamma(\tau, \eta)$ and $\Delta(\tau, \eta)$ are relaxation functions such that

$$\gamma(\tau, \eta) = 0 \text{ for } \tau < 0 \text{ or } \eta < 0, \qquad \Delta(\tau, \eta) = 0 \text{ for } \tau < 0 \text{ or } \eta < 0$$

and

$$\gamma(\tau, \eta) = \gamma(\eta, \tau), \qquad \Delta(\tau, \eta) = \Delta(\eta, \tau).$$

In (8.41), since the material is incompressible, ρ is a constant. The two terms in (8.41) are from the first terms in an expansion of the type (8.28).

Although the form (8.41) is similar to that of (8.30), considered in the reduction to infinitesimal theory, they should not be considered to be interchangeable, since here the strain measure being used is that of an exact nonlinear form, whereas in (8.30) the strains are interpreted in the approximate infinitesimal theory form. The form taken in (8.41), involving histories of rates of deformation rather than the deformation itself, has been used for the reasons of simplicity. The equivalence of these two possible forms was discussed and illustrated in Section 8.1.

By substituting (8.41) into the stress component expressions (8.40), we find them as

$$\sigma_{11}(t) = 2K^2(t) F_1(t) + K(t) F_2(t) + K^2(t) F_3(t),$$
$$\sigma_{22}(t) = F_3(t), \quad (8.42)$$

and

$$\sigma_{12}(t) = 2K(t) F_1(t) + \tfrac{1}{2} F_2(t) + K(t) F_3(t)$$

where

$$F_1(t) = \frac{1}{2} \int_0^t \gamma(t - \tau, 0) \frac{\partial}{\partial \tau} K^2(\tau) \, d\tau,$$
$$F_2(t) = \int_0^t \Delta(t - \tau, 0) \frac{\partial}{\partial \tau} K(\tau) \, d\tau, \quad (8.43)$$

and

$$F_3(t) = \frac{1}{2} \int_0^t \Delta(t - \tau, 0) \frac{\partial}{\partial \tau} K^2(\tau) \, d\tau$$

and where the symmetry relations $\gamma(\tau_1, \tau_2) = \gamma(\tau_2, \tau_1)$ and $\Delta(\tau_1, \tau_2) = \Delta(\tau_2, \tau_1)$ have been used. When the deformation history $K(t)$ is specified, relations (8.42) yield the explicit values of the stresses.

Of particular interest here is the specification of $K(t) = K_0 h(t)$ where K_0 is a constant. This corresponds to the stress relaxation conditions of the infinitesimal theory, and thereby it might be expected to lead to a means of determining mechanical properties. Therefore, we let

$$K(t) = K_0 h(t). \tag{8.44}$$

It must be noted that the deformation field implied by (8.44) violates the continuity assumptions that the strains and their first time derivatives are continuous. We, nevertheless, note that arguments similar to those used in Section 3.2 could be used here to justify this relaxation of the continuity assumptions.

Relations (8.42) and (8.43), subject to (8.44), give

$$\begin{aligned}
\sigma_{11}(t) &= K_0^4 \gamma(t, 0) + K_0^2(1 + K_0^2/2)\, \Delta(t, 0), \\
\sigma_{22}(t) &= (K_0^2/2)\, \Delta(t, 0), \\
\sigma_{12}(t) &= K_0^3 \gamma(t, 0) + (K_0/2)(1 + K_0^2)\, \Delta(t, 0)
\end{aligned} \tag{8.45}$$

and

$$\sigma_{23}(t) = \sigma_{31}(t) = \sigma_{33}(t) = 0.$$

These expressions for the stress components give an explicit display of the relevant relaxation functions $\gamma(t, 0)$ and $\Delta(t, 0)$ and might, therefore, be used in a direct comparison with experimental results to determine these mechanical properties. The situation is expected to be more complicated when more terms are included in the stored energy representation (8.41). It is also expected that it would be necessary to include several more terms in (8.41) for the results to apply over a reasonable range of amplitude parameters K_0. The practical aspects of determining nonlinear theory mechanical properties are considered in Section 9.6.

One simplification, which could be effected in representations as (8.41), would be to assume that the arguments of the relaxation functions occur only in additive form. With regard to the specific representation (8.41) this assumption is given by

$$\gamma(\tau, \eta) = \gamma(\tau + \eta) \qquad \text{and} \qquad \Delta(\tau, \eta) = \Delta(\tau + \eta).$$

A similar situation arose in Chapter 3, where it was shown that the reduced form of the relaxation functions, as above, is a well-motivated and reasonable assumption. In fact, if an assumption of this type is not introduced, relaxation functions, as determined from stress and strain measurements, are not sufficient to characterize the stored energy and the rate of dissipation of energy, as shown by the forms (8.41) and (8.45) in the present example.

Although this nonlinear deformation problem is much more complicated than the comparable infinitesimal theory one, it is nevertheless simple compared with other nonlinear elasticity and viscoelasticity solutions. Since the deformation is homogeneous, the equations of equilibrium are automatically satisfied. Nonhomogeneous deformation problems would necessarily be concerned with these (quasi-static) equilibrium conditions. An example of such a problem is that of the torsion of a viscoelastic right circular cylinder. The exact solution to this problem has been given by Christensen [8.5]. The general analysis there is similar to that used here. The two complications in the torsion problem over the considerations of simple shear deformation are the necessity of solving the equilibrium equations and the notation associated with the use of curvilinear coordinates. In fact, the similarity between these two problems is much more than just superficial. If, in the simple shear solution (8.42), $K(t)$ is replaced by $rK(t)$, this solution compares exactly with the torsion solution given in Ref. [8.5]. This replacement of $K(t)$ by $rK(t)$ is necessitated because in the torsion problem, the amount of distorsion is linearly proportional to the radius coordinate r. Thus, the torsional deformation is locally equivalent to a state of simple shear. This is a well-known result in nonlinear elasticity, and was first pointed out for nonlinear viscoelasticity by Carroll [8.12].

This very brief introduction to nonlinear viscoelastic solids should not be interpreted as being the only possible approach to such problems or as being indicative of the complications one might encounter in solving such nonlinear problems. It is rather intended to show the similarity between some of the fundamental aspects of the linear and nonlinear theories of viscoelasticity and to demonstrate that, as a practical matter, it is possible to solve some nonlinear problems in viscoelasticity. The subject is considered further in Chapter 9.

A different type of thermodynamical derivation has been given by Day [8.13]. Earlier, another theory of nonlinear viscoelasticity was presented by Green and Rivlin [8.14] and Green *et al.* [8.15]. The application of their theory has been considered by many investigators and is considered in Section 9.6. A general method of solving nonlinear viscoelasticity problems by the Green–Rivlin theory has been given by Carroll [8.16], with several such solutions given in Ref. [8.12]. Also, Fosdick [8.17] has considered the means of obtaining solutions of the same types of boundary value problems as are examined in Ref. [8.12]. Wave propagation in solids governed by the nonlinear theory of viscoelasticity has been extensively studied, and is considered in Section 9.2.

8.4. VISCOELASTIC FLUIDS

The derivation of constitutive equations given in Section 8.1 applies to the case of viscoelastic solids. The use of the fixed reference configuration X_K as the preferred configuration was a natural result of the fact that solids have a preferred configuration. The primary results of the preceding sections on viscoelastic solids apply also to viscoelastic fluids, under the proviso that the fixed reference configuration not be interpreted as a preferred configuration. The main implication of this proviso is that for the fluid, material symmetry properties (the isotropy group) cannot be specified relative to a fixed reference configuration.[1] For this reason the preceding results are of only limited usefulness for fluids, and it is more convenient to reformulate a theory directly for viscoelastic fluids. In this theory we will not use measures of deformation relative to a fixed reference configuration but rather to the reference configuration which is the current configuration. It is recalled that the distinction between viscoelastic solids and fluids, on a molecular level, was briefly discussed in Section 1.3.

The derivation to be given here closely follows that given in Section 8.1 for viscoelastic solids, especially with reference to the nature of the fading memory hypothesis. Accordingly, we omit some of the details in the present derivation, when they are essentially the same as in the derivation of Section 8.1.

The stored energy A per unit mass, for a viscoelastic fluid is taken to be a functional of the type

$$A = \underset{s=0}{\overset{\infty}{\varphi}} \left(G_{ij}(t-s), \rho(t)\right) \tag{8.46}$$

where

$$G_{ij}(\tau) = C_{ij}(\tau) - \delta_{ij} \tag{8.47}$$

and

$$C_{ij}(\tau) = \frac{\partial x_k(\tau)}{\partial x_i(t)} \frac{\partial x_k(\tau)}{\partial x_j(t)} . \tag{8.48}$$

$C_{ij}(\tau)$ and $G_{ij}(\tau)$ also have a dependence upon current time t and coordinates $x_i(t)$ which, although not noted explicitly, will always be implied.

We see the deformation (8.47) to be a measure relative to the continuously changing current configuration. The only explicit dependence the stored energy has upon a measure of deformation of the current configuration is through the density $\rho(t)$. If the form (8.46) were allowed to have a more general dependence upon the current configuration deformation than it has through scalar density $\rho(t)$, it would violate the assumption that there is no preferred configuration and correspondingly that all configurations are undistorted.

[1] See Truesdell and Noll [8.1] for the rigorous definitions of simple solids and simple fluids; essentially these may be stated such that a fluid is a material for which the isotropy group is the full unimodular group in all configurations, whereas a solid is a material for which there exists some reference configuration with respect to which the isotropy group is a subgroup of the orthogonal group.

It also follows that viscoelastic fluids of the type considered here are isotropic in all configurations. The stored energy functional (8.46) satisfies the objectivity requirement mentioned in the introduction to this chapter and the isotropy requirement, if the following relation is satisfied:

$$\overset{\infty}{\underset{s=0}{\varphi}}\left(G_{ij}(t-s), \rho(t)\right) = \overset{\infty}{\underset{s=0}{\varphi}}\left(Q_{im}(t)\, Q_{jn}(t)\, G_{mn}(t-s), \rho(t)\right) \tag{8.49}$$

where $Q_{ij}(t)$ is a proper orthogonal tensor. Relation (8.49) simply states the independence of the stored energy from any rigid body rotation of the current configuration, since $G_{ij} = Q_{im}Q_{jn}G_{mn}$ is the tensor transformation law appropriate to the rotation of coordinates.

The time derivative of the stored energy is needed for use in the thermodynamical requirement (8.3). In order to obtain the derivative we proceed in the same manner as was used in the derivation of Section 8.1 for solids. We invoke a fading memory hypothesis relative to the deformation histories $G_{ij}(\tau)$, (8.47), such that there exists the first order Fréchet differential, $\delta\overset{\infty}{\underset{s=0}{\varphi}}(\ )$, of the functional $\overset{\infty}{\underset{s=0}{\varphi}}(\ )$. Thus,

$$\overset{\infty}{\underset{s=0}{\varphi}}\left(G_{ij}(t-s) + \delta G_{ij}(t-s), \rho(t)\right) = \overset{\infty}{\underset{s=0}{\varphi}}\left(G_{ij}(t-s), \rho(t)\right)$$

$$+ \delta\overset{\infty}{\underset{s=0}{\varphi}}\left(G_{ij}(t-s), \rho(t) \mid \delta G_{ij}(t-s)\right) + o\, \| \delta G_{ij}(t-s) \| \tag{8.50}$$

where the first order Fréchet differential $\delta\overset{\infty}{\underset{s=0}{\varphi}}(\ )$ is continuous in all variables and linear in $\delta G_{ij}(t-s)$.

We use (8.50) to get the time derivative of the stored energy A, (8.46), as

$$\dot{A} = \left[\frac{\partial}{\partial\rho(t)} \overset{\infty}{\underset{s=0}{\varphi}}\left(G_{ij}(t-s), \rho(t)\right)\right] \dot{\rho} + \delta\overset{\infty}{\underset{s=0}{\varphi}}\left(G_{ij}(t-s), \rho(t) \,\middle|\, \frac{dG_{ij}(t-s)}{dt}\right) \tag{8.51}$$

where, as in Section 8.1, the dot refers to time differentiation, and, if the time argument is not noted, it is understood to be current time t. In the present context, where the reference configuration is taken to be the current configuration all derivatives with respect to time are understood to be material derivatives, defined through the form

$$\frac{d}{dt}\left[f(t, x_i(t))\right] = \frac{\partial f(t, x_i)}{\partial t} + \frac{\partial f(t, x_i)}{\partial x_i} \frac{dx_i}{dt}.$$

Using (8.47) and (8.48), we can show the term $\dot{G}_{ij}(t-s)$ in (8.51) to be expressed as

$$\frac{d}{dt} G_{ij}(t-s) = -\frac{d}{ds} G_{ij}(t-s) - [C_{mi}(t-s)\, v_{m,j} + C_{mj}(t-s)\, v_{m,i}] \quad (8.52)$$

where $v_i = v_i(t)$ are the velocity components. The velocity gradients in (8.52) are decomposed into the symmetric rate of deformation tensor d_{ij} and the anti-symmetric spin tensor ω_{ij} through

$$v_{i,j} = d_{ij} + \omega_{ij}\,. \quad (8.53)$$

We substitute (8.52) and (8.53) into (8.51) to get

$$\dot{A} = \frac{\partial}{\partial \rho} \underset{s=0}{\overset{\infty}{\varphi}}(\ )\dot{\rho} - \underset{s=0}{\overset{\infty}{\Lambda}}(\ ) - 2\underset{s=0}{\overset{\infty}{\chi_{mj}}}(\ )[d_{mj} + \omega_{mj}] \quad (8.54)$$

where

$$\underset{s=0}{\overset{\infty}{\Lambda}} = \underset{s=0}{\overset{\infty}{\delta\varphi}}\left(G_{ij}(t-s), \rho(t) \,\middle|\, \frac{d}{ds} G_{ij}(t-s)\right) \quad (8.55)$$

and where $\underset{s=0}{\overset{\infty}{\chi_{mj}}}(\ )$ is the tensor valued functional defined by

$$\underset{s=0}{\overset{\infty}{\chi_{mj}}}(\ )\, v_{m,j} = \underset{s=0}{\overset{\infty}{\delta\varphi}}\left(G_{ij}(t-s), \rho(t) \mid C_{mi}(t-s)\, v_{m,j}\right) \quad (8.56)$$

which is valid, since $\underset{s=0}{\overset{\infty}{\delta\varphi}}(\ )$ is linear in the last variable and $v_{m,j} = v_{m,j}(t)$. For $\dot{A}$ to be invariant with respect to rigid body rotation it is necessary for $\underset{s=0}{\overset{\infty}{\chi_{mj}}}(\ )\,\omega_{mj}$ to vanish and it follows, since ω_{mj} is antisymmetric, $\underset{s=0}{\overset{\infty}{\chi_{mj}}}(\ )$ must be symmetric, as

$$\underset{s=0}{\overset{\infty}{\chi_{mj}}}(\ ) = \underset{s=0}{\overset{\infty}{\chi_{jm}}}(\ ) \quad (8.57)$$

where $\underset{s=0}{\overset{\infty}{\chi_{mj}}}(\ )$ is defined by (8.56). Actually, relation (8.57) is satisfied by the objectivity condition (8.49).

One further result is needed. This is the expression of the equation of continuity in terms of the density, through

$$\dot{\rho} + \rho\delta_{ij}\, d_{ij} = 0. \quad (8.58)$$

We substitute $\dot{A}$ from (8.54), subject to (8.57) and (8.58), into the thermodynamic requirement (8.3) obtained from the conservation of energy and the entropy production inequality. This operation gives

$$\sigma_{ij} d_{ij} + \rho^2 \left[\frac{\partial}{\partial \rho} \overset{\infty}{\underset{s=0}{\varphi}} (\ ) \right] \delta_{ij} d_{ij} + \rho \overset{\infty}{\underset{s=0}{\Lambda}} (\ ) + 2\rho \overset{\infty}{\underset{s=0}{\chi_{ij}}} (\ ) \, d_{ij} \geqslant 0. \tag{8.59}$$

A line of reasoning similar to that used in Section 8.1, utilizing the fading memory hypothesis and assuming the instantaneous elasticity effect, shows that, for (8.59) to be satisfied for all processes, the coefficient of d_{ij} must vanish. This results in the stress constitutive equation given by

$$\sigma_{ij} = -\delta_{ij} \rho^2 \frac{\partial}{\partial \rho} \overset{\infty}{\underset{s=0}{\varphi}} (\ ) - 2\rho \overset{\infty}{\underset{s=0}{\chi_{ij}}} (\ ) \tag{8.60}$$

where $\overset{\infty}{\underset{s=0}{\chi_{ij}}} (\ )$ is defined by (8.56) in terms of the first order Fréchet differential of the stored energy functional. With (8.60) the inequality (8.59) becomes simply

$$\rho \overset{\infty}{\underset{s=0}{\Lambda}} (\ ) \geqslant 0 \tag{8.61}$$

which expresses the requirement that the rate of dissipation of energy must be nonnegative.

The stress constitutive relation is written in a more convenient form than that of (8.60) by noting from (8.54), combined with (8.57) and (8.58), that the right-hand side of (8.60) is $\rho(\partial \dot{A}/\partial d_{ij})$. Therefore,

$$\sigma_{ij} = \rho \frac{\partial \dot{A}}{\partial d_{ij}} \tag{8.62}$$

which involves the partial derivatives of $\dot{A}$ with respect to the components of the rate of deformation tensor. Once a representation for the stored energy, A, is specified and $\dot{A}$ is obtained, the simple form (8.62) expresses an operational means of obtaining the stress constitutive relation for a viscoelastic fluid. It is emphasized that the representation for A must satisfy the isotropy and invariance requirement (8.49) and must be consistent with the fading memory hypothesis. A typical representation which does satisfy these requirements is given in the next section, in connection with a specific flow example.

For incompressible materials there must be no local volume change and the stress constitutive relation must include a reactive hydrostatic pressure, as

$$\sigma_{ij} = -p\delta_{ij} + \rho \frac{\partial \dot{A}}{\partial d_{ij}} \tag{8.63}$$

where ρ is now the constant density.

8.5. SIMPLE SHEAR FLOW EXAMPLE

A physical example is now given which illustrates the use of the theory just derived for viscoelastic fluids. The example involves a simple shear deformation, as was considered in Section 8.3, for viscoelastic solids; however, we shall now specifically take a state of unlimited flow in which, relative to a fixed set of coordinates X_K, we specify the state of deformation as being given by

$$\begin{aligned} x_1(\tau) &= X_1 + K(\tau)\, X_2\,, \\ x_2(\tau) &= X_2\,, \\ \text{and} \qquad x_3(\tau) &= X_3 \end{aligned} \tag{8.64}$$

where $K(\tau)$ determines the rate of flow or deformation. For $K(\tau) = \kappa\tau h(\tau)$, these relations specify the start up from rest, of a state of thereafter steady simple shear flow. We seek the transient stress state which must be imposed upon the fluid to give this state of flow. Our constitutive relations use the current configuration as the reference configuration, and we must specify the flow state accordingly. Therefore, we write the forms corresponding to (8.64) for current time t and then eliminate the X_K coordinates between the two sets. This gives

$$\begin{aligned} x_1(\tau) &= x_1(t) + [K(\tau) - K(t)]\, x_2(t), \\ x_2(\tau) &= x_2(t), \\ \text{and} \qquad x_3(\tau) &= x_3(t). \end{aligned} \tag{8.65}$$

The deformation measures from (8.47), which correspond to the flow state (8.65), are given by

$$[G_{ij}(\tau)] = \begin{bmatrix} 0 & [K(\tau) - K(t)] & 0 \\ [K(\tau) - K(t)] & [K(\tau) - K(t)]^2 & 0 \\ 0 & 0 & 0 \end{bmatrix}. \tag{8.66}$$

As an example we consider a particular type of incompressible fluid whereby a specific representation for the stored energy A is utilized. Being guided by the forms used in Section 8.3 for viscoelastic solids, we take

$$\begin{aligned} \rho A = &\int_{-\infty}^{t}\int_{-\infty}^{t} \gamma(t-\tau, t-\eta)\,\frac{\partial G_{ii}(\tau)}{\partial\tau}\,\frac{\partial G_{jj}(\eta)}{\partial\eta}\,d\tau\,d\eta \\ &+ \int_{-\infty}^{t}\int_{-\infty}^{t} \Delta(t-\tau, t-\eta)\,\frac{\partial G_{ij}(\tau)}{\partial\tau}\,\frac{\partial G_{ij}(\eta)}{\partial\eta}\,d\tau\,d\eta \end{aligned} \tag{8.67}$$

where $\gamma(\tau, \eta)$ and $\Delta(\tau, \eta)$ are relaxation functions such that

$$\gamma(\tau, \eta) = 0, \qquad \Delta(\tau, \eta) = 0 \qquad \text{for} \quad \tau < 0 \quad \text{or} \quad \eta < 0 \tag{8.68}$$

and the following symmetry properties are assumed:

$$\gamma(\tau, \eta) = \gamma(\eta, \tau), \qquad \Delta(\tau, \eta) = \Delta(\eta, \tau). \tag{8.69}$$

Also for $\tau > 0$, $\eta > 0$ the relaxation functions are assumed to be positive monotonic decreasing functions of their arguments and they are assumed to be continuous functions with continuous first derivatives. The relaxation functions are further assumed to satisfy the relations

$$\lim_{\tau\to\infty} \frac{\gamma(\tau, \eta)}{e^{-\lambda\tau}} \to 0$$

and (8.70)

$$\lim_{\tau\to\infty} \frac{\Delta(\tau, \eta)}{e^{-\lambda\tau}} \to 0$$

where λ is a sufficiently small, but nonzero, positive constant. Relations (8.70) characterize the fact that the relaxation functions decay to zero exponentially for large arguments. This is entirely consistent with the fading memory assumptions upon which this theory is based. The terms in the representation (8.67) can be considered to be the first terms in an expansion of the type (8.28), derived for solids, if the strain measures $E_{KL}(\tau)$ for the solid are replaced by the deformation measures $G_{ij}(\tau)$ for the fluid.

Using (8.68) and $G_{ij}(\tau)\,|_{\tau=t} = 0$, (8.67) could be integrated by parts to give another form for A, involving histories of $G_{ij}(\tau)$ directly. In this form, it is easy to show that (8.67) satisfies the objectivity and isotropy requirements mentioned earlier, (8.49).

The time derivative of (8.67) is given by

$$\begin{aligned}
\rho\dot{A} &= \left[2\frac{\partial G_{ii}}{\partial\tau}\right]_{\tau=t} \int_{-\infty}^{t} \gamma(t-\tau, 0)\frac{\partial G_{jj}(\tau)}{\partial\tau}\,d\tau \\
&\quad + \int_{-\infty}^{t}\int_{-\infty}^{t} \frac{d}{dt}\left[\gamma(t-\tau, t-\eta)\frac{\partial G_{ii}(\tau)}{\partial\tau}\frac{\partial G_{jj}(\eta)}{\partial\eta}\right] d\tau\, d\eta \\
&\quad + \left[2\frac{\partial G_{ij}}{\partial\tau}\right]_{\tau=t} \int_{-\infty}^{t} \Delta(t-\tau, 0)\frac{\partial G_{ij}(\tau)}{\partial\tau}\,d\tau \\
&\quad + \int_{-\infty}^{t}\int_{-\infty}^{t} \frac{d}{dt}\left[\Delta(t-\tau, t-\eta)\frac{\partial G_{ij}(\tau)}{\partial\tau}\frac{\partial G_{ij}(\eta)}{\partial\eta}\right] d\tau\, d\eta
\end{aligned} \tag{8.71}$$

where ρ is a constant because the fluid is incompressible.

The following kinematical results are needed for use in (8.71):

$$\left.\frac{\partial G_{ij}(\tau)}{\partial \tau}\right|_{\tau=t} = 2d_{ij}$$

and (8.72)

$$\frac{d}{dt}\left[\frac{\partial G_{ij}(\tau)}{\partial \tau}\right] = -\frac{\partial C_{mi}(\tau)}{\partial \tau} v_{m,j}(t) - \frac{\partial C_{mj}(\tau)}{\partial \tau} v_{m,i}(t).$$

We substitute (8.72) into (8.71) which gives $\rho \dot{A}$ as

$$\begin{aligned}
\rho \dot{A} = & \int_{-\infty}^{t} \int_{-\infty}^{t} \left[\frac{d}{dt} \gamma(t-\tau, t-\eta)\right] \frac{\partial G_{ii}(\tau)}{\partial \tau} \frac{\partial G_{jj}(\eta)}{\partial \eta}\, d\tau\, d\eta \\
& - 4\, d_{ij} \int_{-\infty}^{t} \int_{-\infty}^{t} \gamma(t-\tau, t-\eta) \frac{\partial G_{ij}(\tau)}{\partial \tau} \frac{\partial G_{kk}(\eta)}{\partial \eta}\, d\tau\, d\eta \\
& + 4\, \delta_{ij}\, d_{ij} \int_{-\infty}^{t} \gamma(t-\tau, 0) \frac{\partial G_{kk}(\tau)}{\partial \tau}\, d\tau \\
& + \int_{-\infty}^{t} \int_{-\infty}^{t} \left[\frac{d}{dt} \Delta(t-\tau, t-\eta)\right] \frac{\partial G_{ij}(\tau)}{\partial \tau} \frac{\partial G_{ij}(\eta)}{\partial \eta}\, d\tau\, d\eta \\
& - 4\, d_{ij} \int_{-\infty}^{t} \int_{-\infty}^{t} \Delta(t-\tau, t-\eta) \frac{\partial G_{ik}(\tau)}{\partial \tau} \frac{\partial G_{kj}(\eta)}{\partial \eta}\, d\tau\, d\eta \\
& + 4\, d_{ij} \int_{-\infty}^{t} \Delta(t-\tau, 0) \frac{\partial G_{ij}(\tau)}{\partial \tau}\, d\tau
\end{aligned} \tag{8.73}$$

where the decomposition of $v_{i,j}$, according to (8.53), has been used.
We substitute (8.73) into the stress constitutive equation (8.63) to get

$$\begin{aligned}
\sigma_{ij} = & -p\delta_{ij} - 4 \int_{-\infty}^{t} \int_{-\infty}^{t} \gamma(t-\tau, t-\eta) \frac{\partial G_{ij}(\tau)}{\partial \tau} \frac{\partial G_{kk}(\eta)}{\partial \eta}\, d\tau\, d\eta \\
& - 4 \int_{-\infty}^{t} \int_{-\infty}^{t} \Delta(t-\tau, t-\eta) \frac{\partial G_{ik}(\tau)}{\partial \tau} \frac{\partial G_{kj}(\eta)}{\partial \eta}\, d\tau\, d\eta \\
& + 4 \int_{-\infty}^{t} \Delta(t-\tau, 0) \frac{\partial G_{ij}(\tau)}{\partial \tau}\, d\tau
\end{aligned} \tag{8.74}$$

where an integral term proportional to δ_{ij} has been absorbed into p. With an integration by parts of all integrals in (8.74) using (8.70) and $G_{ij}(t) = 0$, the resulting stress constitutive relation corresponds to that given by Coleman and

Noll [8.11] for the second order theory of incompressible viscoelastic fluids, but with one less independent relaxation function owing to the thermodynamical restrictions.

The stress forms given by (8.74) are now specialized to the deformation field characteristics given by (8.65), with the result that

$$\sigma_{11} = -p - 4\int_{-\infty}^{t}\int_{-\infty}^{t} \Delta(t-\tau, t-\eta)\frac{\partial K(\tau)}{\partial \tau}\frac{\partial K(\eta)}{\partial \eta}\,d\tau\,d\eta,$$

$$\begin{aligned}\sigma_{22} = -p &- 4\int_{-\infty}^{t}\int_{-\infty}^{t} \gamma(t-\tau, t-\eta)\frac{\partial}{\partial \tau}[K(\tau)-K(t)]^2\frac{\partial}{\partial \eta}[K(\eta)-K(t)]^2\,d\tau\,d\eta\\ &+ 4\int_{-\infty}^{t} \Delta(t-\tau, 0)\frac{\partial}{\partial \tau}[K(\tau)-K(t)]^2\,d\tau\\ &- 4\int_{-\infty}^{t}\int_{-\infty}^{t} \Delta(t-\tau, t-\eta)\frac{\partial K(\tau)}{\partial \tau}\frac{\partial K(\eta)}{\partial \eta}\,d\tau\,d\eta\\ &- 4\int_{-\infty}^{t}\int_{-\infty}^{t} \Delta(t-\tau, t-\eta)\frac{\partial}{\partial \tau}[K(\tau)-K(t)]^2\frac{\partial}{\partial \eta}[K(\eta)-K(t)]^2\,d\tau\,d\eta,\end{aligned}$$

$$\sigma_{33} = -p,$$

$$\begin{aligned}\sigma_{12} = &-4\int_{-\infty}^{t}\int_{-\infty}^{t} \gamma(t-\tau, t-\eta)\frac{\partial K(\tau)}{d\tau}\frac{\partial}{\partial \eta}[K(\eta)-K(t)]^2\,d\tau\,d\eta\\ &+ 4\int_{-\infty}^{t} \Delta(t-\tau, 0)\frac{\partial K(\tau)}{\partial \tau}\,d\tau\\ &- 4\int_{-\infty}^{t}\int_{-\infty}^{t} \Delta(t-\tau, t-\eta)\frac{\partial K(\tau)}{d\tau}\frac{\partial}{d\eta}[K(\eta)-K(t)]^2\,d\tau\,d\eta,\end{aligned}$$

and

$$\sigma_{23} = \sigma_{31} = 0. \tag{8.75}$$

This represents the solution appropriate to a state of simple shear flow.

The flow is further specialized with the parameter $K(\tau)$ taken as

$$K(\tau) = \kappa\tau h(\tau) \tag{8.76}$$

so that the flow field (8.64) is that of the start up, from rest, of thereafter steady simple shear flow. This flow state and the one to be considered next violates the continuity assumptions of the present derivation. The justification for allowing this follows the arguments given in Section 3.2 for relaxing continuity assumptions. We use $K(\tau)$ from (8.76) in (8.75) which gives the stresses as

$$\sigma_{11} = -p - 4\kappa^2 \int_0^t \int_0^t \Delta(u, v)\, du\, dv,$$

$$\begin{aligned}\sigma_{22} = -p &- 16\kappa^4 \int_0^t \int_0^t \gamma(u, v)uv\, du\, dv \\ &- 8\kappa^2 \int_0^t \Delta(u, 0)u\, du - 4\kappa^2 \int_0^t \int_0^t \Delta(u, v)\, du\, dv \\ &- 16\kappa^4 \int_0^t \int_0^t \Delta(u, v)uv\, du\, dv,\end{aligned}$$

$$\sigma_{33} = -p,$$

$$\begin{aligned}\sigma_{12} = 8\kappa^3 &\int_0^t \int_0^t \gamma(u, v)v\, du\, dv + 4\kappa \int_0^t \Delta(u, 0)\, du \\ &+ 8\kappa^3 \int_0^t \int_0^t \Delta(u, v)v\, du\, dv,\end{aligned}$$

and

$$\sigma_{23} = \sigma_{31} = 0. \tag{8.77}$$

This represents the transient (quasi-static) solution for the startup of steady shear flow. The fact that a steady state is eventually achieved is assured by the decaying nature of the relaxation functions with (8.70).

For a sufficiently slow flow, $\kappa \ll 1$, the solution (8.77) shows that the normal stresses are of higher order than the shear stress. This effect is comparable to that which exists for the nonlinear deformation of a viscoelastic solid, as was shown in Section 8.3. More specially, for $\kappa \ll 1$, the normal stresses can be neglected and the shear stress is given by

$$\sigma_{12} = 4\kappa \int_0^t \Delta(u, 0)\, du. \tag{8.78}$$

In this condition the fluid behaves as a Newtonian viscous fluid, and, under steady flow conditions, after the transient start up effects have disappeared, the effective Newtonian viscosity is given by

$$4 \int_0^\infty \Delta(u, 0)\, du.$$

This situation of a limiting Newtonian type flow is comparable to that which is discussed in Section 1.3 in the much more limited context of the linear theory.

In this problem of simple shear flow, the equations of equilibrium are identically satisfied since the deformation is homogeneous and inertia terms are neglected. Even this rather simple problem demonstrates some of the effects which are inherent in viscoelastic fluid behavior and which cannot be predicted by elastic solid or viscous fluid theory. Moreover, the simple shear flow problem is of more fundamental significance than is obvious. There is a fairly wide class of nonhomogeneous flows, known as viscometric flows, for which the properties of the fluid need only be specified in terms of the corresponding flow properties in a state of simple shear flow. A complete account of such flows is given by Coleman *et al.* [8.18]. With regard to these types of flows, experimental evidence is cited in Ref. [8.18] which implies that in simple shear flow the normal stress component, normal to the direction of flow and in the plane of the flow, must be less than the normal stress component in the direction of flow. In the present notation, this is stated by $\sigma_{22} < \sigma_{11}$. From relations (8.77), we see this inequality to be true for the stresses which result from the representation, (8.67) for the stored energy.

In addition to the flow state just considered, another deformation history of fundamental significance is that of stress relaxation in simple shear deformation. We specify this by taking $K(\tau)$ in (8.64) as

$$K(\tau) = K_1 h(\tau). \tag{8.79}$$

By incorporating this into the stress component expressions (8.75), we get the stresses necessary to maintain the step function in simple shear deformation as

$$\begin{aligned}
\sigma_{11} &= -p - 4K_1{}^2\Delta(t, t),\\
\sigma_{22} &= -p - 4K_1{}^4\gamma(t, t) - 4K_1{}^2\Delta(t, t) - 4K_1{}^2\Delta(t, 0) - 4K_1{}^4\Delta(t, t),\\
\sigma_{33} &= -p,\\
\sigma_{12} &= 4K_1{}^3\gamma(t, t) + 4K_1\Delta(t, 0) + 4K_1{}^3\Delta(t, t),
\end{aligned}$$

and

$$\sigma_{23} = \sigma_{31} = 0. \tag{8.80}$$

In obtaining these forms from (8.75), it is more simple to use the forms obtained by converting, through integration by parts, the expressions in (8.75) to a form involving the history of the deformation than to use the forms involving the history of the rates of deformation.

Considerable simplification occurs in the solutions for the stresses, (8.77) and (8.80), if the relaxation functions are assumed to have the reduced forms

$$\gamma(\tau, \eta) = \gamma(\tau + \eta) \qquad \text{and} \qquad \Delta(\tau, \eta) = \Delta(\tau + \eta).$$

Indeed, if a simplification of this or a similar type is not assumed, then mechanical properties determined from stress relaxation tests, through forms as in (8.80), are not sufficient to characterize the general mechanical properties. That is, as shown in (8.80), the stress relaxation test can be used only to determine $\gamma(t, 0)$ and $\gamma(t, t)$, rather than the more general form $\gamma(\tau, \eta)$, unless it is assumed that $\gamma(\tau, \eta) = \gamma(\tau + \eta)$. Thus, in a manner similar to that discussed in Section 8.3 for viscoelastic solids, it seems reasonable to assume that in a representation for the stored energy of viscoelastic fluids the relaxation functions have the reduced form with additive arguments.

The solutions given here for the simple representation (8.67) of the stored energy cannot be expected to realistically model a viscoelastic fluid's behavior over a wide range of rates of deformation. More realistic types of viscoelastic fluid constitutive relations are considered in Section 9.6.

The viscoelastic fluid theory given here is by no means the only generalization of Newtonian viscous fluid theory. Earlier related but less general theories of non-Newtonian fluid behavior are those of the Reiner–Rivlin fluid, the Rivlin–Erickson fluid, and the Coleman–Noll second order fluid. These theories are briefly summarized and examined by Markovitz [8.19]. There has been a considerable amount of work done on another non-Newtonian fluid theory, which is completely outside the scope of the fluid theory considered here. This theory applies to oriented fluids, sometimes known as liquid crystals. These fluids show some anisotropic effects which are in direct contradiction to the isotropy characteristics of the fluid theory given here. In the corresponding theoretical work on oriented fluids, the stress is not determined solely by the deformation gradient, as was used here, and more general kinematical quantities are involved. A survey of such work has been given by Ericksen [8.20].

Finally, note should be taken of a special approach to viscoelastic fluid characterization whereby forms are directly postulated for the viscoelastic stress constitutive relations. Such forms involve parameters which are adjusted to fit experimental data. This procedure is further considered in Section 9.6 in connection with the determination of mechanical properties.

PROBLEMS

8.1. In Section 8.5 the second order fluid under steady simple shearing flow as shown to have a zero shear rate viscosity function η given by

$$\eta = 4 \int_0^\infty \Delta(u, 0)\, du.$$

The corresponding normal stress differences are defined as $N_1 = (\sigma_{11} - \sigma_{22})/\kappa^2$ and $N_2 = (\sigma_{22} - \sigma_{33})/\kappa^2$. From (8.77) it follows that at zero shear rate

$$N_1 = 8 \int_0^\infty \Delta(u, 0)\, u\, du.$$

Using the definitions of the complex shear modulus given in (1.58) and (1.59), prove that for this η and N_1 we have

$$\eta = \lim_{\omega \to 0} \frac{\mu''(\omega)}{\omega}, \qquad N_1 = \lim_{\omega \to 0} \frac{2\mu'(\omega)}{\omega^2}.$$

8.2. The viscosity function in steady simple shearing flow is defined as the ratio of shear stress to shear rate,

$$\eta(\kappa) = \sigma_{12}/\kappa .$$

Most polymers exhibit a decreasing value of $\eta(\kappa)$ as the shear rate κ increases. This is known as shear thinning. Show that the second order theory of (8.77) is shear thickening. Accordingly, devise a rule providing an effective upper limit for the shear rate in the theory.

8.3. Prove the kinematic relations (8.52) and (8.72) needed in the thermodynamical derivation for viscoelastic fluids.

8.4. Derive the shear flow and shear relaxation results, (8.77) and (8.80), respectively, from the basic relation (8.75).

8.5. In the derivation of viscoelastic solids, include a linear functional term in the energy expression, (8.30). Show that this term contributes only a history independent term to the stress. What is the significance of this term? Show that it is not permissible to use a corresponding linear functional term in the energy form (8.67) for a viscoelastic fluid.

REFERENCES

8.1. Truesdell, C., and W. Noll, *in Handbuch der Physik* (S. Flügge, ed.), Vol. 3, No. 3. Springer, Berlin, 1965.

8.2. Coleman, B. D., "Thermodynamics of Materials with Memory," *Arch. Ration. Mech. Anal.* **17**, 1 (1964).

8.3. Coleman, B. D., "On Thermodynamics, Strain Impulses, and Viscoelasticity," *Arch. Ration. Mech. Anal.* **17**, 230 (1964).

8.4. Noll, W., "A Mathematical Theory of The Mechanical Behavior of Continuous Media," *Arch. Ration. Mech. Anal.* **2**, 197 (1958).

8.5. Christensen, R. M., "On Obtaining Solutions in Nonlinear Viscoelasticity," *J. Appl. Mech.* **35**, 129 (1968).

8.6. Laws, N., "On the Thermodynamics of Certain Materials with Memory," *Int. J. Eng. Sci.* **5**, 427 (1967).

8.7. Coleman, B. D., and V. J. Mizel, "A General Theory of Dissipation in Materials with Memory," *Arch. Ration. Mech. and Anal.* **27**, 255 (1967).

8.8. Wineman, A. S., and A. C. Pipkin, "Material Symmetry Restrictions on Constitutive Equations," *Arch. Ration. Mech. Anal.* **17**, 184 (1964).

8.9. Spencer, A. J. M., and R. S. Rivlin, "Further Results in the Theory of Matrix Polynomials," *Arch. Ration. Mech. Anal.* **4**, 214 (1960).

8.10. Chacon, R. V. S., and R. S. Rivlin, "Representation Theorems in the Mechanics of Materials with Memory," *Z. Angew. Math. Phys.* **15**, 444 (1964).

8.11. Coleman, B. D., and W. Noll, "Foundations of Linear Viscoelasticity," *Rev. Mod. Phys.* **33**, 239 (1961).

8.12. Carroll, M. M., "Finite Deformations of Incompressible Simple Solids I. Isotropic Solids," *Quart. J. Mech. Appl. Math.* **21**, 147 (1968).

8.13. Day, W. A., *The Thermodynamics of Simple Materials with Fading Memory*. Springer–Verlag, Berlin and New York, 1972.

8.14 Green, A. E., and R. S. Rivlin, "The Mechanics of Non Linear Materials with Memory; Part I," *Arch. Ration. Mech. and Anal.* **1**, 1 (1957).

8.15. Green, A. E., R. S. Rivlin, and A. J. M. Spencer, "The Mechanics of Non-Linear Materials with Memory; Part II," *Arch. Ration. Mech. Anal.* **3**, 82 (1959).

8.16. Carroll, M. M., "Controllable Deformations of Incompressible Simple Materials," *Int. J. Eng. Sci.* **5**, 515 (1967).

8.17. Fosdick, R. L., "Dynamically Possible Motions of Incompressible Isotropic Simple Materials," *Arch. Ration. Mech. Anal.* **29**, 272 (1968).

8.18. Coleman, B. D., H. Markovitz, and W. Noll, *in Viscometric Flows of Non-Newtonian Fluids* (C. Truesdell, ed.). Springer, New York, 1966.

8.19. Markovitz, H., "Nonlinear Steady Flow Behavior," *in Rheology* (F. R. Eirich, ed.), Vol. 4, p. 347. Academic Press, New York, 1967.

8.20. Ericksen, J. L., "Continuum Theory of Liquid Crystals," *Appl. Mech. Rev.* **20**, 1029 (1967).

Chapter IX Nonlinear Mechanical Behavior

In Chapter 8 we introduced nonlinear viscoelasticity theory through the use of a thermodynamical derivation. The method served to clarify the underlying relationships among energy, dissipation, stress, strain, and memory effects in the general nonlinear context. The thermodynamical method did not, however, serve to produce concise yet general forms of the stress strain relations that could be employed in solving practical problems. In this final chapter, we focus more explicitly on nonlinear effects in the practical realm of material behavior.

It is tempting to expect that some time in the future simple, universal forms will be discovered for the nonlinear viscoelastic stress strain constitutive relations, comparable to those that exist in the linear theory. Unfortunately, a realistic appraisal of the field indicates that the goal of completely general constitutive forms is unlikely to be realized, either for solids or for fluids. The reasons for this outlook are not simply stated in a few words, but some of the complications will be elucidated in Section 9.6. At this point it suffices to say that the reality of the situation is that the best approach to nonlinear viscoelasticity is to seek limited and special forms of constitutive relations that have well-delineated applicability to particular materials or classes of materials in particular problems or classes of problems. That is the approach we follow here. An approach of this type could go innumerable directions. Accordingly, we do not here attempt a comprehensive treatment of the many different nonlinear viscoelastic stress strain forms. Rather, we use particular examples to illustrate the practicality of approaching nonlinear behavior in the manner described.

We proceed with separate examples applicable to solid and fluid type mechanical behavior. In Section 9.1 we derive what is probably the simplest realistic nonlinear theory of viscoelastic solids. The emphasis is given to the physical requirements necessary to justify the theory and to circumscribe its range of applicability. Specific results are found for the well-defined physical model. In Section 9.2 we turn in a different direction and show that there are some limited but useful nonlinear results that can be obtained without restriction to a particular viscoelastic model. The subject is the propagation of (nonlinear) acceleration waves. Switching to fluids in Section 9.3, we also show that specific practical results can be obtained for the flow of non-Newtonian viscoelastic fluids, without restriction to a particular viscoelastic model. These applications are for viscometric flows. In Sections 9.4 and 9.5, on nonviscometric flows in general and lubrication type flow in particular, we show that meaningful results can be obtained for some very simple fluid models of constitutive behavior. Finally, in Section 9.6 we give an appraisal of the inherent complications in nonlinear behavior and we mention some of the constitutive models in use for viscoelastic solids and fluids.

9.1. A NONLINEAR THEORY OF ELASTOMERIC SOLIDS

In the present work we seek to develop a consistent theory of nonlinear viscoelasticity for application to elastomers. We are not seeking a general theory for which we expect to have wide applicability for all materials. Rather, our approach is to find a simple, physically meaningful nonlinear theory of viscoelastic solids. The range of applicability will be delineated through careful analysis of the underlying physical assumptions of the theory. The work follows that of Christensen [9.1].

Any theory of nonlinear viscoelasticity must admit the limiting case of elastic behavior and therefore must include elasticity theory results. The general continuum theory of nonlinear elasticity, as developed by Cauchy, has been used by Rivlin [9.2] to obtain general solutions to boundary value problems without specifying the strain energy function. Within the continuum context there are no specific derived forms for the stress strain relations. Often the stress strain relations are taken as those resulting from a polynomial expansion of a strain energy function in terms of the invariants of the deformation measure. However, Rivlin and Sawyers [9.3] show that expressing the strain–energy function as an algebraic function is not useful; it is more meaningful simply to use the experimentally derived dependence of the strain energy upon the invariants. The main observation to be made at this point is that from a continuum mechanics point of view there is no simple universal form for the stress strain relations of nonlinear elasticity.

From a molecular point of view, in contrast to the situation just described for continuum mechanics, there is a very specific result. In particular, the kinetic theory of rubber elasticity predicts a specific form for the strain energy function, namely that the strain energy depends linearly on the first invariant of the deformation measure. To be sure, this simple result does not model all experimental data. However, over a limited strain range for many elastomers it does a reasonable job. The overwhelming significance of the kinetic theory of rubber elasticity is that the stress strain relation is derived from a very simple set of reasonable physical assumptions (see Treloar [9.4] for the specifics of the derivation). In general terms, the use of Gaussian statistics leads to the "entropy spring" formulation for the single molecule, which provides the general result when summed over all molecules. The kinetic theory of rubber elasticity is taken here as the guide for the development of a viscoelastic theory at the same level of applicability.

Our purpose is to derive the viscoelastic counterpart of the kinetic theory of rubber elasticity. This problem is approached here not by means molecular theory but rather by seeking the answer in the continuum context. Obtaining the viscoelastic generalization of the kinetic theory of rubber elasticity is not simply a matter of reinterpreting functions as functionals. As we shall see, the problem is far more subtle than that. In fact, we must be very careful in implementing the statement that the viscoelastic theory should reduce to the kinetic elasticity theory. Under what conditions would we expect the viscoelastic theory to model the given elastic behavior? Precise statements of these requirements are given.

Several restrictions on the general theory, in addition to incompressibility and isotropy, will now be delineated. The material will be taken as behaving under isothermal conditions, at a temperature far above that of glass transition. This requirement means that, in the absence of history (time) dependence, the material would respond in the rubbery range.

When time and rate dependence are included, it is not sufficient to say the material is in the rubbery range of behavior. In fact, if the excitation is rapid enough, the material will respond in a manner dictated by a glassy-type elasticity effect, even though the temperature is far above the glass transition temperature. To be precise, when including time and rate dependence we must say that for a sufficiently slow process the material will respond in a rubbery manner. The complete requirements are twofold:

(i) Under a sufficiently slow process, the viscoelastic theory must reduce to the kinetic theory of rubber elasticity, which is given by

$$\sigma_{ij} = x_{i,K}x_{j,L}\frac{\partial W}{\partial E_{KL}}$$

where $W = cI$, with I being the first invariant of the strain measure given by

$$2E_{KL} = C_{KL} - \delta_{KL}$$

where C_{KL} is the right Cauchy–Green tensor.

Although it is intuitively obvious what is meant by a "slow process," this term will be given a mathematical characterization in the developments to follow.

(ii) The viscoelastic theory will be applicable to stress-imposed rather than strain-imposed problems.

A stress-imposed problem is one in which the primary variable of excitation is that of stress (or load) rather than strain (or displacement). Attention will be restricted to these stress-imposed problems rather than strain-imposed problems. In a strain-imposed problem, boundary displacements are the primary input or excitation variables, whereas in stress-imposed problems, boundary stresses are the primary variables. Consider first strain-imposed problems. In a practical context, if an elastomer is required to function in a strain-imposed context, generally speaking the material is required only to "fill the gap," with no concern for its stiffness or response. Such materials are often referred to as fillers and potting compounds. For stress-imposed problems, the polymer is typically required to bear the load, and it deforms by whatever amount is necessary to sustain the load without failure. Examples of stress-imposed problems are the response of rubber tires and the behavior of bond lines in adhesive joints. Clearly the stress-imposed problems pose a greater and much more important challenge than do strain-imposed problems. We also must assume certain smoothness conditions. Specifically, in comparing stress-imposed with strain-imposed histories, it is necessary to assume the same degree of smoothness in each. Thus we compare step functions in strain, for relaxation tests, with step functions in stress, for creep tests. In the present context, it would not be proper to compare, say, a step function in stress with a much smoother history in strain. For the most part, in the following, when speaking of stress and strain-imposed problems, our specific interest will be in the respective creep and relaxation tests.

The starting point for the development of the stress strain relation is the general form of the constitutive relation, given by

$$\sigma_{ij} = -p\,\delta_{ij} + x_{i,K}x_{j,L}\overset{\infty}{\underset{s=0}{\phi_{KL}}}\left(E_{KL}(t-s),\,E_{KL}(t)\right) \tag{9.1}$$

where σ_{ij} is the Cauchy stress, p is the pressure, $x_{i,K}$ denotes the displacement gradients, $\overset{\infty}{\underset{s=0}{\phi_{KL}}}(\ )$ is a tensor-valued functional with a dependence not only on the strain history but also on the current strain, which is defined by

$$2E_{KL} = x_{i,K}x_{i,L} - \delta_{KL}\,. \tag{9.2}$$

The form in (9.1) is similar to that in (8.14), but no assumption is made here of the existence of a strain energy functional.

The next step is to provide a special representation for the functional shown in (9.1). To this end, the Green–Rivlin expansion method used in Section 8.1 will be followed and $\overset{\infty}{\underset{s=0}{\phi_{KL}}}$ will be taken as a polynomial expansion in linear functionals of the type

$$\int_0^t g(t-\tau)\frac{\partial E_{KL}(\tau)}{\partial\tau}\,d\tau.$$

Performing the expansion beginning with the zero order term gives

$$\sigma_{ij} = -p\,\delta_{ij} + x_{i,K}x_{j,L}\left[g_0\,\delta_{KL} + \int_0^t g_1(t-\tau)\frac{\partial E_{KL}(\tau)}{\partial\tau}\,d\tau \right.$$
$$\left. + \delta_{KL}\int_0^t g_2(t-\tau)\frac{\partial E_{JJ}(\tau)}{\partial\tau}\,d\tau + \cdots\right]. \tag{9.3}$$

Pipkin [9.5] has shown that for an incompressible material the requirement of incompressibility gives

$$\operatorname{tr}\mathbf{E} = \operatorname{tr}(\mathbf{E}^2) - [\operatorname{tr}(\mathbf{E})]^2 + \cdots \tag{9.4}$$

where only second and higher order terms are involved on the right-hand side of (9.4). By use of (9.4), the E_{JJ} term in (9.3) can be absorbed into the quadratic integral, and (9.3) can be written as

$$\sigma_{ij} = -p\,\delta_{ij} + x_{i,K}x_{j,L}\left[g_0\,\delta_{KL} + \int_0^t g_1(t-\tau)\frac{\partial E_{KL}(\tau)}{\partial\tau}\,d\tau + \cdots\right]. \tag{9.5}$$

The next term not explicitly shown in (9.5) is that of the quadratic functional. Note that the term involving g_0 in (9.5) corresponds to the kinetic theory of rubber elasticity.

To implement requirement (i) we must analytically specify what is meant by a slow process. To do this, we consider accelerated and retarded processes, as discussed in Section 1.7. For a given deformation history $x_{i,K}(t)$ and $E_{KL}(t)$, accelerated and retarded strain histories are specified by $E_{KL}(\alpha t)$, where $\alpha > 1$ for the accelerated and $\alpha < 1$ for the retarded strain history.

The effect of accelerating the history is the same as compressing the time scale; thus we seek to determine the stress at reduced values of its argument as specified by

$$\sigma_{ij}\left(\frac{t}{\alpha}\right) = -p\,\delta_{ij} + x_{i,K}(t)\,x_{j,L}(t)\left[g_0\,\delta_{KL}\right.$$
$$\left. + \int_0^{t/\alpha} g_1\left(\frac{t}{\alpha}-\tau\right)\frac{\partial E_{KL}(\alpha\tau)}{\partial\tau}\,d\tau + \cdots\right]. \tag{9.6}$$

With a change of variable, (9.6) can be written as

$$\sigma_{ij}\left(\frac{t}{\alpha}\right) = -p\,\delta_{ij} + x_{i,K}(t)\,x_{j,L}(t)\left[g_0\,\delta_{KL} + \int_0^t g_1\left(\frac{t-\eta}{\alpha}\right)\frac{\partial E_{KL}(\eta)}{\partial\eta}\,d\eta + \cdots\right]. \tag{9.7}$$

For infinitely accelerated and retarded histories, it is clear from (9.7) that

$$\sigma_{ij}\left(\frac{t}{\alpha}\right)\bigg|_{\text{accelerated}} = -p\,\delta_{ij} + x_{i,K}(t)\,x_{j,L}(t)[g_0\,\delta_{KL} + g_1(0)\,E_{KL}(t) + \cdots]_{\alpha\to\infty} \tag{9.8}$$

and

$$\sigma_{ij}\left(\frac{t}{\alpha}\right)\bigg|_{\text{retarded}} = -p\,\delta_{ij} + x_{i,K}(t)\,x_{j,L}(t)[g_0\,\delta_{KL} + g_1(\infty)\,E_{KL}(t) + \cdots]_{\alpha\to 0}\,. \tag{9.9}$$

Relations (9.8) and (9.9) express the result of Section 1.7 that for very rapid processes only the initial glassy value of the relaxation function is involved, as with $g_1(0)$ in (9.8), whereas for very slow processes only the long-time asymptote of the relaxation function is involved, as with $g_1(\infty)$ in (9.9).

Now consider requirement (i). For a very slow process the viscoelastic theory is required to give the same result as the kinetic theory of elasticity. Since the kinetic theory is in fact embodied in relations (9.7)–(9.9) by the presence of the term g_0, it follows from (9.9) that $g_1(\infty) \equiv 0$ and that the long-time asymptotes of the relaxation functions in all the higher order terms in (9.7) must vanish. Thus requirement (i) gives

$$g_n(\infty) = 0, \qquad n = 1, 2, \ldots \tag{9.10}$$

where $g_n(t)$ denotes the relaxation functions involved in (9.7). Relation (9.10) is one of basic results sought. Even though the material is a solid, the relaxation functions decay to zero, as in a fluid. The preferred configuration of the solid is remembered by the rubber elasticity term g_0 in (9.7).

Before explicitly considering requirement (ii), we must obtain some background results. We wish to examine the behavior of (9.7) for accelerated processes. To this end, we take the derivative of (9.7) to obtain

$$\frac{\partial \sigma_{ij}}{\partial \alpha} = x_{i,K}(t)\,x_{j,L}(t)\left[\int_0^t \frac{\partial g_1[(t-\eta)/\alpha]}{\partial \alpha}\,\frac{\partial E_{KL}(\eta)}{\partial \eta}\,d\eta + \cdots\right]. \tag{9.11}$$

For a positive, monotone decreasing relaxation function $g_1(t)$ we have

$$\frac{\partial g_1[(t-\eta)/\alpha]}{\partial \alpha} > 0. \tag{9.12}$$

Rewriting (9.11) we obtain

$$\frac{\partial \sigma_{ij}}{\partial \alpha} = x_{i,K}(t)\,x_{j,L}(t)\left[\int_0^t H(t-\eta,\alpha)\,\frac{\partial E_{KL}(\eta)}{\partial \eta}\,d\eta + \cdots\right] \tag{9.13}$$

where

$$H(t-\eta,\alpha) = \frac{\partial g_1[(t-\eta)/\alpha]}{\partial \alpha}\,.$$

From (9.12) and (9.13) we see that as the deformation history is accelerated the stress amplitudes increase. That is, the magnitude of the stress response is increased as the deformation history is accelerated. In arriving at this general conclusion we assume that all the relaxation functions exhibit the type of behavior shown in (9.12).

Up to this point we have been working with a specified deformation history. Our interest in applying requirement (ii) concerns imposed stresses, not strain or deformation. It is not possible to invert (9.7) explicitly to express strain as functionals of stress in terms of those in (9.7). However, for present purposes it is sufficient to consider a particular component of stress at a particular instant of time. Thus, for any one component of σ_{ij} in (9.7), we write

$$\sigma = f(\alpha, E) \tag{9.14}$$

where E is any convenient scalar measure of the amplitude of the given strain history, as specified by

$$E_{KL}(t) = E\,\hat{E}_{KL}(t). \tag{9.15}$$

With E taken as a positive, real number, it will later be used as a strain amplitude scaling factor; we restrict the derivatives of σ in (9.14) as

$$\frac{\partial \mid \sigma \mid}{\partial E} > 0 \tag{9.16}$$

and

$$\frac{\partial \mid \sigma \mid}{\partial \alpha} > 0. \tag{9.17}$$

Requirement (9.16) simply expresses the observation that increasing strain amplitude increases stress magnitude, while (9.17) follows from Eq. (9.12) and (9.13).

Under the conditions specified by requirement (ii) stress is taken to be the control variable, and we write

$$\frac{d\sigma}{d\alpha} = 0. \tag{9.18}$$

This result simply states that as the deformation history, specified by (9.15), is accelerated the stress is required to remain unchanged. Using (9.14), we find (9.18) takes the form

$$\frac{\partial \mid f(\alpha, E) \mid}{\partial \alpha} + \frac{\partial \mid f(\alpha, E) \mid}{\partial E} \frac{dE}{d\alpha} = 0 \tag{9.19}$$

or

$$\frac{dE}{d\alpha} = -\frac{\partial \mid f(\alpha, E) \mid}{\partial \alpha} \Big/ \frac{\partial \mid f(\alpha, E) \mid}{\partial E} = -\frac{\partial \mid \sigma \mid}{\partial \alpha} \Big/ \frac{\partial \mid \sigma \mid}{\partial E}. \tag{9.20}$$

Relations (9.16) and (9.17) then allow (9.20) to be written as

$$\frac{dE}{d\alpha} < 0. \tag{9.21}$$

Relation (9.21) is a basic result we have been seeking; namely, if stress amplitudes are controlled to remain unchanged, then accelerating a deformation history $\hat{E}_{KL}(t)$ in (9.15) must diminish the amplitude of the deformation history. This result is useful in truncating expansion (9.5), which is considered next.

To collect our observations, for slow processes the integral terms in (9.5) are negligible compared with the g_0 term because of requirement (9.10). The g_0 term is just the contribution of the kinetic theory of rubber elasticity. Therefore the viscoelastic theory must be subject to the same restrictions as the elasticity theory. In this latter case, for example, extension ratios greater than about 1.5 cannot be accomodated in natural rubber. Indeed, in any material the strain range of applicability of the kinetic theory is limited because of the assumption of Gaussian statistics, as well as other possible complications. Thus the strain levels will be restricted to small strains in the sense of the possible ultimate strains in elastomers. This does not, however, mean that the strain level is infinitesimal; we retain a fully nonlinear treatment of the kinematics of deformation.

We next consider the effect of rapid processes. For a very rapid process (in stress control problems) we can show with (9.21) that the strain amplitude must be small. With small strains the quadratic and higher functionals in (9.5) are of higher order than the linear functional in (9.5). Thus there is a physical justification for truncation of (9.5) under very fast process conditions. With such a justification for truncation under both very fast and very slow processes, we in fact truncate (9.5) explicitly, to give

$$\sigma_{ij} = -p\,\delta_{ij} + x_{i,K}(t)\,x_{j,L}(t)\left[g_0\,\delta_{KL} + \int_0^t g_1(t-\tau)\,\frac{\partial E_{KL}(\tau)}{\partial \tau}\,d\tau\right]. \tag{9.22}$$

This result, subject to condition (9.10), is the form sought, the simple, physically meaningful generalization of the kinetic theory of rubber elasticity to model viscoelastic effects. Note that obtaining this viscoelastic form was not simply a case of reinterpreting the elastic constant g_0 as a relaxation function and inserting a hereditary effect.

It can be shown that the shear relaxation function of the linear theory is related to the properties herein by

$$2\mu(t) = 2g_0 + g_1(t). \tag{9.23}$$

Thus the nonlinear theory contains the same material properties involved in the infinitesimal theory. This result lends special utility to the nonlinear form(9.22). That is, the properties in the nonlinear theory are accessible through infinitesimal testing.

As a general remark here, we consider the types of applications for which (9.22) might be expected to apply. Requirement (ii) was a central influence in the derivation. This requirement concerned the applicability to stress- rather than strain-imposed problems. The standard creep test is certainly a stress-imposed problem, but the relaxation test is strain imposed. Thus we expect that at the simplest level the result (9.22) of this derivation is more likely successfully to model creep conditions than relaxation conditions. This observation can be generalized to include any type of history, with the present results being more applicable to stress- than to strain-imposed histories. Of course, this distinction between strain- and stress-imposed histories disappears as infinitesimal deformation conditions are approached.

Creep Inversion

With respect to stress-imposed problems, it is logical to invert the stress strain relation (9.22) such that strain is the variable specified in terms of stress history. Contract (9.22) with $X_{M,i}X_{N,j}$ to obtain

$$s_{KL} + pX_{K,i}X_{L,i} - g_0\,\delta_{KL} = \int_0^t g_1(t-\tau)\,\frac{\partial E_{KL}(\tau)}{\partial \tau}\,d\tau \tag{9.24}$$

where s_{KL} is the symmetric Piola stress,

$$s_{KL} = \sigma_{ij}X_{K,i}X_{L,j}\,.$$

The form (9.24) may be inverted to give

$$E_{KL}(t) = -J_1(t)\,g_0\,\delta_{KL} + \int_0^t J_1(t-\tau)\,\frac{\partial}{\partial \tau}\,[s_{KL}(\tau) + p(\tau)\,X_{K,i}(\tau)\,X_{L,i}(\tau)]\,d\tau \tag{9.25}$$

where the creep function $J_1(t)$ is defined through the relation

$$\int_0^t J_1(t-\tau)\,\frac{dg_1(\tau)}{d\tau}\,d\tau = t. \tag{9.26}$$

While the integral equation (9.25) does involve stress history in the integrand, unfortunately the deformation gradient is also involved in the integrand. Thus it is not possible to specify stress, perform the integration, and thereby determine strain response. Rather, the integral equation must be evaluated indirectly, which, in general, is a very difficult matter. Thus, although the stress strain relation (9.22) can be formally inverted to give relation (9.25), there is no practical utility in doing so. For application to creep test conditions, some other means must be found to invert (9.22). This will be presented in later developments.

Simple Extension

Next, we consider simple extension conditions. The deformation is specified by

$$\begin{aligned} x_1 &= \lambda X_1 \\ x_2 &= (1/\lambda)^{1/2} X_2 \\ x_3 &= (1/\lambda)^{1/2} X_3 \,. \end{aligned} \tag{9.27}$$

The strain tensor is given by

$$[2E_{KL}] = \begin{bmatrix} \lambda^2 - 1 & 0 & 0 \\ 0 & 1/\lambda - 1 & 0 \\ 0 & 0 & 1/\lambda - 1 \end{bmatrix} . \tag{9.28}$$

Using the condition $\sigma_{22} = \sigma_{33} = 0$ to evaluate p, it is found from (9.22) that

$$\sigma_{11} = g_0 \left(\lambda^2 - \frac{1}{\lambda}\right) - \frac{1}{2\lambda} \int_0^t g_1(t-\tau) \frac{d}{d\tau}\left(\frac{1}{\lambda}\right) d\tau + \frac{\lambda^2}{2} \int_0^t g_1(t-\tau) \frac{d\lambda^2}{d\tau}\, d\tau. \tag{9.29}$$

Relation (9.29) will be evaluated first for stress relaxation conditions specified by

$$\lambda = (\lambda_0 - 1) h(t) + 1 \tag{9.30}$$

where $h(t)$ is the unit step function. Before we substitute (9.30) into (9.29), it is useful to integrate (9.29) by parts. It is found that

$$\sigma_{11} = \left(\lambda_0^2 - \frac{1}{\lambda_0}\right) g_0 + \frac{1}{2}\left(\lambda_0^4 - \lambda_0^2 + \frac{1}{\lambda_0} - \frac{1}{\lambda_0^2}\right) g_1(t). \tag{9.31}$$

The difficulties in dealing with stress-imposed problems were discussed earlier. Here we determine a general method to deal with creep-imposed conditions, but only in the context of simple extension, which is by far the most important test state. In pursuing this objective of inverting (9.22) for creep conditions, we shall restrict the results to second order deformation conditions. That is, we write

$$\lambda = 1 + \epsilon \tag{9.32}$$

where ϵ is the strain, and we retain only first and second order effects in ϵ. This is not a serious restriction in the strain range of intended application. Using (9.32) and (9.29) and neglecting third and higher order terms gives

$$\sigma_{11} = 3 g_0 \epsilon(t) + \frac{3}{2}[1 + \epsilon(t)] \int_0^t g_1(t-\tau) \frac{d\epsilon(\tau)}{d\tau}\, d\tau. \tag{9.33}$$

It is interesting to note there is no second order contribution to the g_0 term of rubber elasticity.

Common creep tests use imposed load rather than imposed stress. To deal with this case we introduce the engineering stress p_{11}, defined by

$$p_{11} = \sigma_{11}/\lambda \tag{9.34}$$

which is the stress per unit initial area. Substituting (9.34) into (9.33), using (9.32), and retaining up to second order terms gives

$$p_{11} = 3g_0\epsilon(t)[1 - \epsilon(t)] + \frac{3}{2}\int_0^t g_1(t-\tau)\frac{d\epsilon(\tau)}{d\tau}\,d\tau. \tag{9.35}$$

Let

$$3g_0 + \tfrac{3}{2}g_1(t) = E(t). \tag{9.36}$$

Then with (9.36), equation (9.35) becomes

$$p_{11} + 3g_0\epsilon^2 = \int_0^t E(t-\tau)\frac{d\epsilon(\tau)}{d\tau}\,d\tau. \tag{9.37}$$

The form of (9.37) can be inverted directly to give

$$\epsilon(t) = \int_0^t J(t-\tau)\frac{d}{d\tau}[p_{11}(\tau) + 3g_0\epsilon^2(\tau)]\,d\tau. \tag{9.38}$$

We are finally to the point of being able to specify the conditions of a creep test involving a step change in load given by

$$p_{11} = Th(t) \tag{9.39}$$

where T is the load level. Inserting (9.39) into (9.38) gives

$$\epsilon(t) = J(t)\,T + 3g_0\int_0^t J(t-\tau)\frac{d\epsilon^2(\tau)}{d\tau}\,d\tau. \tag{9.40}$$

This form is still an integral equation and must be solved for the creep function $J(t)$. To obtain the solution, let

$$J(t) = \epsilon(t)/T + f(t)/T \tag{9.41}$$

where $J(t) = \epsilon(t)/T$ is the linear theory result and the term involving $f(t)$ represents the correction to the linear result so that nonlinear effects can be modeled. Substituting (9.41) into (9.40) gives

$$f(t) + \frac{3g_0}{T}\int_0^t f(t-\tau)\frac{d\epsilon^2(\tau)}{d\tau}\,d\tau = \frac{-3g_0}{T}\int_0^t \epsilon(t-\tau)\frac{d\epsilon^2(\tau)}{d\tau}\,d\tau. \tag{9.42}$$

Note that if the second term on the left-hand side of (9.42) is neglected, then $f(t)$ is $O(\epsilon^3)$ in $\epsilon(t)$. If $f(t)$ is taken to be $O(\epsilon^3)$, then the term that was neglected is $O(\epsilon^5)$ and thus of higher order than the remaining terms. Therefore, it is legitimate to neglect the integral term involving $f(t)$ in (9.42) provided the coefficient $3g_0/T$ is less than $O(\epsilon^{-1})$. It will be assumed that the coefficient $3g_0/T$ is such that this term can be neglected, and it will be necessary to check this assumption in particular applications. With this order, assessment (9.42) reduces to

$$f(t) \simeq -\frac{3g_0}{T}\int_0^t \epsilon(t-\tau)\frac{d\epsilon^2(\tau)}{d\tau}\,d\tau. \tag{9.43}$$

Now (9.41) can be written as

$$J(t) = \frac{\epsilon(t)}{T} - \frac{3g_0}{T^2}\int_0^t \epsilon(t-\tau)\frac{d\epsilon^2(\tau)}{d\tau}\,d\tau. \tag{9.44}$$

Equation (9.44) is the final form to be used to deduce the creep function from the nonlinear creep test. The last term in (9.44) represents the nonlinear correction to the linear theory result $\epsilon(t)/T$. It is fortunate that the present theory adapts so easily to the specification of constant load rather than constant Cauchy stress, since the constant load case is the expedient test method.

The general solution (9.22) is the form sought consistent with the requirements stated in its derivation. More complicated theories which reduce to (9.22) in special cases are mentioned in Section 9.6. The special result (9.44) has been evaluated by Christensen [9.1] with application to a particular elastomer. The complications we encountered in deriving this simple nonlinear theory suggest the underlying difficulties in attempting to derive more complicated and more general nonlinear theories of viscoelasticity. Some of the complications are discussed in Section 9.6.

9.2. NONLINEAR ACCELERATION WAVES

We now proceed with an example to show that specific, important results can be obtained without restricting the analysis to any particular form of constitutive relation. In particular, we consider the propagation of one dimensional acceleration waves in nonlinear viscoelastic solids. Although there have been many contributions to this field, the original work was principally by Coleman and Gurtin; we follow the account of Chen [9.6], who was also one of the basic contributors.

By *acceleration waves* we mean waves that possess discontinuities in the first derivatives of stresses and deformation gradient, and a discontinuity in particle acceleration. Taking X as the reference configuration, let the stress constitutive relation be expressed as a nonlinear functional

$$\sigma(X, t) = \mathop{\psi}_{s=0}^{\infty}\left(F(t-s), F(t)\right) \tag{9.45}$$

where

$$x = x(X, t) \qquad \text{and} \qquad F = \partial x / X . \tag{9.46}$$

Under conditions of assumed continuity, the time derivative of (9.45) is given by

$$\dot{\sigma} = \frac{\partial}{\partial F} \overset{\infty}{\underset{s=0}{\psi}} (F(t-s), F(t)) \dot{F}(t) + \overset{\infty}{\underset{s=0}{\delta\psi}} (F(t-s), F(t) \mid \dot{F}(t-s)) \tag{9.47}$$

where the Fréchet derivative, defined in Section 7.4 and used extensively in Chapter 8, is given by

$$\overset{\infty}{\underset{s=0}{\delta\psi}} (F(t-s), F(t) \mid f(t-s)) = \frac{d}{d\alpha} \overset{\infty}{\underset{s=0}{\psi}} (F(t-s) + \alpha f(t-s), F(t)) \big|_{\alpha=0} . \tag{9.48}$$

The acceleration waves have discontinuities in $\dot{\sigma}$, $\ddot{x}$, and $\dot{F}$ but not in the lower order derivatives. Take the waves as having the discontinuity across

$$X = Y(t) \tag{9.49}$$

and use the following notation for the "jump" in a field variable across the discontinuity

$$[\![f]\!] = f^{-} - f^{+}. \tag{9.50}$$

Note that this convention is opposite to that employed in Section 6.1. Now at an acceleration wave, from the balance of momentum

$$[\![\dot{\sigma}]\!] = - \rho_R v [\![\ddot{x}]\!] \tag{9.51}$$

$$[\![\partial\dot{\sigma}/\partial X]\!] = \rho_R [\![\ddot{x}]\!] \tag{9.52}$$

and

$$[\![\ddot{x}]\!] = - v [\![\dot{F}]\!] \tag{9.53}$$

where ρ_R is the position independent density in the reference configuration and v is the speed of propagation of the acceleration wave. We wish to determine the rate at which the amplitude of the acceleration wave grows or decays.

First, define the kinematical conditions for the rate of change of the "jump" of a variable. From Coleman and Gurtin [9.7] we have by definition of the time derivative

$$\frac{d[\![f]\!]}{dt} = \left[\!\!\left[\frac{\partial f}{\partial t} \right]\!\!\right] + v \left[\!\!\left[\frac{\partial f}{\partial X} \right]\!\!\right]. \tag{9.54}$$

Now let

$$a = [\![\ddot{x}]\!] \qquad \text{at} \quad X = Y(t) \tag{9.55}$$

Then from (9.54)

$$\frac{da}{dt} = [\![\dddot{x}]\!] + v \left[\!\!\left[\frac{\partial \ddot{x}}{\partial X} \right]\!\!\right] \tag{9.56}$$

or from (9.53)

$$\frac{da}{dt} = [\![\ddot{x}]\!] + v[\![\dot{F}]\!]. \tag{9.57}$$

Next, find $[\![\ddot{x}]\!]$ for use in (9.57). Using $f = \dot{\sigma}$ in (9.54), we have

$$\frac{d[\![\dot{\sigma}]\!]}{dt} = [\![\ddot{\sigma}]\!] + v\left[\!\!\left[\frac{\partial\dot{\sigma}}{\partial X}\right]\!\!\right]. \tag{9.58}$$

Solving this for $[\![\partial\dot{\sigma}/\partial X]\!]$ and substituting into (9.52) gives

$$[\![\ddot{x}]\!] = \frac{1}{\rho_R v}\frac{d[\![\dot{\sigma}]\!]}{dt} - \frac{1}{\rho_R v}[\![\ddot{\sigma}]\!]. \tag{9.59}$$

Now $[\![\ddot{x}]\!]$ can be substituted into (9.57) to give, after using the derivative of (9.51)

$$2\frac{da}{dt} = -\frac{a}{v}\frac{dv}{dt} - \frac{1}{\rho_R v}[\![\ddot{\sigma}]\!] + v[\![\dot{F}]\!]. \tag{9.60}$$

It is da/dt for which we are seeking an explicit solution. To this end, we need to evaluate the various terms in (9.60). Begin with an evaluation of $[\![\ddot{\sigma}]\!]$. By definition, $\delta\psi_{s=0}^{\infty}(\ )$, (9.47), is linear in $\dot{F}(t-s)$, and we can write it as a linear functional,

$$\underset{s=0}{\overset{\infty}{\delta\psi}}(\ ) = \int_0^\infty \frac{d}{ds}G_t(s)\dot{F}(t-s)\,ds. \tag{9.61}$$

The memory function $G_t(s)$ in (9.61) is the stress relaxation function corresponding to the history $F(t-s)$. That is to say, since we have general nonlinear behavior, $G_t(s)$ must depend upon the history $F(t-s)$. Let

$$E_t = \frac{d}{dF}\underset{s=0}{\overset{\infty}{\psi}}(F(t-s), F(t)). \tag{9.62}$$

E_t is called the instantaneous tangent modulus, and we take the integral of $G_t(s)$, in (9.61), such that

$$G_t(0) = E_t\,. \tag{9.63}$$

Obviously E_t also depends on the history of deformation and thereby on X and t. E_t is a measure of the initial slope of the instantaneous stress strain curve superimposed on the past history.

Define $\tilde{E}_t$ as

$$\tilde{E}_t = \frac{d^2}{dF^2}\underset{s=0}{\overset{\infty}{\psi}}(F(t-s), F(t)) \tag{9.64}$$

and refer to $\tilde{E}_t$ as the instantaneous second order tangent modulus. With this terminology, we can first find $[\![\dot{\sigma}]\!]$. From (9.47) we have

$$\dot{\sigma} = E_t \dot{F} + \underset{s=0}{\overset{\infty}{\delta\psi}}\,(\;) \tag{9.65}$$

and thus

$$[\![\dot{\sigma}]\!] = E_t [\![\dot{F}]\!] \tag{9.66}$$

where the "jump" of the last term in (9.65) vanishes because $\delta\psi_{s=0}^{\infty}\,(\;)$ is independent of the current value of the deformation, it depends only on the history of the deformation. Using (9.53) in (9.66), we get

$$[\![\dot{\sigma}]\!] = -\frac{1}{v} E_t [\![\ddot{x}]\!] \tag{9.67}$$

and on comparison with (9.51) it follows that

$$v^2 = E_t / \rho_R \,. \tag{9.68}$$

This is our first basic result. We have determined the speed of propagation of the acceleration wave in terms of the density and the instantaneous tangent modulus.

Differentiate (9.47), but not at a discontinuity, to get

$$\begin{aligned}\ddot{\sigma} = {} & \underset{s=0}{\overset{\infty}{\delta^2\psi}}\,(\;) + 2\frac{d}{dF}\underset{s=0}{\overset{\infty}{\delta\psi}}(\;)\dot{F} + \frac{d^2}{dF^2}\underset{s=0}{\overset{\infty}{\psi}}\,(\;)\dot{F}^2 \\ & + \frac{d}{dF}\underset{s=0}{\overset{\infty}{\psi}}\,(\;)\ddot{F} + \frac{\partial}{\partial\tau}\underset{s=0}{\overset{\infty}{\delta\psi}}\left(F(t-s), F(t)\,\Big|\,\frac{\partial F(\tau-s)}{\partial\tau}\right)\Big|_{\tau=t}\end{aligned} \tag{9.69}$$

where the $\delta^2\psi_{s=0}^{\infty}\,(\;)$ term is the second order Fréchet derivative. At the discontinuity, (9.69) becomes

$$[\![\ddot{\sigma}]\!] = 2I_t[\![\dot{F}]\!] + \tilde{E}_{tY}[\![\dot{F}^2]\!] + E_{tY}[\![\ddot{F}]\!] + \left[\!\!\left[\frac{\partial}{\partial\tau}\underset{s=0}{\overset{\infty}{\delta\psi}}\left(F(t-s), F(t)\,\Big|\,\frac{\partial F(\tau-s)}{\partial\tau}\right)\Big|_{\tau=t}\right]\!\!\right] \tag{9.70}$$

where

$$\begin{aligned} I_t &= \frac{d}{dF}\underset{s=0}{\overset{\infty}{\delta\psi}}\,(\;)\,|_{X=Y(t)} \\ \tilde{E}_t &= \tilde{E}_t\,|_{X=Y(t)} \\ E_t &= E_t\,|_{X=Y(t)}\,. \end{aligned} \tag{9.71}$$

There is no contribution of $\delta^2\psi_{s=0}^{\infty}\,(\;)$ to the jump for the reason explained in connection with (9.66).

Now, away from the "jump," where $X \neq Y(t)$, we have by differentiating (9.61)

$$\frac{\partial}{\partial \tau} \underset{s=0}{\overset{\infty}{\delta\psi}} \left(F(t-s), F(t) \,\Big|\, \frac{\partial F(\tau - s)}{\partial \tau} \right) \Big|_{\tau=t} = \left[\frac{\partial}{\partial \tau} \int_0^\infty G_t'(s) \frac{\partial F(\tau - s)}{\partial \tau} \, ds \right]_{\tau=t}. \tag{9.72}$$

Integrate (9.72) by parts to obtain the form

$$G_t'(0)\, \dot{F} + \int_0^\infty G_t''(s)\, \dot{F}(t-s)\, ds.$$

Using this form in (9.72) and evaluating the "jump" gives

$$\left[\!\!\left[\frac{\partial}{\partial \tau} \underset{s=0}{\overset{\infty}{\delta\psi}} \left(F(t-s), F(t) \,\Big|\, \frac{\partial f(\tau - s)}{\partial \tau} \right) \Big|_{\tau=t} \right]\!\!\right] = G_t'(0) \,|_{X=Y(t)} [\![\dot{F}]\!]. \tag{9.73}$$

Substituting (9.73) into (9.70) and the result therefrom into (9.60) gives

$$2 \frac{da}{dt} = - \frac{a}{v} \frac{dv}{dt} - \frac{1}{\rho_R v} \{[2I_t + G_t'(0)][\![\dot{F}]\!] + \tilde{E}_t [\![\dot{F}^2]\!]\} \tag{9.74}$$

where $G_t(0) = G_t(0) \mid_{X=Y(t)}$ and (9.68) has been used.

From (9.53) we write

$$[\![\dot{F}]\!] = - a/v. \tag{9.75}$$

Form $[\![\dot{F}^2]\!]$ as

$$[\![\dot{F}^2]\!] = (\dot{F}^-)^2 - (\dot{F}^+)^2$$

which can be written as

$$[\![\dot{F}^2]\!] = [\![\dot{F}]\!]^2 + 2\dot{F}^+ [\![\dot{F}]\!].$$

Using (9.75) in this gives

$$[\![\dot{F}^2]\!] = \frac{a^2}{v} - 2\dot{F}^+ \frac{a}{v}. \tag{9.76}$$

Substituting (9.75) and (9.76) into (9.74) and using (9.68) gives the final result

$$\frac{da}{dt} = -\mu a + \beta a^2 \tag{9.77}$$

where

$$\mu = - \frac{1}{4E_t} \left[4I_t + 2G_t'(0) + 4\tilde{E}_t \dot{F}^+ - \frac{dE_t}{dt} \right]$$

and

$$\beta = -\tilde{E}_t / 2E_t v. \tag{9.78}$$

The result (9.77) will now be further specialized to a particular deformation field, in order to illustrate basic physical effects.

Accleration Waves Entering a Homogeneously Deformed Medium

Take for $X \geqslant Y(t)$

$$x(X, t) = F_0 X + X_0 \tag{9.79}$$

where F_0 and X_0 are constants referring to the uniform deformation ahead of wave, and a possible rigid body displacement. In this situation, ahead of the wave, we have

$$\dot{F}^+ = 0 \tag{9.80}$$

and

$$\left.\begin{aligned} E_t &= G_t(0) = G_0 \\ \tilde{E}_t &= \tilde{E}_0 \\ G_t'(0) &= G_0' \end{aligned}\right\} \quad \text{(constants).} \tag{9.81}$$

From (9.68), the velocity of the acceleration wave is given by

$$v^2 = G_0/\rho_R \qquad \text{(constant).} \tag{9.82}$$

From (9.79) we see $\dot{F}(t - s) = 0$ for $X \geqslant Y(t)$; thus $\delta\psi_{s=0}^{\infty}(\ )$, from (9.47), vanishes and from (9.71)

$$I_t = 0.$$

Collecting these results into (9.78), μ and β reduce to

$$\left.\begin{aligned} \mu &= -G_0'/2G_0 = \mu_0 \\ \beta &= -\tilde{E}_0/2G_0 v = \beta_0 \end{aligned}\right\} \quad \text{(constants).} \tag{9.83}$$

For μ_0 we have $\mu_0 \geqslant 0$, since $G_0' \leqslant 0$ implies positive dissipation, while β_0 remains algebraic.

From (9.77) we have, by integrating, the solution for the "jump" in acceleration at the wave

$$a(t) = \frac{\lambda_0}{[\lambda_0/a(0) - 1]\, e^{\mu_0 t} + 1} \tag{9.84}$$

where

$$\lambda_0 = \mu_0/\beta_0 = G_0' v/\tilde{E}_0 \tag{9.85}$$

and $a(0)$ is $a(t)\,|_{t=0}$. Note that μ_0 and λ_0 depend only upon F_0 and the properties of the material. Physically realistic conditions are

$$G_0 > 0, \qquad G_0' \leqslant 0, \qquad \tilde{E}_0 \neq 0.$$

Consider various cases:

(i) If sgn $a(0) =$ sgn $\tilde{E}_0$ or if sgn $a(0) = -$ sgn $\tilde{E}_0$ and $|a(0)| < |\lambda_0|$, then we have $a(t) \to 0$ monotonically as $t \to \infty$.

(ii) If $a(0) = \lambda_0$, then we have $a(t) = a(0)$, a constant.

(iii) If sgn $a(0) = -$ sgn $\tilde{E}_0$ and $|a(0)| > |\lambda_0|$, then we have $|a(t)| \to \infty$ monotonically in a finite time given by

$$t_\infty = -\frac{1}{\mu_0} \ln \left[1 - \frac{\lambda_0}{a(0)}\right].$$

Thus we have the unexpected result from (iii) that acceleration waves can grow in strength as they propagate. In fact, this suggest that acceleration waves can grow into shock waves. This growth possibility is in sharp contrast to the situation in linear theory. In Sections 6.1 and 6.3 we found that linear waves always decay due to the inherent dissipation in the material. In the more general nonlinear theory, however, we find according to (iii) that if the acceleration wave is sufficiently strong, the nonlinear effect can overwhelm the dissipative effect, and the wave can grow as it propagates. This growth condition only occurs if the sign of the acceleration wave is opposite the sign of the instantaneous second order tangent modulus.

For large amplitude waves with

$$|\lambda_0/a(0)| \ll 1$$

we have

$$ln[1 - \lambda_0/a(0)] \simeq -\lambda_0/a(0)$$

and condition (iii) gives

$$t_\infty \simeq -2G_0 v/\tilde{E}_0 a(0)$$

which is the same as the corresponding nonlinear elasticity theory result. Thus, if the amplitude of the wave is very large relative to λ_0, then the dissipative effects are negligible.

For small amplitude waves with

$$|\lambda_0/a(0)| \gg 1$$

then from (9.84) the solution is

$$a(t) \simeq a(0)e^{-\mu_0 t}.$$

This result is consistent with the linear theory result of Section 6.1, whereby wave disturbances experience an exponential decay with distance traveled.

These amazingly powerful results were found without appeal to any particular constitutive relation. Rather, we merely used the properties of nonlinear constitutive functionals that allow Fréchet differentiation. The success we experienced here with acceleration waves encourages us to attempt a similar treatment for shock waves, which have discontinuities in σ, F, and $\dot{x}$. Unfortunately, it is found that the deformation behind the shock wave cannot be predicted only in terms of material properties and conditions ahead of the shock wave. However, it can at least be proven that growing shock waves can exist. See Chen [9.6] for the details.

9.3. VISCOMETRIC FLOWS

We turn now to problems of viscoelastic fluid flow. We follow the motivation of the preceding section on acceleration waves in solids to show that practical problems can be solved without restricting attention to particular constitutive models.

To introduce viscometric flows, let us first discuss simple shearing flow. For the simple shearing flow specified by (8.64), the steady state stresses are normally expressed as

$$\eta = \frac{\sigma_{12}}{\kappa}$$

$$N_1 = \frac{\sigma_{11} - \sigma_{22}}{\kappa^2} \tag{9.86}$$

$$N_2 = \frac{\sigma_{22} - \sigma_{33}}{\kappa^2}$$

where η is the viscosity function, N_1 and N_2 are said to be the first and second normal stress differences, respectively, and κ is the shear rate. These give the state of stress to within the hydrostatic pressure for an incompressible fluid. N_1 is always measured to be positive, while N_2 is usually considered to be negative and small in magnitude compared with N_1. The forms η, N_1, and N_2 are said to be the viscometric coefficients and characterize the material response. The viscometric coefficients have greater applicability than just to steady, simple shearing flow. As will be shown, they apply to any viscometric flow.

Following the description by Pipkin [9.8], a viscometric flow is one which can be visualized as the relative sliding of inextensible material surfaces over each other. The motions need not be rectilinear and the surfaces need not be planar. It can be seen that motions of this type involve only relative shearing of the material, and their material response has been shown by Coleman and Noll [9.9] to be governed by the viscometric coefficients when the shear rate κ is constant in time at a given particle. The direction of shear flow is taken as x_1, with the shearing motion occurring in the x_1, x_2 tangent plane. This establishes the directions of σ_{11}, σ_{22}, and σ_{12}. The relative sliding of the inextensible material surfaces is a necessary but not a sufficient condition for viscometric flow. Sufficiency requires a steady shear rate and satisfaction of the balance of momentum relations for the flow to exist. In the present context, inertia term effects are neglected.

For example,

(i) steady flow in a circular, uniform pipe is viscometric, and

(ii) steady flow in a noncircular uniform pipe is not viscometric.

The latter type of flow, taken as relative motion of inextensible surfaces, does not satisfy balance of momentum relations or the boundary conditions.

Poiseuille Flow

Consider first Poiseuille flow, that is, steady flow through a uniform cirular pipe of radius a. In cylindrical coordinates the shear rate is given by

$$\kappa = du/dr = u'$$

where u is the velocity in the z direction and axial symmetry is used. The momentum balance equations are

$$\begin{aligned} \frac{\partial \sigma_{rr}}{\partial r} + \frac{\sigma_{rr} - \sigma_{\theta\theta}}{r} &= 0 \\ \frac{\partial \sigma_{zz}}{\partial z} + \frac{1}{r}\frac{\partial}{\partial r}(r\sigma_{rz}) &= 0 \qquad (9.87) \\ \frac{\partial \sigma_{zz}}{\partial z} &= -\frac{\partial p}{\partial z} \end{aligned}$$

where only the pressure part of σ_{zz} is assumed to vary with z.

In the present notation

$$\begin{aligned} \kappa\eta &= \sigma_{rz} \\ \kappa^2 N_1 &= \sigma_{zz} - \sigma_{rr} \\ \kappa^2 N_2 &= \sigma_{rr} - \sigma_{\theta\theta} \,. \end{aligned}$$

Using the last of (9.87) in the second equation gives

$$\frac{\partial}{\partial r}(r\kappa\eta) = r\frac{\partial p}{\partial z}. \qquad (9.88)$$

We assume $\partial p/\partial z$ is independent of r and integrate (9.88) to find

$$r\kappa\eta = \frac{r^2}{2}\frac{\partial p}{\partial z} \qquad (9.89)$$

which can be written as

$$\eta \frac{\partial u}{\partial r} = \frac{r}{2} \frac{\partial p}{\partial z}. \tag{9.90}$$

Knowing from material behavior the relation $\eta(\kappa)$ versus $\kappa = \partial u / \partial r$, we can integrate (9.90) to find the velocity profile

$$u = u(r).$$

The flow rate is given by

$$Q = 2\pi \int_0^a u(r)\, r\, dr$$

or by integrating by parts and using $u(a) = 0$:

$$Q = -\pi \int_0^a r^2 \frac{\partial u(r)}{\partial r}\, dr. \tag{9.91}$$

From (9.90) we find $\partial u / \partial r$ and substitute in (9.91) to get

$$Q = -\frac{\pi}{2} \frac{\partial p}{\partial z} \int_0^a \frac{r^3\, dr}{\eta}. \tag{9.92}$$

With known $\eta = \eta(r)$ from the velocity profile, (9.92) gives the flow rate. However, we follow an inverse procedure and seek to determine $\eta = \eta(\kappa)$ from flow rate measurements.

Using $\sigma_{rz} = \kappa\eta$ and (9.89) we have

$$\sigma_{rz} = \frac{r}{2} \frac{\partial p}{\partial z} \tag{9.93}$$

where $\partial p / \partial z$ is a constant. Thus, we can write

$$\sigma_{rz} / \tau_a = r / a \tag{9.94}$$

where

$$\tau_a = \sigma_{rz}|_{r=a}.$$

We next solve (9.93) for r, evaluate dr, and substitute into (9.92) to get

$$Q = -\frac{\pi a^3}{\tau_a^3} \int_0^{\tau_a} \sigma_{rz}^2 \frac{\partial u}{\partial r}\, d\sigma_{rz} \tag{9.95}$$

and take the derivative with respect to τ_a of (9.95) to get

$$\left.\frac{\partial u}{\partial r}\right|_{\sigma_{rz}=\tau_a} = -\frac{1}{\pi a^3 \tau_a^2}\frac{\partial}{\partial \tau_a}(\tau_a^3 Q). \tag{9.96}$$

Combining (9.93) and (9.94) at $r = a$ gives

$$\tau_a = \frac{a}{2}\frac{\partial p}{\partial z}. \tag{9.97}$$

Relations (9.96) and (9.97) provide the basic forms to be used to determine $\eta = \eta(\kappa)$. Specifically, knowing the pressure gradient $\partial p/\partial z$, from (9.97) we know τ_a. Then by varying the pressure gradient, and correspondingly τ_a, we vary the flow rate Q. With $(\partial/\partial\tau_a)(\tau_a^3 Q)$ known from measurements, relation (9.96) gives the shear rate at the wall. Knowledge of τ_a versus κ gives

$$\eta = \eta(\kappa) = \frac{\tau_a}{(\partial u/\partial r)\,|_{\sigma_{rz}=\tau_a}}.$$

We have therefore found a means by which flow measurements in a pipe can be used to determine the viscoemetric coefficient η. As shown by Walters [9.10], this derivation can be taken further, using the first of the momentum balance equations (9.87) to gain information on the normal stress differences. The problem with determining properties this way is one of sensitivity, involving derivatives. A better method is direct measurements in fields involving a homogeneous shear rate. This method is followed next. The present analysis of Poiseuille flow shows that knowledge of the viscometric coefficients allows us to solve the associated flow problem completely.

Cone and Plate Flow

Consider flow in the annular space between a coaxially rotating cone and fixed plate, as shown in Fig. 9.1. Following Walters [9.10], the momentum balance equations in spherical coordinates are

$$\frac{1}{r^2}\frac{\partial}{\partial r}(r^2\sigma_{rr}) + \frac{1}{r\sin\theta}\frac{\partial}{\partial\theta}(\sin\theta\,\sigma_{r\theta}) - \frac{\sigma_{\theta\theta}+\sigma_{\phi\phi}}{r} = 0 \tag{9.98}$$

$$\frac{1}{r}\frac{\partial}{\partial r}(r^2\sigma_{r\theta}) + \frac{1}{r\sin\theta}\frac{\partial}{\partial\theta}(\sin\theta\,\sigma_{\theta\theta}) + \frac{\sigma_{r\theta}}{r} - \frac{\cot\theta}{r}\sigma_{\phi\phi} = 0 \tag{9.99}$$

and

$$\frac{1}{r^2}\frac{\partial}{\partial r}(r^2\sigma_{r\phi}) + \frac{1}{r}\frac{\partial\sigma_{\theta\phi}}{\partial\theta} + \frac{\sigma_{r\phi}}{r} + \frac{2\cot\theta}{r}\sigma_{\theta\phi} = 0. \tag{9.100}$$

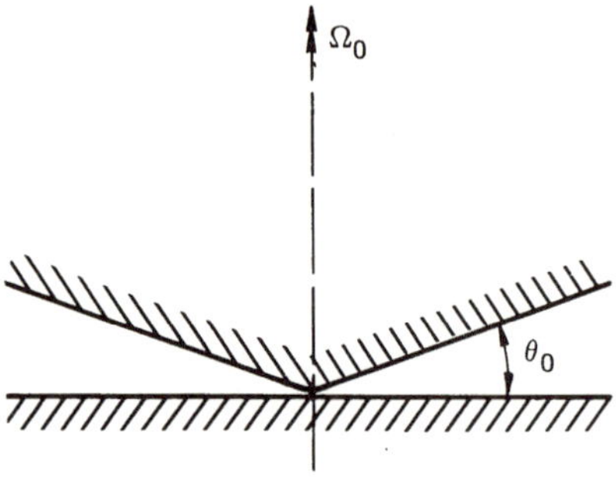

Fig. 9.1. Cone and plate geometry.

The equation of continuity is

$$\frac{1}{r^2}\frac{\partial}{\partial r}(r^2 u_r) + \frac{1}{r \sin\theta}\frac{\partial}{\partial\theta}(\sin\theta\, u_\theta) = 0. \tag{9.101}$$

Assume the form of solution

$$\begin{aligned} u_r &= 0 \\ u_\theta &= 0 \\ u_\phi &= r \sin\theta\, \Omega(\theta). \end{aligned} \tag{9.102}$$

The boundary conditions are that

$$\begin{aligned} \Omega &= 0 \qquad \text{at} \quad \theta = \pi/2 \\ \Omega &= \Omega_0 \qquad \text{at} \quad \theta = (\pi/2 - \theta_0). \end{aligned} \tag{9.103}$$

For the velocity field in (9.102) the shear rate is

$$\kappa = \sin\theta \frac{d\Omega}{d\theta} \tag{9.104}$$

and the stresses are

$$\begin{aligned} \sigma_{rr} = -p, \qquad \sigma_{\theta\theta} &= -p + \kappa^2 N_2, \qquad \sigma_{\phi\phi} = -p + \kappa^2(N_1 + N_2) \\ \sigma_{r\theta} = 0, \qquad \sigma_{r\phi} &= 0, \qquad \sigma_{\theta\phi} = \kappa\eta(\kappa). \end{aligned} \tag{9.105}$$

The equation of continuity is satisfied by (9.102), and relations (9.98)–(9.100) become

$$\begin{aligned} -\frac{1}{r^2}\frac{\partial}{\partial r}(r^2 p) - \frac{\sigma_{\theta\theta} + \sigma_{\phi\phi}}{r} &= 0 \\ \frac{1}{r \sin\theta}\frac{\partial}{\partial\theta}(\sin\theta\, \sigma_{\theta\theta}) - \frac{\cot\theta}{r}\sigma_{\phi\phi} &= 0 \\ \frac{1}{r}\frac{\partial\sigma_{\theta\phi}}{\partial\theta} + \frac{2\cot\theta}{r}\sigma_{\theta\phi} &= 0. \end{aligned} \tag{9.106}$$

Consider the case of a small angular gap between the cone and plate and take θ_0 small to reduce (9.106) to

$$-\frac{1}{r^2}\frac{\partial}{\partial r}(r^2 p) - \frac{\sigma_{\theta\theta} + \sigma_{\phi\phi}}{r} = 0$$

$$\frac{\partial \sigma_{\theta\theta}}{\partial \theta} = 0 \qquad (9.107)$$

$$\frac{\partial \sigma_{\theta\phi}}{\partial \theta} = 0.$$

The shear rate becomes

$$\kappa = \Omega_0/\theta_0 . \qquad (9.108)$$

The first of (9.107), after using (9.105), becomes

$$\frac{\partial p}{\partial r} + \frac{\kappa^2}{r}(N_1 + 2N_2) = 0. \qquad (9.109)$$

Integrating (9.109) gives

$$p = -\kappa^2(N_1 + 2N_2)\ln r + \text{const.} \qquad (9.110)$$

Thus, p has logarithmic dependence on r.

The boundary condition at the free surface is given by

$$\sigma_{rr} = -p = 0 \qquad \text{at} \quad r = a. \qquad (9.111)$$

Thus (9.110) becomes

$$p = \kappa^2(N_1 + 2N_2)\ln(a/r). \qquad (9.112)$$

The last two of the momentum equations (9.107) are now satisfied.

To find total thrust on the cone, defined by

$$F = 2\pi \int_0^a \sigma_{\theta\theta} r \, dr \qquad (9.113)$$

we use (9.112) in (9.105) and that result in (9.113) to get

$$F = -(\pi/2)a^2\kappa^2 N_1 . \qquad (9.114)$$

The total torque is defined by

$$T = 2\pi \int_0^a \sigma_{\theta\phi} r^2 \, dr$$

and it is found that

$$T = \tfrac{2}{3}\pi a^3 \kappa \eta(\kappa). \qquad (9.115)$$

The axial force, (9.114), and the torque, (9.115), solutions permit the use of direct measurements to determine the viscometric coefficients $\eta(\kappa)$ and $N_1(\kappa)$. The determination of the second normal stress difference N_2 is a far more complicated problem. The solutions given here show the ease with which viscometric flows can be treated.

9.4. NONVISCOMETRIC FLOWS

Complementary to the definition of viscoemetric flows, all other flows are those that cannot be described as being locally a steady simple shearing flow. Extensional flow is an obvious example of a nonviscometric flow. We shall give a special means of treating such general flows under slow conditions.

Consider the strain history measure for a fluid, (8.47), repeated here:

$$G_{ij}(\tau) = \frac{\partial x_k(\tau)}{\partial x_i(t)} \frac{\partial x_k(\tau)}{\partial x_j(t)} - \delta_{ij} \,. \tag{9.116}$$

As in (8.46), we would write the stored energy functional as

$$A = \mathop{\phi}_{s=0}^{\infty} (G_{ij}(t-s))$$

with a similar expression for stress. Examine the deformation measure, $G_{ij}(t-s)$. Following Pipkin [9.8], for slow motion expand the relative strain $\mathbf{G}$ in powers of s as

$$\mathbf{G}(t-s) = \sum_{n=1}^{\infty} \frac{1}{n!} \mathbf{A}_n(t)(-s)^n \tag{9.117}$$

where

$$\mathbf{A}_n(t) = \frac{d^n}{d\tau^n} \mathbf{G}(\tau) \big|_{\tau=t} \,. \tag{9.118}$$

The $\mathbf{A}_n$ are said to be the Rivlin–Ericksen tensors.

Obtain an explicit formula for the Rivlin–Ericksen tensors. Let $\mathbf{x}(\tau)$ be the position of a particle, which at time t is $\mathbf{x}(t)$. Then from (9.116), taking the derivative with respect to τ gives

$$\frac{d}{d\tau} G_{ij}(\tau) = x_{k,i}\dot{x}_{k,j} + x_{k,j}\dot{x}_{k,i}$$

where (9.119)

$$\dot{x}_{k,j} = \frac{\partial \dot{x}_k(\tau)}{\partial x_j(t)}$$

and $\dot{x}_k(\tau)$ is the velocity at time τ. Similar to the case in (9.119), derivatives of any order can be taken with the general result

$$\frac{d^n}{d\tau^n} G_{ij}(\tau) = x_{k,i}x_{k,j}^{(n)} + x_{k,j}x_{k,i}^{(n)} + \sum_{m=1}^{n-1} \binom{n}{m} x_{k,i}^{(m)} x_{k,j}^{(n-m)} \tag{9.120}$$

where $\binom{n}{m}$ is the binomial coefficient, and the order of the time derivatives are as shown. Evaluate (9.120) at time $\tau = t$. Then from (9.118) we have

$$A_{ij}^{(n)} = [x_i^{(n)}]_{,j} + [x_j^{(n)}]_{,i} + \sum_{m=1}^{n-1} \binom{n}{m} [x_k^{(m)}]_{,i} [x_k^{(n-m)}]_{,j} \,. \tag{9.121}$$

In terms of velocity components u_i,

$$A_{ij}^{(1)} = u_{i,j} + u_{j,i}$$

and (9.122)

$$A_{ij}^{(2)} = \left(\frac{du_i}{dt}\right)_{,j} + \left(\frac{du_j}{dt}\right)_{,i} + 2u_{k,i}u_{k,j} \,.$$

Thus $\mathbf{A}^{(2)}$ involves not only particle acceleration but velocity gradients as well. The following recursion relation can be proved:

$$A_{ij}^{(n+1)} = \frac{d}{dt} A_{ij}^{(n)} + A_{ik}^{(n)} u_{k,j} + A_{jk}^{(n)} u_{k,i} \,. \tag{9.123}$$

For cases in which the flow is slow and slowly changing we see that we can truncate the expansion (9.117) because higher $\mathbf{A}_n$ are expected to be of higher order than lower $\mathbf{A}_n$. Substitute $\mathbf{G}(t - s)$, from (9.117) into the integral expansion form of the constitutive functional. There will be constant coefficients of the $\mathbf{A}_n$ that involve the integrals of the relaxation functions times the various powers of s. To formalize the end result of the process, write

$$\boldsymbol{\sigma} = -p\mathbf{I} + \mathbf{S}_1 + \mathbf{S}_2 + \cdots$$

where $\mathbf{S}_i$ are terms of the various orders. Take $\mathbf{S}_i$ as isotropic polynomial functions of $\mathbf{A}_n$. Thus $\mathbf{S}_1$ involves $\mathbf{A}_1$ and $\mathbf{I}$ tr $\mathbf{A}_1$; $\mathbf{S}_2$ involves $\mathbf{A}_1^2$, $\mathbf{A}_2$, and $\mathbf{A}_1$ tr $\mathbf{A}_1$, etc. For an incompressible material tr $\mathbf{A}_1 = 0$, leaving

$$S_1 = \eta\mathbf{A}_1 \,, \qquad S_2 = \alpha_1\mathbf{A}_2 + \alpha_2\mathbf{A}_1^2 \,.$$

Other tensor polynomial identities such as the Cayley–Hamilton theorem can be used to reduce the number of independent terms. For example,

$$S_3 = c_1\mathbf{A}_3 + c_2\mathbf{A}_1 \operatorname{tr} \mathbf{A}_2 + c_2(\mathbf{A}_1\mathbf{A}_2 + \mathbf{A}_2\mathbf{A}_1).$$

Writing the stress constitutive relation explicitly through second order gives

$$\boldsymbol{\sigma} = -p\mathbf{I} + \eta\mathbf{A}_1 + \alpha_1\mathbf{A}_2 + \alpha_2\mathbf{A}_1^2 + \cdots. \tag{9.124}$$

Examine the Rivlin–Erickson tensors in the case of steady simple shearing flow. Using (9.122) it is found that for

$$u_1 = \kappa X_2, \qquad u_2 = u_3 = 0$$

then

$$\left.\begin{aligned} A_{12}^{(1)} &= A_{21}^{(1)} = \kappa \\ A_{22}^{(2)} &= 2\kappa^2 \end{aligned}\right\} \qquad \text{(all other components vanishing).}$$

Higher order Rivlin–Ericksen tensors follow from the recursion relation (9.123). It is found that all of these vanish:

$$A_{ij}^{(n)} = 0, \qquad n \geqslant 3.$$

More generally, in steady flow, the $\mathbf{A}_n$ are of higher order as n increases. However, in unsteady flow there is no assurance of ordering in the various terms. Note that the coefficients in (9.124) are constants, independent of flow conditions, in contrast to viscometric flows. As shown by Pipkin [9.8], for viscometric flow we can write

$$\boldsymbol{\sigma} = -p\mathbf{I} + \eta\mathbf{A}_1 - \tfrac{1}{2}N_1\mathbf{A}_2 + (N_1 + N_2)\,\mathbf{A}_1^2$$

but here in the context of Rivlin–Ericksen tensors we must have η, N_1, and N_2 independent of shear rate. Thus the utility of the Rivlin–Ericksen tensor formulation is in application to nonviscometric flows.

We see that the restrictions for the use of (9.124) are that the flow must be both slow and slowly changing. However, to be quantitative about the number of terms we would need to retain, we require complete experimental information concerning the response of the fluid as a function of rate of flow and also concerning the spectrum of relaxation times inherent in the fluid. We seldom have such complete information.

The method of derivation of (9.124) suggests that problems should be solved using perturbation techniques, with the lowest order form being that of Newtonian flow. The second order solution would introduce normal stress effects and so on. This method has been highly developed and applied by Joseph and Fosdick [9.11] in an example of the rod climbing effect, to be considered in the following example.

Rod Climbing Effect Example

Before considering the Weissenburg, or rod climbing, effect, we will first discuss the related problem of Couette flow. Consider the fluid filled annular space between an infinite rotating rod and a concentric fixed outer cylinder. Using polar cylindrical coordinates, as a trial solution take the velocity field in the fluid as

$$u_r = 0, \qquad u_\theta = f(r), \qquad u_z = 0. \tag{9.125}$$

We shall treat the fluid as being modeled by the Rivlin–Ericksen tensor formalism, even though it is a viscometric flow. The reason for this will be apparent shortly.

The momentum balance equations are given by

$$\frac{1}{r}\frac{\partial}{\partial r}(r\sigma_{rr}) - \frac{\sigma_{\theta\theta}}{r} = 0$$

and (9.126)

$$\frac{\partial}{\partial r}(r^2\sigma_{r\theta}) = 0.$$

The first equation in (9.126) can be written

$$r\frac{\partial \sigma_{rr}}{\partial r} - \kappa^2 N_1 = 0 \tag{9.127}$$

where

$$\kappa^2 N_1 = \sigma_{\theta\theta} - \sigma_{rr}$$

$$\kappa^2 N_2 = \sigma_{rr} - \sigma_{zz} \tag{9.128}$$

and

$$\kappa = \partial u_\theta / \partial r.$$

The second equation in (9.126) can be written as

$$\frac{\partial}{\partial r}(r^2 \eta\kappa) = 0. \tag{9.129}$$

If we were treating this problem as a viscometric flow we would need to account for the shear rate dependence of $\eta = \eta(\kappa)$. However, with the Rivlin–Ericksen tensor form (9.124), η is a constant and drops out of (9.129), leaving

$$r^2\kappa = C_1 .$$

Thus using (9.128) these results

$$\frac{\partial u_\theta}{\partial r} = \frac{C_1}{r^2}$$

which when integrated gives

$$u_\theta = -C_1/r + C_2. \tag{9.130}$$

The boundary conditions are

$$\begin{aligned} u_\theta &= a\omega \quad &\text{at} \quad r = a \\ u_\theta &= 0 \quad &\text{at} \quad r = b \end{aligned} \tag{9.131}$$

where ω is the rate of rotation.

Combining (9.130) and (9.131) allows determination of the constants, giving

$$C_1 = \frac{a^2\omega}{a/b - 1} \quad \text{and} \quad C_2 = \frac{a\omega}{1 - b/a}. \tag{9.132}$$

Then κ follows from (9.128) as

$$\kappa = \frac{a^2\omega}{r^2(a/b - 1)}. \tag{9.133}$$

From (9.127) we have

$$\frac{\partial \sigma_{rr}}{\partial r} = \frac{\kappa^2 N_1}{r}. \tag{9.134}$$

Combining (9.133) and (9.134)

$$\frac{\partial \sigma_{rr}}{\partial r} = \left(\frac{a^2\omega}{1 - a/b}\right)^2 \frac{N_1}{r^5}. \tag{9.135}$$

Integrating (9.135), we get the solution

$$\sigma_{rr} = -\frac{1}{4}\left(\frac{a^2\omega}{1 - a/b}\right)^2 \frac{N_1}{r^4} + \text{const.} \tag{9.136}$$

The velocity distribution (9.130) and (9.132), along with the stress distribution (9.136), completes the solution of the Couette flow problem.

Now we adapt the Couette flow solution to model the rod climbing problem. This problem is posed as being of the Couette flow type but with a free surface that is aligned vertically with a gravity field, as shown in Fig. 9.2. The free surface condition makes this problem one of nonviscometric flow. The free surface of the viscoelastic fluid climbs the rotating rod. This is known as the Weissenburg effect. A qualitative explanation of the rod climbing will be given now following the suggestion of Coleman *et al.* [9.12] and Pipkin [9.8]. Take

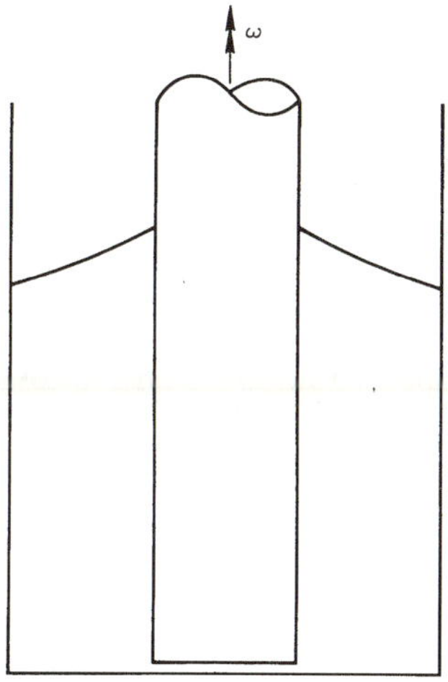

Fig. 9.2. Rod climbing effect.

the Couette flow solution (9.136), combine with N_2 in (9.128), and add the gravity field effect to get

$$\sigma_{zz} = -p_0 + \rho g z - \frac{1}{r^4}\left(\frac{a^2\omega}{1 - a/b}\right)^2 \left(\frac{N_1}{4} + N_2\right) \tag{9.137}$$

where p_0 is the atmospheric pressure at $z = 0$ for $\omega = 0$, and g is the gravitational constant. Set the fluid stress σ_{zz} equal to $-p_0$ to approximate the shape of the free surface from (9.137) as

$$z = \frac{1}{\rho g}\left(\frac{a^2\omega}{1 - a/b}\right)^2 \frac{(N_1/4 + N_2)}{r^4}. \tag{9.138}$$

Because $N_1 \geqslant 0$ and large compared with the magnitude of N_2, (9.138) predicts the rod climbing effect. This procedure satisfies only the z component of the free surface condition, but this is acceptable for small changes of the free surface from the horizontal plane. Thus, for small rates of rotation ω, relation (9.138) should be realistic when combined with the centrifugal force effect. It is emphasized that N_1 and N_2 here are not the rate dependent viscometric coefficients; rather they have the zero shear rate values of the corresponding quantities.

Joseph and Fosdick [9.11] have solved the rod climbing problem using Rivlin–Ericksen tensors through fourth order. They use a perturbation procedure involving a perturbation upon a state of rest, with the small parameter being rotation rate ω. They also include surface tension and centrifugal force effects.

Motions with Constant Stretch History

In viscometric flows the stress response is independent of current time and depends only on the past history relative to the state at current time. This is exemplified by the fact that η, N_1, and N_2 involve only integrals of the relaxation function.

The question arises, is there a larger class of flows than just viscometric flows in which the stress functional depends not on the current time but only on the past history? There is such a classification, called motions with constant stretch history (see Truesdell and Noll [9.13]). An example is steady extensional flow in which the components d_{ij} of the rate of deformation tensor are independent of time. Motions with constant stretch history include viscometric flows as a subclass.

Motions with constant stretch history do not have the utility of properties representation that the subclass of viscometric flows has. For example, the flow response properties in steady extensional flow are different from those in viscometric flow. The properties in steady biaxial flow are different from those in steady uniaxial flow, and so on. The only possible unification of all these different responses must be in a general constitutive relation which applies to transient as well as steady state flow response.

In steady extensional flow we find that for many viscoelastic models at some rate of deformation the stress becomes unbounded; see, for example, Walters [9.10]. However, this situation involves the assumption of steady state. More generally we should examine the transient problem of "start up," to see if the solution admits a steady state. That is the analysis which we shall do next. We find that above a certain rate of deformation no steady state exists. The stresses build up exponentially to infinite values.

Extensional Flow

We specify extensional flow by the coordinate description

$$\begin{aligned} x_1(t) &= [(e^{\lambda t} - 1)\, h(t) + 1]\, X_1 \\ x_2(t) &= [(e^{-\lambda t/2} - 1)\, h(t) + 1]\, X_2 \\ x_3(t) &= [(e^{-\lambda t/2} - 1)\, h(t) + 1]\, X_3 \,. \end{aligned} \tag{9.139}$$

Eliminate the X_i coordinates between $x_i(t)$ and $x_i(\tau)$ to get

$$\begin{aligned} x_1(\tau) &= \left[\frac{(e^{\lambda t} - 1)\, h(\tau) + 1}{(e^{\lambda t} - 1)\, h(t) + 1}\right] x_1(t) \\ x_2(\tau) &= \left[\frac{(e^{-\lambda \tau/2} - 1)\, h(\tau) + 1}{(e^{-\lambda t/2} - 1)\, h(t) + 1}\right] x_2(t) \\ x_3(\tau) &= \left[\frac{(e^{-\lambda \tau/2} - 1)\, h(\tau) + 1}{(e^{-\lambda t/2} - 1)\, h(t) + 1}\right] x_3(t) \end{aligned} \tag{9.140}$$

The deformation measure G_{ij}, (9.116), is then found to be

$$G_{ij} = \begin{bmatrix} \left[\dfrac{(e^{\lambda\tau}-1)\,h(\tau)+1}{(e^{\lambda t}-1)\,h(t)+1}\right]^2 & 0 & 0 \\ 0 & \left[\dfrac{(e^{-\lambda\tau/2}-1)\,h(\tau)+1}{(e^{-\lambda t/2}-1)\,h(t)+1}\right]^2 & 0 \\ 0 & 0 & \left[\dfrac{(e^{-\lambda\tau/2}-1)\,h(\tau)+1}{(e^{-\lambda t/2}-1)\,h(t)+1}\right]^2 \end{bmatrix}. \tag{9.141}$$

Using

$$d_{ij} = \left.\frac{\partial G_{ij}}{\partial \tau}\right|_{\tau=t}$$

we find

$$d_{11} = \frac{2\lambda e^{\lambda t} h(t)}{(e^{\lambda t}-1)\,h(t)+1}.$$

Thus

$$d_{11} = \begin{cases} 0 & \text{for} \quad t < 0 \\ 2\lambda & \text{for} \quad t \geqslant 0. \end{cases}$$

Parameter λ determines the rate of constant extension. Similarly,

$$d_{22} = -\lambda, \quad d_{33} = -\lambda, \qquad \text{for} \quad t \geqslant 0$$

and thus $d_{ii} = 0$, satisfying the incompressibility condition.

To proceed further, we must employ a particular constitutive relation. We take the very simple lowest order form of (8.74), written here as

$$\sigma_{ij} = -p\delta_{ij} + 4\int_{-\infty}^{t} \Delta(t-\tau, 0)\,\frac{\partial G_{ij}(\tau)}{\partial \tau}\,d\tau. \tag{9.142}$$

In component form (9.142) is

$$\begin{aligned} \sigma_{11} &= -p + \int_{-\infty}^{t} \tilde{E}(t-\tau)\,\dot{G}_{11}(\tau)\,d\tau \\ \sigma_{22} &= -p + \int_{-\infty}^{t} \tilde{E}(t-\tau)\,\dot{G}_{22}(\tau)\,d\tau \end{aligned} \tag{9.143}$$

where

$$\tilde{E}(t-\tau) = 4\Delta(t-\tau, 0).$$

For stress-free lateral surfaces, $\sigma_{22} = 0$, we evaluate p from (9.143) to obtain σ_{11} as

$$\sigma_{11} = \int_{-\infty}^{t} \tilde{E}(t-\tau)[\dot{G}_{11}(\tau) - \dot{G}_{22}(\tau)]\, d\tau. \tag{9.144}$$

Using (9.141) in (9.144) we find for $t \geqslant 0$

$$\sigma_{11} = \frac{\dot{\lambda}}{e^{-\dot{\lambda}t}} \int_{0}^{t} \tilde{E}(t-\tau)\, e^{-\dot{\lambda}\tau}\, d\tau + \frac{2\dot{\lambda}}{e^{2\dot{\lambda}t}} \int_{0}^{t} \tilde{E}(t-\tau)\, e^{2\dot{\lambda}\tau}\, d\tau. \tag{9.145}$$

To illustrate effects take the simple relaxation function

$$\tilde{E}(t) = E_0 e^{-t/\tau_0}. \tag{9.146}$$

Substituting (9.146) in (9.145) gives the response form

$$\frac{\sigma_{11}}{E_0} = \frac{2\dot{\lambda}}{2\dot{\lambda} + 1/\tau_0}\left[1 - e^{-(2\dot{\lambda}+1/\tau_0)t}\right] + \frac{\dot{\lambda}}{\dot{\lambda} - 1/\tau_0}\left[-1 + e^{(\dot{\lambda}-1/\tau_0)t}\right]. \tag{9.147}$$

The solution (9.147) is unstable for $\dot{\lambda} > 1/\tau_0$, and unlimited stress growth occurs. This suggests unstable flow behavior and breakage when filaments are drawn at excess rates in polymer processing operations. For $\dot{\lambda} < 1/\tau_0$ the steady state solution is found from (9.147) to be

$$\sigma_{11} = \frac{3E_0\tau_0\dot{\lambda}}{(1 + 2\dot{\lambda}\tau_0)(1 - \dot{\lambda}\tau_0)}.$$

For very slow extension

$$\sigma_{11} = 3E_0\tau_0\dot{\lambda}$$

which is typically written as $\sigma_{11} = 3\eta\dot{\lambda}$ where $\eta = E_0\tau$ is the shear viscosity and 3η, called the Trouton viscosity, is the viscosity governing extensional flow.

The experimental data of Meissner [9.14] appear to support an unstable behavior at sufficiently high rates of extension for a polyethylene melt. Many viscoelastic fluid models predict an instability of the type shown here. Conversely, many models have been devised that do not predict an instability. The assessment of experimental data for a variety of polymers is inconclusive on the existence of an instability [9.15]. This state of affairs is typical of problems in polymer technology, yet this very situation provides the source for a rich field of research problems. The difficulties with extensional flow are discussed further in Section 9.6.

9.5. VISCOELASTIC LUBRICATION

Lubricating oils often contain long chain polymer additives, and these are said to increase the load bearing capacity of pad and journal bearings. There could be several reasons for this effect. We detail one possible source here that relates to the non-Newtonian nature of the fluid.

The problem we solve is that of a slider bearing configuration, shown in Fig. 9.3. Specifically, the lower surface in Fig. 9.3 undergoes a motion as

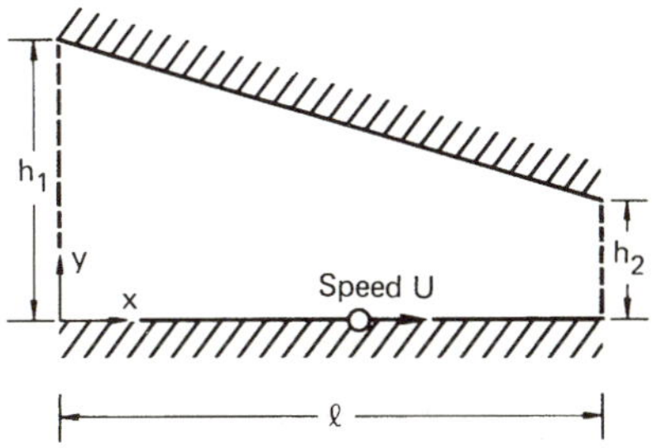

Fig. 9.3. Slider bearing.

shown, while the upper surface remains fixed. The problem is taken to be two dimensional, the fluid entering from the left in Fig. 9.3 and exiting from the right. We seek to determine the load that the bearing can support as characterized by the resultant force in the y direction. This problem was first solved for a Rivlin–Ericksen fluid by Tanner [9.16]. We here follow the analysis of Christensen and Saibel [9.17], which uses a different fluid model.

The starting point of the analysis is the incompressible fluid characterization through the Coleman–Noll second order form, (8.74), rewritten here as

$$\begin{aligned}\sigma_{ij} = -p\,\delta_{ij} &+ \int_{-\infty}^{t} \Delta(t-\tau, 0)\,\frac{\partial G_{ij}(\tau)}{\partial \tau}\,d\tau \\ &- \int_{-\infty}^{t}\int_{-\infty}^{t} \Delta(t-\tau, t-\eta)\,\frac{\partial G_{ik}(\tau)}{\partial \tau}\,\frac{\partial G_{kj}(\eta)}{\partial \eta}\,d\tau\,d\eta \\ &- \int_{-\infty}^{t}\int_{-\infty}^{t} \gamma(t-\tau, t-\eta)\,\frac{\partial G_{ij}(\tau)}{\partial \tau}\,\frac{\partial G_{kk}(\eta)}{\partial \eta}\,d\tau\,d\eta \qquad (9.148)\end{aligned}$$

where the relaxation functions $\Delta(\ )$ and $\gamma(\ )$ are taken to be symmetric in their arguments, and $G_{ij}(\tau)$ is given by (9.116). Following Section 8.5 and restricting

(9.148) to simple shearing flow gives the following results, comparable to (8.77) with only lowest order terms in the shear rate κ being retained:

$$\begin{aligned}
\sigma_{11} &= -p - \kappa^2 \int_0^\infty \int_0^\infty \Delta(u, v)\, du\, dv \\
\sigma_{22} &= -p - 2\kappa^2 \int_0^\infty \Delta(u, 0)\, u\, du - \kappa^2 \int_0^\infty \int_0^\infty \Delta(u, v)\, du\, dv \qquad (9.149) \\
\sigma_{12} &= \kappa \int_0^\infty \Delta(u, 0)\, du.
\end{aligned}$$

The simplicity of (9.149), involving only a single relaxation function $\Delta(u, v)$, will be of utility to us later. It should be emphasized that the reduction of the Coleman–Noll second order theory form results from the thermodynamical restriction that (9.149) be derived from a free energy functional. Note from (9.149) that in terms of the normal stress differences $N_1 = \sigma_{11} - \sigma_{22}$ and $N_2 = \sigma_{22} - \sigma_{33}$ there results $| N_2 | > N_1$, which is at variance with much experimental evidence. However, because our purpose is merely to assess the magnitude of normal stress effects, the form (9.149) has the overwhelming advantage of involving only a single mechanical property $\Delta(u, v)$, as well as being derived in a rational theoretical framework. Also, there is evidence that in approaching a low rate asymptotic limit $| N_2 |$ and N_1 are of the same size, and the smallness generally attributed to $| N_2 |$ compared with N_1 results from thinning effects, such as shear thinning with high rate. This matter is discussed at some length by Christensen and Saibel [9.17]. The main point is that the forms of (9.149) result from a consistent theoretical derivation, but we employ them with caution outside the range of their strict validity, as is common practice. Because we are only looking for the size of normal stress relative to shear stress, rather than the particular value of each, the application of (9.149) at high rates of deformation may be justifiable on an order of magnitude basis over a certain range of rates.

Next we characterize the relaxation function that appears in (9.149). Specifically, we take

$$\Delta(t) = \beta_0 e^{-t/\tau_0} + \beta_a e^{-t/\tau_a} \qquad (9.150)$$

where two relaxation times, τ_0 and τ_a, are employed. The first term in (9.150) is taken to be that of the base oil, while the second term is associated with the effect of the polymer additive. The charactesistic relaxation time of the base oil is very short and normally of no concern in lubrication problems. The relaxation time τ_a associated with the interaction between the additive molecules and the base oil is generally taken to be in the range of about 10^{-5}–10^{-3} sec. The steady state shear flow viscosity associated with (9.150) is given by

$$\eta = \beta_0 \tau_0 + \beta_a \tau_a . \qquad (9.151)$$

The relative magnitudes of the β_0 and β_a amplitudes in (9.150) and (9.151) will be considered later. The form (9.151) will be shown to separate conveniently effects of the base oil and the additive. Although we loosely speak of the second term in (9.151) as being due to the additive, it is, of course, actually a result of the interaction of the additive and the base oil.

Under simple shear flow conditions, the stress relations (9.149), together with the relaxation function (9.150), take the form

$$\begin{aligned} \sigma_{xy} &= (\beta_0\tau_0 + \beta_a\tau_a)\frac{\partial u}{\partial y} \\ \sigma_{xx} &= -p - (\beta_0\tau_0^2 + \beta_a\tau_a^2)\left(\frac{\partial u}{\partial y}\right)^2 \\ \sigma_{yy} &= -p - 3(\beta_0\tau_0^2 + \beta_a\tau_a^2)\left(\frac{\partial u}{\partial y}\right)^2 \end{aligned} \tag{9.152}$$

where a coordinate convention (as in the slider bearing of Fig. 9.3) is used, u being the velocity in the x direction. The steady state stress constitutive equations (9.152) will be applied to model the flow in the slider bearing of Fig. 9.3. Thus an assumption of quasi-steady flow is implied, with all transient effects in the non-Newtonian fluid being neglected.

Under ordinary conditions, normal stress effects are of no importance for base oils in lubrication because of the extreme smallness of the relaxation time τ_0. Thus the $\beta_0\tau_0^2$ terms in (9.152) will be neglected compared with the $\beta_a\tau_a^2$ terms, which involve a much longer relaxation time. A similar reduction in the shear stress term in (9.152) cannot be made, because the additive is known to have a strong effect on viscosity, and relaxation times enter viscosity to a different order than they enter the normal stress terms. The reduction leaves (9.152) in the form

$$\begin{aligned} \sigma_{xy} &= (\beta_0\tau_0 + \beta_a\tau_a)\frac{\partial u}{\partial y} \\ \sigma_{xx} &= -p - \beta_a\tau_a^2\left(\frac{\partial u}{\partial y}\right)^2 \\ \sigma_{yy} &= -p - 3\beta_a\tau_a^2\left(\frac{\partial u}{\partial y}\right)^2. \end{aligned} \tag{9.153}$$

Our objective is to assess the effect of the normal stress terms in the slider bearing configuration. We proceed by the usual analysis of slider bearing lubrication. Specifically, following the lubrication approximation, we consider only the momentum equation,

$$\frac{\partial \sigma_{xx}}{\partial x} + \frac{\partial \sigma_{xy}}{\partial y} = 0 \tag{9.154}$$

taking the other equation to be of higher order. Substituting (9.153) into (9.154) gives

$$\eta \frac{\partial^2 u}{\partial y^2} - \beta_a \tau_a^2 \frac{\partial}{\partial x} \left(\frac{\partial u}{\partial y}\right)^2 = \frac{\partial p}{\partial x} \tag{9.155}$$

where η is given by (9.151). Let the velocity u be given by

$$u = u_0 + u_1 y + u_2 y^2 + \cdots \tag{9.156}$$

where truncation will be taken at the explicit level shown, and (9.155) becomes

$$2\eta u_2 - \beta_a \tau_a^2 \frac{\partial}{\partial x} [u_1 + 2u_2 y]^2 = \frac{\partial p}{\partial x}. \tag{9.157}$$

The boundary conditions are of the type $u = U$ at $y = 0$, and $u = 0$ at $y = h$. Satisfying these conditions as well as (9.157) to lowest order consistent with the lubrication approximation gives u as

$$u = \left(\frac{-U}{h} - \frac{h}{2\eta} \frac{\partial p}{\partial x}\right) y + \frac{1}{2\eta} \frac{\partial p}{\partial x} y^2 + U. \tag{9.158}$$

We see that as a natural consequence of the lubrication approximation the non-Newtonian effects do not appear in (9.158).

The continuity equation is

$$\partial q_x / \partial x = 0 \tag{9.159}$$

where

$$q_x = \int_0^h u \, dy \tag{9.160}$$

and (9.159) takes the form

$$6U\eta \frac{\partial h}{\partial x} = \frac{\partial}{\partial x} \left(h^3 \frac{\partial p}{\partial x}\right) \tag{9.161}$$

where $\partial p / \partial x$ is given by

$$\frac{\partial p}{\partial x} = 6\eta \left(\frac{U}{h^2} - \frac{2q_x}{h^3}\right). \tag{9.162}$$

We shall use the conventions of Fig. 9.3, as in the derivation of Batchelor [9.18], with

$$h = h_1 - \alpha x. \tag{9.163}$$

Integrating (9.161) gives

$$p - p_0 = \frac{6\eta}{\alpha} \left[U \left(\frac{1}{h} - \frac{1}{h_1}\right) - q_x \left(\frac{1}{h^2} - \frac{1}{h_1^2}\right)\right]. \tag{9.164}$$

At $h = h_2$, we take $p = p_0$ such that the pressure is the same at each end of the slider bearing. This procedure gives

$$q_x = U\left(\frac{h_1 h_2}{h_1 + h_2}\right) \tag{9.165}$$

leaving (9.164) as

$$p - p_0 = \left(\frac{6\eta U}{\alpha}\right) \frac{(h_1 - h)(h - h_2)}{h^2(h_1 + h_2)} . \tag{9.166}$$

We shall have use for the integral of (9.166), which is

$$\int_0^l (p - p_0)\, dx = \frac{6\eta U}{\alpha^2}\left[\ln \frac{h_1}{h_2} - 2\left(\frac{h_1 - h_2}{h_1 + h_2}\right)\right]. \tag{9.167}$$

Our next objective is to determine the total load that can be supported by the bearing. The stress integral is decomposed into the parts

$$\int_0^l \sigma_{yy}\, dx = \int_0^l (-p)\, dx + \int_0^l \hat{\sigma}_{yy|y=0}\, dx \tag{9.168}$$

where the first term on the right-hand side of (9.168) is given by (9.167). From (9.153), $\hat{\sigma}_{yy|y=0}$ in the last term is given by

$$\hat{\sigma}_{yy|y=0} = -3\beta_a \tau_a^2 (\partial u/\partial y\,|_{y=0})^2. \tag{9.169}$$

Using the preceding results to evaluate the derivative in (9.169) gives

$$\hat{\sigma}_{yy|y=0} = -12\beta_a \tau_a^2 \left(-2 + \frac{3h_1 h_2}{h(h_1 + h_2)}\right)^2 \frac{U^2}{h^2} . \tag{9.170}$$

When $h = h(x)$, given by (9.163), is substituted into (9.170) and then integrated, we obtain

$$\int_0^l \hat{\sigma}_{yy}\, dx = -\frac{12\beta_a \tau_a^2 U^2}{\alpha}\left[\frac{2(h_1 - h_2)}{h_1 h_2} + \frac{3(h_1^3 - h_2^3)}{h_1 h_2 (h_1 + h_2)^2}\right]. \tag{9.171}$$

Now, substituting (9.167) and (9.171) into (9.168) and rearranging the algebraic forms gives

$$-\int_0^l \sigma_{yy}\, dx = p_0 l + \frac{6\eta U}{\alpha^2}\left\{\left[\ln\left(\frac{h_1}{h_2}\right) - 2\frac{h_1/h_2 - 1}{h_1/h_2 + 1}\right]\right.$$
$$\left. + \frac{2\beta_a \tau_a^2 U}{\eta l}\left[\frac{h_2}{h_1} \frac{(h_1/h_2 - 1)^2 (h_1^2/h_2^2 - h_1/h_2 + 1)}{(h_1/h_2 + 1)^2}\right]\right\}. \tag{9.172}$$

This expression gives the total load that can be supported by the bearing. The first bracketed term in (9.172) is just due to the wedging effect of the bearing, and it involves only Newtonian flow characteristics. This term is the usual Reynolds solution for the lubrication effect with a Newtonian fluid. The last term in (9.172) is the non-Newtonian effect resulting from the normal stresses.

We next examine the general characteristics of the non-Newtonian term in (9.172). Note first that this term involves no materials parameters independent of those that determine the viscosity coefficient η, given by (9.151). Thus, as we shall see, we can evaluate the size of the non-Newtonian term without appealing to independent measurements of the normal stress, measurements which are particularly difficult to determine.

The order of the coefficient of the non-Newtonian term inside the braces in (9.172) can be written as

$$O(\text{non-Newtonian coefficient}) = \beta_a \tau_a^2 U/\eta l. \qquad (9.173)$$

Because in (9.151) the additive makes a significant contribution to the viscosity so that $\beta_0 \tau_0$ and $\beta_a \tau_a$ are of the same order, (9.173) reduces to

$$O(\text{non-Newtonian coefficient}) = \tau_a U/l.$$

Thus, we see that the strength of the non-Newtonian effect relative to the Newtonian effect is not governed by the Deborah number, which would be expressed as $\tau_a U/h$. Of course, the abolute (not relative) magnitude of the non-Newtonian term is directly related to the magnitude of the normal stress effect.

Examples have been worked by Christensen and Saibel [9.17] showing that the present theory predicts a non-Newtonian effect that is at a threshold significance in the range of practical bearing and fluid characteristics. These characteristics include the geometry of the bearing, the speed, the relaxation time of the additive in solution, and the contribution of the additive to the total viscosity of the fluid. The examples show that the load bearing capability of the non-Newtonian effect is significant for certain practical ranges of these parameters but is insignificant for others. In general, the effect must be evaluated explicitly from the result (9.172) to determine its possible significance. The added effect of shear thinning is discussed below. Shear thinning is the decrease of the viscosity $\eta(\kappa)$ with increasing shear rate κ.

It is an important feature of the present method that only the relaxation time of the additive and its effect on the viscosity are needed to evaluate the load capacity of the bearing. This is in contrast to many second order theories, in which an independent normal stress coefficient is needed to evaluate the magnitude of the non-Newtonian effect. Of course, the present use of the Coleman–Noll second order theory limits the range of applicable rates, as does any second order theory. However, in the present application only the relative effect of Newtonian and non-Newtonian sources is sought. Of course, the second order theory does not model shear and normal stress thinning effects. If equivalent-magnitude thinning effects were to occur in shear and normal stress terms, then the results could have relative validity beyond the range of the usual second order theory. Typically, however, the normal stresses show a stronger thinning effect than does the shear stress. The present method shows that in fluids where this is the case, the additive gives negligible load support through the normal stress contribution. Thus the thinning effect is of importance to the lubrication applications. Certainly, the results show the strength of the non-Newtonian effect in the shear rate range of the inception of non-Newtonian effects, and these results provide a basic solution in that respect.

9.6. NONLINEAR THEORY MECHANICAL PROPERTIES

We can see from Chapter 4 that the general approach and conceptual procedures for determining mechanical properties appropriate to the linear theory are well established and understood. The primary difficulty involved in determining linear theory mechanical properties resides in the actual experimental implementation of such procedures. The situation with regard to nonlinear theories of viscoelasticity is much less straightforward. Not only is it more difficult in general to determine mechanical properties for a nonlinear theory than for a linear theory, but in the particular case of viscoelasticity there is not even common agreement as to which of many possible nonlinear theories should be used. Accordingly, we shall not be too much concerned here with detailed results for specific materials; rather, we shall just broadly outline a few typical approaches.

We find it convenient to subdivide the general area of nonlinear viscoelastic mechanical properties determination into procedures appropriate to solids and fluids. Even this distinction is not as clear cut as it might seem. For example, materials that are at a temperature far below their glass transition temperature have a glassy, solidlike behavior even though they may in fact be fluids. In or near the glassy state it is very difficult to mechanically ascertain whether a preferred configuration exists corresponding to solid behavior. The relaxation times can be so long in this case that it is not practical to wait long enough to see if the material has a long-time nonzero asymptote for the relaxation or creep functions. However, elevated temperature testing can be used to make the solid versus fluid determination. Also, knowledge of the molecular structure serves to establish the material type. Nevertheless, near the glassy state it may not even be relevant to distinguish between solidlike or fluidlike behavior. As will be seen in the following discussion, fluid-type theories have been used to model the behavior of solids and the converse. Of course, from a mathematical point of view, the treatments of solids or fluids are very different in the nonlinear range of behavior, since completely different types of kinematics are involved in the two cases, as discussed in Sections 8.1 and 8.4.

The fact that the distinction between solids and fluids is blurry in some applications reveals the complicated nature of the field of modeling nonlinear viscoelastic behavior. Once the guiding principles and postulates of mechanics are satisfied, there is freedom to proceed by any means possible. The wide variety of existing nonlinear methods and models is testimony to the expedient nature of the field. Until rather recently, efforts were made based on the expectation that very general nonlinear constitutive equations ultimately would be obtained for modeling broad classes of material behavior. That expectation has not been realized. A more realistic approach involves the delineation of a variety of nonlinear constitutive relations, each of which may have only limited but well-understood ranges of application. The key objective in seeking such constitutive relations is to provide physical realism in the underlying models and to have the relations in forms that are not so complicated that they cannot admit practical application.

Solids

A common approach in determining the mechanical properties for viscoelastic solids is based upon the use of the Green–Rivlin theory [9.19]. This theory involves a representation in which stress is expressed as a polynomial expansion in linear functionals of strain history. The procedure is similar to that outlined in Section 8.1 where the stored energy was expressed by a polynomial expansion in linear functionals of strain history. With the series truncated at some level, the number of integrating functions, or kernel functions, to be determined corresponds to the number of terms retained. Although this theory is of a three-dimensional character, generally properties relevant to the three-dimensional form have not been determined. Rather, most mechanical properties which have been determined are appropriate to the one-dimensional specialization of the theory. This, of course, severely limits the general usefulness of such mechanical properties. However, the methods and techniques which are found to be useful in the one dimensional case certainly suggest the direction to be followed in the more general case.

The one-dimensional uniaxial stress strain constitutive relation, from the Green–Rivlin theory has the form

$$\begin{aligned}\sigma(t) = &\int_{-\infty}^{t} G_1(t-\tau)\frac{dE(\tau)}{d\tau}\,d\tau \\ &+ \int_{-\infty}^{t}\int_{-\infty}^{t}\int_{-\infty}^{t} G_3(t-\tau_1, t-\tau_2, t-\tau_3) \\ &\times \frac{dE(\tau_1)}{d\tau_1}\frac{dE(\tau_2)}{d\tau_2}\frac{dE(\tau_3)}{d\tau_3}\,d\tau_1\,d\tau_2\,d\tau_3 + \cdots \end{aligned} \tag{9.174}$$

where $E(\tau)$ is the nonlinear strain measure defined by the one-dimensional counterpart of (8.2). It should be noted that constitutive relation (9.174) is perfectly consistent with the viscoelastic solid theory of Section 8.1. In fact, using the method of derivation of Section 8.1, the form (9.174) follows directly from a comparable one-dimensional representation of the stored energy involving only even ordered terms in the polynomial expansion. The omission of even ordered terms in (9.174) follows directly from the omission of odd ordered terms in the comparable stored energy form; otherwise a nonnegative character of the stored energy could not be assumed. The integrating functions in (9.174) such as

$$G_1(\tau_1), \quad G_3(\tau_1, \tau_2, \tau_3), \quad G_5(\tau_1, \tau_2, \tau_3, \tau_4, \tau_5), \ldots, \qquad \tau_i \geqslant 0$$

are the relaxation function type properties to be determined for each different material of interest. These functions are not intrinsic mechanical properties of the material since it is the nature of the approximation theorem upon which this representation is based that the form of these functions depends upon the level at which the series is truncated. Also, the multidimensional nature of these integrating functions renders their determination to be a rather difficult matter. For example, the determination of $G_3(\tau_1, \tau_2, \tau_3)$ requires the characterization of a surface in the four-space comprised of $G_3, \tau_1, \tau_2, \tau_3$. The difficulties involved in the determination of the integrating functions which come from the higher order terms becomes obvious. In a particular case, Lockett [9.20] considers the number of independent tests required to characterize these properties. Despite these difficulties some progress has been made in determining these properties or similar ones related to comparable creep integral formulations. This alternative creep integral formulation is next stated and after that some actual methods of determining the properties are considered.

We cannot analytically invert relation (9.174) to express strain as a functional of stress history, but by formally applying the same expansion procedure that lead to (9.174), with the roles of stress and strain interchanged, we arrive at

$$\begin{aligned} E(t) = & \int_{-\infty}^{t} J_1(t-\tau) \frac{d\sigma(\tau)}{d\tau} \, d\tau \\ & + \int_{-\infty}^{t} \int_{-\infty}^{t} \int_{-\infty}^{t} J_3(t-\tau_1, t-\tau_2, t-\tau_3) \\ & \cdot \frac{d\sigma(\tau_1)}{d\tau_1} \frac{d\sigma(\tau_2)}{d\tau_2} \frac{d\sigma(\tau_3)}{d\tau_3} \, d\tau_1 \, d\tau_2 \, d\tau_3 + \cdots \end{aligned} \tag{9.175}$$

where the integrating functions $J_i(\ )$ are the creep-type properties which characterize the material. Most of the experiments which have been conducted to determine mechanical properties utilize (9.175) rather than (9.174) since the conditions of creep at constant stress are experimentally easier to realize than conditions of relaxation at constant strain. The levels of deformation in these tests are sufficiently low that it is not necessary to distinguish between constant-stress and constant-load type tests.

A typical test result of this type is given by the data of Ward and Wolfe [9.21]. This work utilizes a creep integral constitutive relation which is truncated at the cubic functional term, and is based upon the previous work of Ward and Onat [9.22]. Under conditions of a single step creep test, the form assumed by (9.175) is given by

$$E(t) = J_1(t)\sigma_0 + J_3(t, t, t)\,\sigma_0^3$$

where

$$\sigma(t) = 0, \qquad t < 0$$
$$\sigma(t) = \sigma_0, \qquad t \geqslant 0.$$

Thus, single step creep tests at two different levels of stress can be used to determine the functions $J_1(t)$ and $J_3(t, t, t)$. In Ref. [9.21] it is shown that more general forms of J_3, namely $J_3(t - t_1, t - t_1, t)$ and $J_3(t - t_1, t, t)$ where $t > t_1$ can be determined from the creep and recovery in a single step creep loading program along wih the creep response in a loading program involving the superposition of two identical loads. Since only two terms are retained in the stress constitutive relation, test data can be fit exactly only at two separate levels of stress, which limits the amount and/or type of nonlinearity which can be treated. The mechanical properties determined from these types of creep tests are used in Ref. [9.21] to predict the response to two consecutive step changes in unequal stress level, with satisfactory results. More extensive comparisons of this type have been given by Findley and his coworkers in references to be cited.

A similar but more general type of mechanical properties determination procedure has been given by Neis and Sackman [9.23]. Also a similar type of study has been given by Lifschitz and Kolsky [9.24]. Findley and his coworkers have made several mechanical properties studies using multiple integral expansions. In some of these works the integrating functions, as for example in relation (9.175), are assumed to have the special product form given by

$$J_i(\tau_1, \tau_2, \ldots, \tau_i) = \hat{J}_i(\tau_1)\, \hat{J}_i(\tau_2) \cdots \hat{J}_i(\tau_i). \tag{9.176}$$

Thus rather than having to determine i dimensional functions $J_1(\tau_1, \tau_2, \ldots, \tau_i)$, the problem is reduced to determining one dimensional functions $\hat{J}_i(\tau)$. An approach of this type has been given by Findley and Onaran [9.25]. Gottenberg *et al.* [9.26] have employed a somewhat similar means of simplifying the procedure for determining mechanical properties in their study of stress relaxation experiments. However, rather than taking a product form of the integrating functions, as in (9.176), they assume

$$G_i(\tau_1, \tau_2, \ldots, \tau_i) = \tilde{G}_i(\tau_1 + \tau_2 + \cdots + \tau_i)$$

which again reduces the multidimensional function to be determined to a one-dimensional function. Yet other forms for the integrating functions have been assumed by Smart and Williams [9.27].

A completely different type of mechanical properties determination procedure, known as the modified superposition principle, has been explored by Findley and coworkers. This procedure, originated by Leaderman [9.28], as described by Findley and Khosla [9.29], involves an assumption that the strain response to a series of N step changes in stress level can be expressed as

$$E(t) = \sum_{i=1}^{N} [F(\sigma_i, t - t_i) - F(\sigma_{i-1}, t - t_i)] \tag{9.177}$$

where

$$F = F(\sigma_i, t - t_i) \tag{9.178}$$

is the creep-type response beginning at time $t = t_i$ for a previously unloaded material being subjected to a suddenly applied stress level σ_i. In the linear approximation, F would be proportional to σ_i; in the nonlinear case there is allowed a general nonlinear dependence of F on σ_i. We see from formula (9.177) that the strain response to a series of suddenly applied changes of stress level assumes that the response is due to a sequence of applied and removed stress levels, considered as acting upon a previously unloaded specimen. This assumed type of behavior does not have any theoretical basis, in the sense that (9.174) and (9.175) do; however, it has been shown to be a useful relation in some cases. Lai and Findley [9.30] have expressed this modified superposition principle in the alternate form involving the stress response to a series of step changes in strain, and have compared certain resulting predictions therefrom with those obtained from the direct application of (9.175) which uses the reduced forms of the generating functions (9.176). Both procedures were shown to be satisfactory. It may be noted that this modified superposition principle procedure is in accordance with behavior of the linear theory of viscoelasticity therefore the usual linear theory behavior can be recovered as a limiting case of this type of nonlinear behavior. Pipkin and Rogers [9.31] have proposed a generalization of the procedure involved in the modified superposition principle.

Mechanical properties relative to the two-dimensional behavior of non-linear viscoelastic solids have been considered, as for example by Findley and Lai [9.32] and Lai and Findley [9.33]. However these multidimensional formulations are somewhat arbitrary in as much as they are not deduced as special cases of a more general theory, but rather they are essentially one-dimensional formulations with additional cross coupling terms added. Findley *et al.* [9.34] have provided a general compilation of such procedures.

Stafford [9.35] reviewed several Green–Rivlin type procedures, as well as others. Lockett [9.36] provided an extensive discussion of nonlinear models for solids, with the emphasis on Green–Rivlin procedures. In connection with the general use of Green–Rivlin type expansions, Gradowczyk [9.37] has shown that the kernels in these types of expansions are in general very sensitive to experimental errors and this sensitivity to errors increases with the number of terms in the expansion. Despite this restriction, the results just cited seem to indicate reasonably successful procedures for determining mechanical properties of nonlinear viscoelastic solids. However, in most of these experimental results one or more of the following limitations are present: (1) the deformations are sufficiently small that the infinitesimal definition of strain may be used, (2) in the range of deformations considered the material is only weakly nonlinear, or (3) the material is only slightly viscoelastic. Any of these limitations provides a considerable simplification in the complexity of the problem. There is very little experimental work on the determination of three-dimensional theory mechanical properties for strongly viscoelastic materials in the range of strongly nonlinear deformations.

One work which is outside the scope of the limitations just mentioned is that of Lianis and colleagues in application of the special theory known as finite linear viscoelasticity. This theory, mentioned in Section 8.2, was developed by Coleman and Noll [9.38]. Following the methodology of Sections 8.1 and 8.4, the basic kinematics of the problem can be expressed as a dependence on the current configuration, measured relative to the reference configuration, and the past history deformation, measured relative to the current configuration. Accordingly, the norm of the history can be made small if the deformation of the material is slowly changing, even though the deformation levels themselves may be arbitrarily large. The smallness of the norm follows from (i) the fading memory character of the material, (ii) the fact that the kinematical history is expressed relative to the current configuration, and (iii) the history is slowly changing relative to the current configuration. With a small norm, the representation theorem employed in Section 8.1 can be used to express stress in terms of a linear functional of the history. In contrast to the infinitesimal theory, however, the integrating functions in the functional must include a dependence on the deformation measure of the current configuration relative to the reference configuration. In fact, this dependence is the means by which the solid material remembers its preferred configuration. The theory then takes the form

$$\sigma_{ij}(t) = \sigma_{ij}^{\infty}(\mathbf{C}) + x_{i,K}(t)\, x_{j,L}(t) \int_{-\infty}^{t} g_{KLmn}(\mathbf{C}, t-\tau) \frac{\partial G_{mn}(\tau)}{\partial \tau}\, d\tau \quad (9.179)$$

where $x_{i,K}(t)$ is the deformation gradient relative to the reference configuration, and $\mathbf{C}$ is the right Cauchy–Green deformation tensor relative to the reference configuration,

$$C_{KL} = \frac{\partial x_k(t)}{\partial X_K} \frac{\partial x_k(t)}{\partial X_L}.$$

$G_{mn}(\tau)$ is the history dependent deformation measure, (8.47), relative to the current configuration and $\overset{\infty}{\sigma}_{ij}$ is the equilibrium or long-time stress, which depends upon $\mathbf{C}$. The mechanical properties $g_{KLmn}(\ )$ are time dependent relaxation functions which also depend upon the current state of deformation. According to Lianis [9.39], material isotropy provides restrictions such that $\overset{\infty}{\sigma}_{ij}$ can be expressed in terms of three scalar functions of $\mathbf{C}$ and $g_{KLmn}(\mathbf{C}, t)$ can be represented by nine scalar functions of $\mathbf{C}$ and t. Thus there is a total of 12 independent material properties, nine of which are time dependent. In a series of references [9.40–9.43], Lianis, DeHoff, Goldberg, McGuirt, and Ting have argued and assumed certain simplifications in the 12 properties and have successfully modeled a variety of behaviors for certain polymers, including elastomers.

An entirely different approach to nonlinear modeling has been given by Schapery [9.44, 9.45]. Proceeding from a thermodynamical basis involving internal variables, and the division of effects into "forces" and "fluxes," he arrived at a general nonlinear constitutive relation. The resulting expressions have a parallel formalism, whether expressed in relaxation integral or creep integral forms. Using infinitesimal strain ϵ, the uniaxial form in terms of a relaxation function is given by

$$\sigma = \sigma_e + h_1 \int_0^t \Delta G(\xi - \xi') \frac{d\tilde{\epsilon}(\tau)}{\partial \tau}\, d\tau \tag{9.180}$$

where

$$\xi = \int_0^t \frac{d\tau}{a_\epsilon(\tau)}$$

and σ_e, h_1 and $\tilde{\epsilon}$ are quantities which have a nonlinear function-type dependence on the strain ϵ. The relaxation function ΔG is the linear theory form according to the decomposition

$$\Delta G(t) = G(t) - G(t)\mid_{t\to\infty}$$

but the shift factor $a_\epsilon(\tau)$ has a general nonlinear dependence on the strain level. Equation (9.180) is obviously a generalization of the linear theory form. A specialization of (9.180) to the case where the nonlinearity enters only through strain ϵ and equilibrium stress σ_e corresponds to the modified superposition principle mentioned earlier. Specifically, the relaxation function is taken to have a nonlinear dependence on the strain level through a scalar multiplier.

Yet another widely used theory is that of the BKZ model, stated in the next section. Actually this theory was derived explicitly for fluids. Nevertheless, it has been used to model solids. All these theories have been applied to model polymers in the small strain range where the distinction between solids and fluids is unimportant. Smart and Williams [9.46] have compared the use of the modified superposition principle, the Schapery theory, and the BKZ theory in strain ranges up to 5 % for two different polymers. The distinctions between the different models were not sharp.

Mechanical models have sometimes been used to obtain nonlinear forms. Specifically, the springs or the dashpots can be endowed with a nonlinear behavior that then produces a nonlinear constitutive relation. We do not find a procedure such as this to be particularly helpful or enlightening; the situation is much the same as that stated in connection with the linear theory in Section 1.5. It also may be noted that arguments have been made in favor of inserting masses into mechanical models. This, of course, could be done in either the linear or the nonlinear context. Our position is that mass belongs in the balance relations, i.e., balance of momentum and conservation of mass. To insert mass effects arbitrarily into the constitutive relations through mechanical models does not enhance basic physical understanding.

Other theories include that of Peng *et al.* [9.47], whereby the free energy is assumed to be a separable function of the three principle stretch ratios and a set of internal variables. Theories of viscoelasticity that include nonlinear plasticity effects have been discussed by Murch and Naghdi [9.48] and Robotnov [9.49]. A hereditary-type theory, known as endocronic theory, has been devised by Valanis [9.50] to model inelastic but rate independent effects. Applications have been primarily to the behavior of metals and concrete. Section 9.1 gave an example of a simple theory derived explicitly for elastomers in a particular range of deformation. The situation with respect to polymers in the range of glassy behavior is more complicated, as we next discuss.

Consider for a moment the distinction between solid polymers in or near the glassy range and polymers in or near the rubbery range. Materials appropriate to the latter range of behavior, elastomers, are obviously describable by nonlinear elasticity theory for slow processes and by nonlinear viscoelasticity for processes which are not slow. The rate effect brings out the dissipative properties of the material. A rather simple nonlinear theory would be reasonably expected to model time and rate effects if the state of deformation is not too extreme. For glassy polymers, on the other hand, the situation is much more complicated in the nonlinear range of behavior. Such materials are observed to undergo local failure, even at moderate strain levels. Local instabilities, in the form of crazing and shear banding, often dominate behavior. A broad distribution of local regions of high deformation certainly contributes overwhelmingly to the overall material behavior. Macroscopic theories that do not account for local failure or "damage" amount to little more than curve fitting. Another class of materials which would be expected to possess a complicated behavior is the semicrystalline polymers. The crystalline domains may be strain level dependent, they may partially act as effective cross links for the molecular structure, or they can act as an anisotropic filler phase. Nonlinear moldeling for such materials must be very complicated.

One last technical complication should be mentioned. Nonlinear properties are often deduced indiscriminately from either creep tests or relaxation tests. The test to be used is usually selected on a basis of experimental expediency. In the linear theory it makes no difference which is used; the resulting properties are interconvertible. In the nonlinear range of deformation, however, the results are not interchangeable. This nonlinear complication is easily reasoned as follows. Consider an elastomer which is tested by the use of both a relaxation test and a creep test. Furthermore, take the deformation and stress level in both tests to be equivalent at the long-time equilibrium value. Using uniaxial linear theory results, it can easily be shown that the initial stress $\sigma_R(0)$ in the relaxation test is related to the initial strain $\epsilon_C(0)$ in the creep test through

$$\sigma_R(0) = [G(0)/G(\infty)]\, G(0)\epsilon_C(0)$$

where $G(t)$ is the relaxation function. This result shows the initial state of stress and strain in the relaxation test is magnified by the factor $G(0)/G(\infty)$ over the initial state of stress and strain in the creep test. This magnification is typically by many orders of magnitude. If nonlinear effects are present, obviously the relaxation test provides a far greater penetration of the nonlinear range of behavior than does the creep test when the final equilibrium states are identical in the two tests. More generally, displacement control of material deformation can provide much different types of nonlinear behavior than would be caused by load control. This distinction appears to have been largely ignored.Another way of saying this is that in nonlinear viscoelasticity the type and degree of nonlinearity depend not only on the current value of stress and deformation but also on the entire past history. This complication is discussed somewhat further by Christensen [9.1].

Finally, a note of caution should be conveyed about modeling nonlinear effects in viscoelasticity. Most theories contain many parameters or functions which are to be adjusted to fit data. Often, one more additional effect is modeled by inserting yet one more parameter. A better approach would be to minimize the number of unknown parameters or functions by maximizing the physical basis or content of the underlying theory. Fitting every corner and inflection point of experimental data may be misleading. It is more rational to recognize the essential aspects of mechanical behavior and relegate the other facets to secondary roles.

Fluids

The flow behavior of non-Newtonian fluids has been a subject of extensive study. Early efforts attempted to account for nonlinear effects by simply allowing the coefficient of viscosity in a Newtonian fluid constitutive relation to be a function of the rate of deformation. This approach is recognized to be inadequate, because, among other things, it does not account for the normal stress effects which are observed in the simple shear flow of viscoelastic fluids. All other attempts to formulate a theory of viscoelastic fluids less general than functional theories of the type considered in this book have exhibited various shortcomings. For example, the Reiner–Rivlin formalism takes the stress to be a function of tensor products of the rate of deformation tensor. However, the resulting forms do not give the general features exhibited by viscoelastic fluids.

Special theories, however, are applicable in certain situations. The steady viscometric flows discussed in Section 9.3 are the best examples of flows for which a nonfunctional but function-type theory is valid. These flows have the characteristic that everywhere the local deformation is that of simple shear flow, and therefore, in a fairly wide class of steady flow problems, the flow characteristics are governed by the shear and normal stress functions (functions of the rate of deformation), determined from a steady simple shear experiment. This point of view is fully explored by Coleman *et al.* [9.12]. Another useful function-type theory is that involving the Rivlin–Ericksen tensors, as discussed in Section 9.4. Henceforth, however, we shall not be concerned with special theories of this type, but rather we shall examine mechanical properties in the light of theories that attempt to characterize the general behavior of viscoelastic fluids. Our attention is restricted to incompressible fluids.

Begin by reconsidering the Maxwell model, written in the differential form

$$\sigma_{ij} + \tau \frac{d\sigma_{ij}}{dt} = \eta\, d_{ij} \tag{9.181}$$

where τ is the relaxation time, η is the steady state viscosity, d_{ij} is the rate of deformation tensor, (8.4), and σ_{ij} here is the nonhydrostatic part of the stress tensor. The form (9.181) is appropriate only to infinitesimal deformation conditions even though we have replaced the infinitesimal strain rate tensor $\dot{\epsilon}_{ij}$ by d_{ij}. This restriction is due to the presence of the derivative with respect to time in (9.181). Oldroyd [9.51] was the first to recognize the need for a special derivative operation in order to satisfy the principle of material frame indifference, or objectivity, as discussed in the introduction to Chapter 8. There is no unique definition of such a frame indifferent derivative operation. Written in rectangular Cartesian components, the three most common derivative forms are

$$\frac{\mathfrak{d}\sigma_{ij}}{\mathfrak{d}t} = \frac{\partial\sigma_{ij}}{\partial t} + v_k\sigma_{ij,k} - v_{i,k}\sigma_{kj} - v_{j,k}\sigma_{ik} \tag{9.182}$$

$$\frac{\mathfrak{d}\sigma_{ij}}{\mathfrak{d}t} = \frac{\partial\sigma_{ij}}{\partial t} + v_k\sigma_{ij,k} + v_{k,i}\sigma_{kj} + v_{k,j}\sigma_{ik} \tag{9.183}$$

$$\frac{\mathscr{D}\sigma_{ij}}{\mathscr{D}t} = \frac{\partial\sigma_{ij}}{\partial t} + v_k\sigma_{ij,k} - \omega_{ik}\sigma_{kj} - \omega_{jk}\sigma_{ik} \tag{9.184}$$

Unfortunately, the same symbolic designation is usually used for (9.182) and (9.183) even though they are completely different forms. In curvilinear coordinates (9.182) is usually referred to as the upper or contravariant convected derivative, while (9.183) is the lower or covariant convected derivative. The form (9.184) is known as the Jaumann or corotational derivative. Formally, the convected derivatives differ from the Jaumann derivative in that they involve velocity gradients $v_{i,k}$ whereas the Jaumann derivative involves the vorticity tensor ω_{ik}.

Any of the three forms (9.182)–(9.184) when used as a replacement for the time derivative in (9.181) renders the Maxwell model appropriate to unrestricted flow. The physical interpretation of the derivatives (9.182)–(9.184) is meaningful. As discussed by Schowalter [9.52], by Astarita and Marrucci [9.53], and by Huilgol [9.54], the form (9.182) results from the definition of a time derivative as seen by an observer moving with coordinates which are embedded in or convected with the medium and for which the unit vector basis is in coordinate directions. Alternatively, (9.183) depends on the reciprocal basis, which involves unit vectors normal to coordinate planes. Both bases translate, rotate, and deform with the material as it flows. However, a linear combination of these two bases is possible for which the resulting basis rotates but does not deform with the material. This type of basis is the source of the Jaumann or corotating derivative, (9.184). The Maxwell model, (9.181) with (9.182)–(9.184), can be generalized to a spectrum of relaxation times, and corresponding hereditary integral forms can be found. Many such generalizations are extensively discussed by Bird *et al.* [9.55]. However, no such simple models have been found to give a general and satisfactory model of viscoelastic fluid behavior. Other models have been proposed which mathematically resemble the combination of a Maxwell and a Kelvin model, but these too have not met with general success For practical applications to polymer melt processing problems, Middleman [9.56] recommends the use of the upper convected derivative (9.182) in the Maxwell model as providing a balance between physical reality and the simplicity necessary for the numerical solution of practical problems. Sometimes the shear viscosity in the Maxwell model is taken to be shear rate dependent.

Bernstein *et al.* [9.57] have proposed the model which is commonly called the BKZ model. The constitutive relation for viscoelastic fluids is in the form

$$\sigma_{ij} = -p\,\delta_{ij} + \int_{-\infty}^{t} [\mu_1(t-\tau, \mathrm{I}, \mathrm{II})\, C_{ij}(\tau) + \mu_2(t-\tau, \mathrm{I}, \mathrm{II})\, C_{ij}^{-1}(\tau)]\, d\tau \tag{9.185}$$

where

$$C_{ij}(\tau) = \frac{\partial x_k(\tau)}{\partial x_i(t)} \frac{\partial x_k(\tau)}{\partial x_j(t)} \tag{9.186}$$

and the inverse tensor $C_{ij}^{-1}(\tau)$ is defined through

$$C_{ik}^{-1}(\tau)\, C_{kj}(\tau) = \delta_{ij}\,, \qquad C_{ij}^{-1}(\tau) = \frac{\partial x_i(t)}{\partial x_k(\tau)} \frac{\partial x_j(t)}{\partial x_k(\tau)}\,. \tag{9.187}$$

The hydrostatic pressure is designated p. The relaxation functions μ_1 and μ_2 in (9.185) depend on the invariants I and II of the $C_{ij}^{-1}(\tau)$ Finger tensor and are derived from a potential function U through

$$\mu_1(t-\tau, \mathrm{I}, \mathrm{II}) = -2\,\frac{\partial U(t-\tau, \mathrm{I}_{C_{ij}^{-1}}(\tau), \mathrm{II}_{C_{ij}^{-1}}(\tau))}{\partial \mathrm{II}_{C_{ij}^{-1}}(\tau)}$$

and

$$\mu_2(t-\tau, \mathrm{I}, \mathrm{II}) = 2\,\frac{\partial U(t-\tau, \mathrm{I}_{C_{ij}^{-1}}(\tau), \mathrm{II}_{C_{ij}^{-1}}(\tau))}{\partial \mathrm{I}_{C_{ij}^{-1}}(\tau)}\,.$$

The constitutive relation (9.185) characterizes the stress as determined by linear functionals of the deformation history, $C_{ij}(\tau)$ and $C_{ij}^{-1}(\ )$, but with nonlinearity introduced through the deformation measures and the dependence of the relaxation functions μ_1 and μ_2 on the deformation history relative to the current configuration.

A somewhat different type of theory is that known as the Bird–Carreau model [9.58]. The model takes the form

$$\sigma_{ij} = -p\,\delta_{ij} + \int_{-\infty}^{t} \mu(t-\tau, \mathrm{II})[(1+\tfrac{1}{2}\epsilon)\, C_{ij}^{-1}(\tau) - \tfrac{1}{2}\epsilon C_{ij}(\tau)]d\tau \tag{9.188}$$

where C_{ij} and C_{ij}^{-1} are defined by (9.186) and (9.187), respectively, or by forms that differ from these only by δ_{ij}, and ϵ is an empirical parameter to be determined. The relaxation function μ is taken to have a dependence on the rate of deformation tensor through the invariant

$$\mathrm{II}(\tau) = d_{ij}(\tau) d_{ij}(\tau)$$

where d_{ij} is the rate of deformation tensor given by (8.4). A specific form for the relaxation function μ in (9.188), by Bird and Carreau [9.58], involves many parameters to be determined from experimental data. A further generalization, given by Carreau [9.59], has the same form for stress, (9.188), but a different relaxation function $M(t-\tau, \text{II})$. The essential difference between the two constitutive relations (9.185) and (9.188) is that in the former the relaxation functions include a dependence on the history of the deformation relative to the current configuration, whereas in the latter form the relaxation function is taken to depend on the history of the rate of deformation. This is certainly a fundamental difference, and we shall consider it further. It should be noted that neither of the theories of behavior represented by (9.185) and (9.188) are straightforward resulting forms of the general theory of viscoelastic fluids given in Chapter 8. Nevertheless, they remain of value if they can be demonstrated to model realistically the behavior of typical viscoelastic fluids. A different type of thermodynamical derivation from that given in Chapter 8 has been given by Bernstein *et al.* [9.60] in support of the constitutive relation (9.185).

Many viscoelastic fluid theories are derived from so-called network considerations. In such theories the molecular composition is idealized as an interconnecting network. The first theory of this type was that of Lodge [9.61], as applicable to a rubberlike liquid. The two theories mentioned in connection with (9.188) were derived from network considerations; another network theory is that of Liu *et al.* [9.62]. The primary feature of network theories is a balance or lack of balance between the creation and destruction of network junctions, thereby modeling the state of entanglements between the molecules. Network theories have been proposed that distinguish between the kinematics of the macroscopic continuum and the kinematics of the network itself. That is,network junctions are not taken to move in an affine manner relative to the continuum. The net effect is the introduction of additional kinematical variables. Such theories are those of Johnson and Segalman [9.63], Phan-Thien and Tanner [9.64], and Phan-Thien [9.65]. The theory of Phan-Thien takes a form reminiscent of a generalized Maxwell model,

$$\sigma_{ij} = \sum_n \sigma_{ij}^{(n)} \tag{9.189}$$

where

$$\lambda_n \frac{D}{Dt} \sigma_{ij}^{(n)} + K\sigma_{ij}^{(n)} = G_n \lambda_n d_{ij} \tag{9.190}$$

with λ_n and G_n being specified by the infinitesimal theory relaxation function. D/Dt is the derivative

$$\frac{D}{Dt} \sigma_{ij} = \frac{\partial \sigma_{ij}}{\partial t} + v_k \sigma_{ij,k} - \mathscr{L}_{ik}\sigma_{kj} - \mathscr{L}_{jk}\sigma_{ik} \tag{9.191}$$

where

$$\mathscr{L}_{ij} = v_{i,j} - \xi\, d_{ij} \tag{9.192}$$

which is written in terms of the velocity gradient and the rate of deformation tensor. The parameter ξ is a shear rate dependent adjustable quantity, and K depends on the trace of the stress tensor and is related to the rate of destruction of network junctions. The variable $\mathscr{L}_{ij}$, (9.192), is the effective gradient written in terms of *both* the velocity gradient and the rate of deformation tensor, rather than in terms of either as in the derivative definitions (9.182)–(9.184). Lau and Schowalter [9.66] have provided a conceptual framework which includes the above theory by introducing time derivatives which are a linear combination of the convected and Jaumann time derivatives in (9.182) and (9.184).

In a completely different approach, Doi and Edwards [9.67] and de Gennes [9.68] have prescribed a molecular model which replaces the network model having entanglements by a more general conception of molecular interaction. Specifically, molecular interactions are considered to occur not at points but as sliding along the contours of the molecules. The constraint of molecular configuration is provided by a "tube" of constraint along which the molecule moves. The model has had considerable success in providing correlations with molecular parameters [9.69].[1] Many other theories and models of various types are available; a discussion of these is given by Petrie [9.15]. These models include the eight constant model of Oldroyd and those of Chang *et al.*, Ward and Jenkins, Bogue and White, Tanner, Metzner and colleagues, Marrucci and colleagues, and Acierno.

There is no all-inclusive experimental procedure for determining viscoelastic fluid mechanical properties such as the relaxation functions in (9.185) and (9.188). Certainly the standard relaxation test procedure involving the application of a step change in the deformation state could be used. However, other types of experiments suggest more critical tests of a fluid theory. One such procedure involves the transient stress growth from a state of rest of suddenly imposed, steady simple shear flow and stress decay due to the immediate cessation of a state of steady simple shear flow. It is not clear, however, that mechanical properties for the types of theories considered here as determined from any one of these procedures could also be expected to apply perfectly generally. The reasons for this uncertainty may be seen from a consideration of some typical experimental data. Vinogradov and Belkin [9.70] have obtained data for the initiation from a state of rest of a suddenly imposed, steady state of simple shear flow. Actually, the flow state is that imposed by the flow in the space between a flat plate and a nearly flat rotating cone, as in Section 9.3. The resulting transient shear stress for polystyrene at elevated temperatures is as shown schematically

[1] I am appreciative to Dr. D. J. Meier for pointing out these references to me.

in Fig. 9.4. At low rates of deformation the stress increases monotonically with time to the asymptotic steady state value. But as the step function in the rate of deformation is increased, the stress exhibits an "overshoot" characteristic, whereby at short times, during the transient response, the shear stress actually reaches a larger value than the ultimate long-time steady state value. As the rate of deformation is increased in succeeding tests, the amount of the overshoot also increases; the time at which an effective steady state is reached also diminishes with the increasing rate of deformation. This overshoot type of behavior has not been satisfactorily modeled by the constitutive relation(9.185). The Carreau form (9.188) does fairly successfully model the overshoot, as do most of the network theories mentioned earlier.

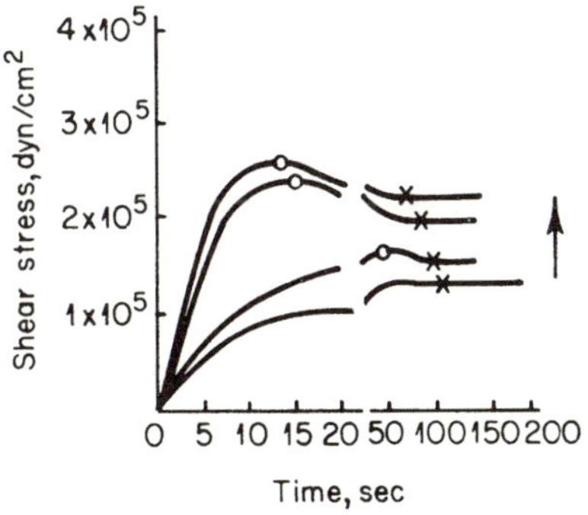

Fig. 9.4. Initiation of steady simple shear flow for polystyrene at 160° C (after Vinogradov and Belkin [9.70]). The arrow indicates increasing deformation rate.

The overshoot behavior is rather unexpected. With regard to a deformation measure taken with respect to the initial stress free undeformed position, the type of behavior shown in Fig. 9.4 describes a situation in which the stress, for a certain length of time, is decreasing even though this particular measure of deformation is continuously increasing. This behavior could have a number of possible mechanisms; for example, it could conceivably represent a dynamic response of the fluid speciemen, a combined dynamic interaction between the fluid speciemen and the testing machine, or perhaps some type of instability mechanism. However, a careful examination of typical experimental procedures and apparatuses reveals that none of these mechanisms are likely explanations. It appears that this anomalous behavior is due to a type of failure mechanism in the fluid. The term "failure" has a different meaning for a fluid from the usual implications in solids of crack formation and propagation. Rather, failure here, in the context of viscoelastic fluids, implies some changes on the scale of molecular dimensions which alter the basic constitutive nature of the material. For example, if molecular bonds and/or entanglements are broken and are not immediately reformed, then a behavior of the type considered here would be consistent. In contrast to failure in solids, the apparent failure in viscoelastic fluids is a reversible process, since testing results are repeatable on the same fluid specimen. Thus, it could appear that this possible mechanism of failure in viscoelastic fluids does not violate time translation invariance, as is implied by failure in solids.

The start up of extensional flow provides an even more complicated situation than that just considered for simple shearing flow. Consider first the model prediction from the analysis in Section 9.4. The very simple single integral model used there predicted the existence of unlimited stress growth at sufficiently high rates of extension. If the extension rate was greater than the reciprocal of the relaxation time, the instability would occur. Do real materials behave in this manner? Early data of Meissner [9.14] seemed to indicate so. More recent data, such as from Meissner *et al.* [9.71], appear to indicate that a stress maximum is reached, followed by a stress decline. It is uncertain whether a steady state stress is ultimately reached. The material used in these tests was low density polyethylene. This overshoot situation in extensional flow is similar to that in shear, but the uncertainty as to the existence of a steady state for extensional flow at high rates is a striking difference from the situation in shear flow. Especially if no steady state exists, the overshoot effect could be an indication of a type of material failure whereby the material tried to respond in the unlimited stress growth manner predicted by simple theories, reached a limit of response, and failed on a molecular scale. The limit of response could be a maximum stored energy, just as for elastic solids. More recent network theories, such as by Phan-Thien [9.65], do not predict the unstable stress growth of the simpler theories. Certainly, the possible instability in extensional flow and the overshoot behavior in extensional and shearing flows is a compelling indication of the difficulty in determining mechanical properties for viscoelastic fluid theories.

Another obvious approach to viscoelastic fluid theory would be a polynomial representation of the stress constitutive relation, similar to that employed in the case of viscoelastic solids. However, an examination of viscoelastic fluid data reveals that in typical ranges of interest the response is sufficiently nonlinear that a great many terms would be needed in a polynomial expansion to model the behavior adequately. Similar considerations, of course, also apply to viscoelastic solids; if only two or three terms are retained in a polynomial expansion, then in general there are strong limitations on the amount or type of nonlinearity which can be modeled. If the polynomial expansion method is considered impractical in the case of viscoelastic fluids, then alternative means must be sought, and some of these have already been mentioned. The starting point of another possible theory is given by part of the stress constitutive equation (8.74).

This is

$$\sigma_{ij} = -p\,\delta_{ij} + 4\int_{-\infty}^{t} \mu(t-\tau)\,\frac{\partial G_{ij}(\tau)}{d\tau}\,d\tau$$
$$- 4\int_{-\infty}^{t}\int_{-\infty}^{t} \mu(2t-\tau-\eta)\,\frac{\partial G_{ik}(\tau)}{\partial\tau}\,\frac{\partial G_{kj}(\eta)}{\partial\eta}\,d\tau\,d\eta \qquad (9.193)$$

where $\Delta(\tau_1, \tau_2)$ in (8.74) is taken in the reduced form

$$\Delta(\tau_1, \tau_2) = \mu(\tau_1 + \tau_2)$$

and where $G_{ij}(\tau)$ is defined by (8.47). This partial form of the stress strain relation (8.74), derived in Chapter 8 from a stored energy functional, is that part of a polynomial expansion form which reduces to the usual Newtonian viscous fluid behavior under sufficiently slow steady flow conditions. The relaxation function $\mu(t)$ is the linear theory relaxation function and can be determined in the usual way. We hypothesize that one primary cause of the nonlinear behavior of viscoelastic fluids is that under certain flow conditions the constitutive character of the material is changed, with these changes being fully accounted for by changes in the relaxation function $\mu(t)$. This hypothesis is motivated by the aforementioned apparent changes or temporary failure mechanism on the molecular scale. We confront a major problem, however, when we consider how to account for the dependence of the relaxation function on some characteristic or characteristics of the flow field. Visualizing these changes as a result of the temporary breakage of bonds and entanglements on the molecular scale, we see that stress is the field variable which most readily admits an interpretation on the molecular level and which could most conceivably be related to a failure criterion on that scale.

Accordingly, we postulate a nonlinear stress constitutive relation for viscoelastic fluids in the form

$$\sigma_{ij}(t) = -p\,\delta_{ij} + 4\int_{-\infty}^{t} \mu(t-\tau, J_2(\tau), J_2(\tau))\,\frac{\partial G_{ij}(\tau)}{\partial\tau}\,d\tau$$
$$- 4\int_{-\infty}^{t}\int_{-\infty}^{t} \mu(2t-\tau-\eta, J_2(\tau), J_2(\eta))\,\frac{\partial G_{ik}(\tau)}{\partial\tau}\,\frac{\partial G_{kj}(\eta)}{\partial\eta}\,d\tau\,d\eta \qquad (9.194)$$

where J_2 is the invariant of the deviatoric stress tensor s_{ij} defined through

$$J_2(\tau) = s_{ij}(\tau)s_{ij}(\tau) \qquad \text{where} \quad s_{ij}(\tau) = \sigma_{ij}(\tau) - 1/3\delta_{ij}\sigma_{kk}(\tau). \qquad (9.195)$$

The deviatoric stress tensor, rather than σ_{ij}, is employed here, but more general forms could include any type of dependence on stress. It should be noted that the dependence on the invariant J_2 is similar to the nature of the Mises yield criteria in the theory of inviscid plasticity. In fact, there are interesting parallels between the specifications of constitutive changes as influenced by stress here for fluids and in plasticity theory. The essence of the above specification is that the relaxation function in the general theory is taken to depend on the stress history, through the invariant J_2 of the stress deviation tensor.

It is informative to contrast the type of behavior represented by the constitutive relation (9.194) with the two previously mentioned constitutive relations(9.185) and (9.188). Although the three forms are different in several important respects, one essential difference is that the relaxation functions in (9.194), (9.188), and (9.185) depend respectively on the stress history, the rate of deformation history, and the deformation history, measured with respect to the current configuration. These represent apparently fundamental differences among the three types of theories. A theory involving the stress dependence of the relaxation function has been introduced by Bernstein and Shokooh [9.72]. Specifically, they have generalized the BKZ theory to include a dependence of the relaxation function on a shift parameter which is stress level dependent. The procedure is similar to that used by Schapery, which was discussed in connection with viscoelastic solids. Also, the theory of Phan–Thien[9.65] mentioned previously has a stress dependence of the relaxation function.

The discussion of these theories is intended to reveal the diversity of approaches in this field. It should now be evident that viscoelastic fluid theory is not at the point where there is a generally accepted constitutive relation, with associated standard procedures for determining the mechanical properties. The same is true for nonlinear viscoelastic solids. Many more nonlinear constitutive relations for viscoelastic solids and fluids have been proposed and studied than have been mentioned here. Those mentioned here, however, should be sufficient to reveal the nature of the difficulties with the topic, but should also serve to highlight the opportunities for further development and application.

PROBLEMS

9.1. Formulate reasons why it is expected to be "easier" and more successful to develop nonlinear theories of viscoelasticity for elastomers than for glassy polymers.

9.2. Compile a list of stress and strain imposed problems in the sense that these two terms are used in Section 9.1.

9.3. Examine Ref. [9.6] to deduce why shock waves are not amenable to predictive behavior in the manner that acceleration waves are found to be in Section 9.2.

9.4. Hardening materials are those for which the nonlinear stress strain curve is concave upward. Softening materials are those for which the stress strain curve is concave downward. Interpreting these terms for the instantaneous stress strain curve, relate the instability criterion for acceleration waves in Section 9.2 to these two types of material behavior.

9.5. Verify the recursion relation (9.123) for the second order Rivlin–Ericksen tensor $A_{ij}^{(2)}$.

9.6. Find the analytical means by which viscometric normal stress information can be extracted from Poiseuille flow. Discuss the expected difficulties.

9.7. In the cone and plate analysis of Section 9.3 we made the assumption of a small angle gap. Would the flow be viscometric if the gap angle were not small?

9.8. Discuss the complication of operating the cone and plate rheological device at large rates of rotation. Examine the literature on viscometry to see if N_2 can be deduced from the cone and plate viscometer.

9.9 Formulate the Couette flow solution, treating it as a viscometric flow rather than through the use of the Rivlin–Ericksen tensors, as was done in Section 9.4.

9.10. Compare the solution given by (9.138) for the free surface profile in the rod climbing problem, with the various order solutions given by Joseph and Fosdick [9.11]. Include the centrifugal force effect in the rod climbing solution of Section 9.4.

9.11. Discuss possible means by which shear thinning and normal stress thinning effects can be incorporated into the lubrication analysis of Section 9.5.

REFERENCES

9.1. Christensen, R. M., "A Nonlinear Theory of Viscoelasticity for Application to Elastomers," *J. Appl. Mech.* **47**, 762 (1980).
9.2. Rivlin, R. S., "Large Elastic Deformations of Isotropic Materials. IV. Further Developments of the General Theory," *Philos. Trans. R. Soc. London, Ser. A* **241**, 379 (1948).
9.3. Rivlin, R. S., and K. N. Sawyers, "The Strain-Energy Function for Elastomers," *Trans. Soc. Rheol.* **20**, 545 (1976).
9.4. Treolar, L. R. G., *The Physics of Rubber Elasticity*, 3rd ed. Oxford Univ. Press, London and New York, 1975.
9.5. Pipkin, A. C., "Small Finite Deformation of Viscoelastic Solids," *Rev. Mod. Phys.* **36**, 1034 (1964).
9.6. Chen, P. J., "Growth and Decay of Waves in Solids," *in Handbuch der Physik* (C. Truesdell, ed.), Vol. 6a, Part 3. Springer–Verlag, Berlin and New York, 1973.
9.7. Coleman, B. D., and M. E. Gurtin, "Waves in Materials with Memory. II. On the Growth and Decay of One-Dimensional Acceleration Waves," *Arch. Ration. Mech. Anal.* **19**, 239 (1965).

9.8. Pipkin, A. C., *Lectures on Viscoelasticity Theory*. Springer–Verlag, Berlin and New York, 1972.

9.9. Coleman, B. D., "Kinematical Concepts with Applications in the Mechanics and Thermodynamics of Incompressible Viscoelastic Fluids," *Arch. Ration. Mech. Anal.* **9**, 273 (1962).

9.10. Walters, K., *Rheometry*. Chapman & Hall, London, 1975.

9.11. Joseph, D. D., and R. L. Fosdick, "The Free Surface on a Liquid between Cylinders Rotating at Different Speeds," *Arch. Ration. Mech. Anal.* **49**, 321 (1973).

9.12. Coleman, B. D., H. Markovitz, and W. Noll, *Viscometric Flows of Non-Newtonian Fluids*. Springer–Verlag, Berlin and New York, 1966.

9.13. Truesdell, C., and W. Noll, *in Handbuch der Physik* (S. Flügge, ed.), Vol. 3, No. 3. Springer–Verlag, Berlin and New York, 1965.

9.14. Meissner, V. J., "Dehnungsverhalten von Polyäthylen-Schmelzen," *Rheol. Acta* **10**, 320 (1971).

9.15. Petrie, J. S., *Elongational Flows*. Pitman, London, 1979.

9.16. Tanner, R. I., "Increase of Bearing Loads Due to Large Normal Stress Differences in Viscoelastic Lubricants," *J. Appl. Mech.* **36**, 634 (1969).

9.17. Christensen, R. M., and E. A. Saibel, "Normal Stress Effects in Viscoelastic Fluid Lubrication," *J. Non-Newtonian Fluid Mech.* **7**, 63 (1980).

9.18. Batchelor, G. K., *An Introduction to Fluid Dynamics*. Cambridge Univ. Press, London and New York, 1967.

9.19. Green, A. E., and R. S. Rivlin, "The Mechanics of Non-Linear Materials with Memory. I," *Arch. Ration. Mech. Anal.* **1**, 1 (1957).

9.20. Lockett, F. J., "Creep and Stress Relaxation Experiments for Non-Linear Materials," *Int. J. Eng. Sci.* **3**, 59 (1965).

9.21. Ward, I. M., and J. M. Wolfe, "The Non-Linear Mechanical Behavior of Polypropylene Fibres under Complex Loading Programs," *J. Mech. Phys. Solids* **14**, 131 (1966).

9.22. Ward, I. M., and E. T. Onat, "Non-Linear Mechanical Behavior of Oriented Polypropylene," *J. Mech. Phys. Solids* **11**, 217 (1963).

9.23. Neis, V. V., and J. L. Sackman, "An Experimental Study of a Nonlinear Material with Memory," *Trans. Soc. Rheol.* **11**, 307 (1967).

9.24. Lifschitz, J. M., and H. Kolsky, "Non-Linear Viscoelastic Behavior of Polyethylene," *Int. J. Solids Struct.* **3**, 383 (1967).

9.25. Findley, W. N., and K. Onaran, "Product Form of Kernel Functions for Nonlinear Viscoelasticity of PVC under Constant Rate Stressing," *Trans. Soc. Rheol.* **12**, 217 (1968).

9.26. Gottenberg, W. G., J. O. Bird, and G. L. Agrawal, "An Experimental Study of a Nonlinear Viscoelastic Solid in Uniaxial Tension," *J. Appl. Mech.* **36**, 558 (1969).

9.27. Smart, J., and J. G. Williams, "A Power-Law Model for the Multiple-Integral Theory of Non-Linear Viscoelasticity," *J. Mech. Phys. Solids* **20**, 325 (1972).

9.28. Leaderman, H., *Elastic and Creep Properties of Filamentous Materials*. Textile Foundation, Washington, D. C., 1943.

9.29. Findley, W. N., and G. Khosla, "Application of the Superposition Principle and Theories of Mechanical Equation of State, Strain, and Time Hardening to Creep of Plastics Under Changing Loads," *J. Appl. Phys.* **26**, 821 (1955).

9.30. Lai, J. S. Y., and W. N. Findley, "Stress Relaxation of Nonlinear Viscoelastic Material under Uniaxial Strain," *Trans. Soc. Rheol.* **12**, 259 (1968).

9.31. Pipkin, A. C., and T. G. Rogers, "A Non-Linear Integral Representation for Viscoelastic Behavior," *J. Mech. Phys. Solids* **16**, 59 (1968).

9.32. Findley, W. N., and J. S. Y. Lai, "A Modified Superposition Principle Applied to Creep of Nonlinear Viscoelastic Material under Abrupt Changes in State of Combined Stress," *Trans. Soc. Rheol.* **11**, 361 (1967).
9.33. Lai, J. S. Y., and W. N. Findley, "Behavior of Nonlinear Viscoelastic Material under Simultaneous Stress Relaxation in Tension and Creep in Torsion," *J. Appl. Mech.* **36**, 22 (1969).
9.34. Findley, W. N., J. S. Lai, and K. Onaran, *Creep and Relaxation of Nonlinear Viscoelastic Materials*. North–Holland Publ., Amsterdam, 1976.
9.35. Stafford, R. O., "On Mathematical Forms for the Material Functions in Nonlinear Viscoelasticity," *J. Mech. Phys. Solids* **17**, 339 (1969).
9.36. Lockett, F. J., *Nonlinear Viscoelastic Solids*. Academic Press, New York, 1972.
9.37. Gradowczyk, M. H., "On the Accuracy of the Green-Rivlin Representation for Viscoelastic Materials," *Int. J. Solids Struct.* **5**, 873 (1969).
9.38. Coleman, B. D., and W. Noll, "Foundations of Linear Viscoelasticity," *Rev. Mod. Phys.* **33**, 239 (1961).
9.39. Lianis, G., "Time Reversal and Symmetry Relations in Finite Linear Viscoelasticity," *Int. J. Eng. Sci.* **18**, 1349 (1980).
9.40. DeHoff, P. H., G. Lianis, and W. Goldberg, "An Experimental Program for Finite Linear Viscoelasticity," *Trans. Soc. Rheol.* **10**, 385 (1966).
9.41. Goldberg, W., and G. Lianis, "Behavior of Viscoelastic Media under Small Sinusoidal Oscillations Superposed on Finite Strain," *J. Appl. Mech.* **35**, 433 (1968).
9.42. McGuirt, C. W., and G. Lianis, "Constitutive Equations for Viscoelastic Solids under Finite Uniaxial and Biaxial Deformations," *Trans. Soc. Rheol.* **14**, 117 (1970).
9.43. Ting, E. C., and G. Lianis, "Stress Analysis for a Nonlinear Viscoelastic Rubberlike Material," *Int. J. Solids Struct.* **8**, 999 (1972).
9.44. Schapery, R. A., "Application of Thermodynamics to Thermomechanical, Fracture, and Birefringent Phenomena in Viscoelastic Media," *J. Appl. Phys.* **35**, 1451 (1964).
9.45. Schapery, R. A., "On the Characterization of Nonlinear Viscoelastic Materials," *Pol. Eng. Sci.* **9**, 295 (1969).
9.46. Smart, J., and J. G. Williams, "A Comparison of Single-Integral Non-Linear Viscoelasticity Theories," *J. Mech. Phys. Solids* **20**, 313 (1972).
9.47. Peng, S. T. J., K. C. Valanis, and R. F. Landel, "Nonlinear Viscoelasticity and Relaxation Phenomena of Polymer Solids," *Acta Mech.* **25**, 229 (1977).
9.48. Murch, S. A., and P. M. Naghdi, "On the Mechanical Behaviour of Viscoelastic/Plastic Solids," *J. Appl. Mech.* **30**, 321 (1963).
9.49. Robotnov, Y. N., *Elements of Hereditary Solid Mechanics*. Mir, Moscow, 1980.
9.50. Valanis, K. C., "On the Foundations of the Endocronic Theory of Plasticity," *Arch. Mech.* **27**, 857 (1975).
9.51. Oldroyd, J. G., "On the Formulation of Rheological Equations of State," *Proc. R. Soc., London, Ser. A* **200**, 523 (1950).
9.52. Schowalter, W. R., *Mechanics of Non-Newtonian Fluids*. Pergamon, Oxford, 1978.
9.53. Astarita, G., and G. Marrucci, *Principles of Non-Newtonian Fluid Mechanics*. McGraw–Hill, New York, 1974.
9.54. Huilgol, R. R., *Continuum Mechanics of Viscoelastic Fluids*. Wiley, New York, 1975.
9.55. Bird, R. B., R. C. Armstrong, and O. Hassager, *Macromolecular Hydrodynamics*, Vol. 1. Wiley, New York, 1977.
9.56. Middleman, S., *Fundamentals of Polymer Processing*. McGraw–Hill, New York, 1977.
9.57. Bernstein, B., E. A. Kearsley, and L. J. Zapas, "A Study of Stress Relaxation with Finite Strain," *Trans. Soc. Rheol.* **7**, 391 (1963).

9.58. Bird, R. B., and P. J. Carreau, "A Nonlinear Viscoelastic Model for Polymer Solutions and Melts; I," *Chem. Eng. Sci.* **23**, 427 (1968).
9.59. Carreau, P. J., "Rheological Equation from Molecular Network Theories," *Trans. Soc. Rheol.* **16**, 99 (1972).
9.60. Bernstein, B., E. A. Kearsley, and L. J. Zapas, "Thermodynamics of Perfect Elastic Fluids," *J. Res. Natl. Bur. Stand., Sect. B* **68**, 103 (1964).
9.61. Lodge, A. S., *Elastic Liquids*. Academic Press, New York, 1964.
9.62. Liu, T. Y., D. S. Soong, and M. C. Williams, "Time-Dependent Rheological Properties and Transient Structural States of Entangled Polymeric Liquids—A Kinetic Network Model," *Polymer Engr. and Science* **21**, 675 (1981).
9.63. Johnson, M. W., Jr., and D. Segalman, "A Model for Viscoelastic Fluid Behavior Which Allows Non-Affine Deformation," *J. Non-Newtonian Fluid Mech.* **2**, 255 (1977).
9.64. Phan-Thien, N., and R. I. Tanner, "A New Constitutive Equation Derived from Network Theory," *J. Non-Newtonian Fluid Mech.* **2**, 353 (1977).
9.65. Phan-Thien, N., "A Nonlinear Network Viscoelastic Model," *J. Rheol.* **22**, 259 (1978).
9.66. Lau, H. C., and W. R. Schowalter, "On the Use of Mixed Corotational and Codeformational Properties in Constitutive Equations," *J. Rheol.* **24**, 507 (1980).
9.67. Doi, M., and S. F. Edwards, "Dynamics of Concentrated Polymer Systems Part 4—Rheological Properties," *J. Chem. Soc., Faraday Trans. 2* **75**, 38 (1979).
9.68. de Gennes, P. G., "Reptation of a Polymer Chain in the Presence of Fixed Obstacles," *J. Chem. Phys.* **55**, 572 (1971).
9.69. Doi, M., "Molecular Interpretation for the Linear Viscoelasticity of Concentrated Polymer Solutions and Melts—Explanation for the 3.4 Power Law," *Polym. Prepr., Am. Chem. Soc., Div. Polym. Chem.* **22**, 100 (1981).
9.70. Vinogradov, G. V., and I. M. Belkin, "Elastic, Strength, and Viscous Properties of Polymer (Polyethylene and Polystyrene) Melts," *J. Polym. Sci., Part A* **3**, 917 (1965).
9.71. Meissner, J., T. Raible, and S. E. Stephenson, "Rotary Clamp in Uniaxial and Biaxial Extensional Rheometry of Polymer Melts," *J. Rheol.* **25**, 1 (1981).
9.72. Bernstein, B., and A. Shokooh, "The Stress Clock Function in Viscoelasticity," *J. Rheol.* **24**, 189 (1980).

Appendixes

A. STEP FUNCTIONS AND DELTA FUNCTIONS

The unit step function, or Heaviside unit step function, is defined by

$$
\begin{aligned}
h(t) &= 0 \qquad \text{for} \quad t < 0 \\
&= 1 \qquad \text{for} \quad t > 1
\end{aligned}
\tag{A.1}
$$

and may be thought of as shown in Fig. A.1. A function $h(t - a)$ merely shifts the discontinuity in Fig. A.1, either to the right or left, depending upon the sign of a.

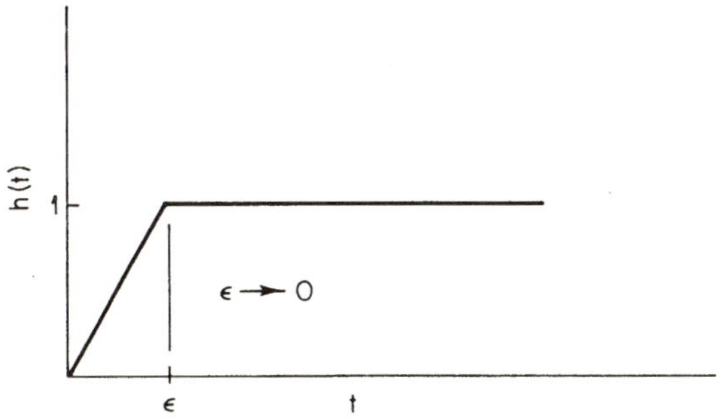

Fig. A.1. Unit step function.

The delta function, or Dirac delta function, $\delta(t - a)$ is defined as having the properties

$$
\delta(t - a) = 0 \quad \text{for} \quad t \neq a, \qquad \int_{-\infty}^{\infty} \delta(t - a)\, dt = 1. \tag{A.2}
$$

These properties can be associated with the function shown in Fig. A.1 as $\epsilon \to 0$. An important property of the delta function is that, for $f(t)$ being a continuous function of t in the neighborhood of $t = a$,

$$
\int_{-\infty}^{\infty} f(t)\, \delta(t - a)\, dt = f(a).
$$

The relationship between unit step functions and delta functions can easily be established. Consistent with the form shown in Fig. A.1, define $h(t)$ as

$$
\lim_{\epsilon \to 0} h(t) \begin{cases} = 0 & \text{for} \quad t \leqslant 0 \\ = t/\epsilon & \text{for} \quad 0 \leqslant t \leqslant \epsilon \\ = 1 & \text{for} \quad t \geqslant \epsilon. \end{cases} \tag{A.3}
$$

This, of course, corresponds to the definition of $h(t)$ given by (A.1). From (A.2) and (A.3), it can be seen that

$$dh(t)/dt = \delta(t) \tag{A.4}$$

where the derivative has been taken before $\epsilon \to 0$.

It is to be noted that these step function and delta function characteristics are obtained here in a nonrigorous, but formally correct, manner. The rigorous justification of these results must be obtained from distribution theory.

B. LAPLACE TRANSFORMATION PROPERTIES

Several properties of the Laplace transformation will be recalled here. For the proofs and further applications of these results, reference may be made to any one of the many books on the subject, such as Churchill [B.1] or Carslaw and Jaeger [B.2] or the more rigorous treatments of Doetsch [B.3] or Widder [B.4].

Definition

The Laplace transform, $\mathscr{L}f(t)$ or $\bar{f}(s)$, of a function $f(t)$ is defined by

$$\mathscr{L}f(t) = \bar{f}(s) = \int_0^\infty f(t)\, e^{-st}\, dt \tag{B.1}$$

where the transform parameter s may in general be complex. For the Laplace transform $\bar{f}(s)$ of $f(t)$ to exist, it is sufficient that $f(t)$ is sectionally continuous in every finite interval for $t \geqslant 0$, and that $f(t)$ is of exponential order as $t \to \infty$. This last requirement means that some constant α exists such that

$$\lim_{t\to\infty} e^{-\alpha t} f(t) = 0. \tag{B.2}$$

Thus, $f(t)$ must not grow more rapidly than at an exponential rate as $t \to \infty$.

Transform of Derivatives

Let the function $f(t)$ and its first $(n - 1)$ derivatives be continuous. Then the Laplace transform of the nth derivative of $f(t)$ is given by

$$\begin{aligned}\mathscr{L}\frac{d^n f(t)}{dt^n} &= s^n \bar{f}(s) - s^{n-1} f(0) - s^{n-2} f^{(1)}(0) \\ &\quad - \cdots - s f^{(n-2)}(0) - f^{(n-1)}(0)\end{aligned} \tag{B.3}$$

where $f^{(k)}(0)$ designates $d^k f(t)/dt^k$ evaluated at $t = 0$.

Transform of Convolution Integrals

The convolution of two functions $f(t)$ and $g(t)$, which are sectionally continuous, is defined by

$$\int_0^t f(t)\, g(t-\tau)\, d\tau.$$

The Laplace transform of this convolution integral is given by

$$\mathscr{L}\int_0^t f(t)\, g(t-\tau)\, d\tau = \bar{f}(s)\, \bar{g}(s).$$

It follows from $\bar{f}(s)\,\bar{g}(s) = \bar{g}(s)\,\bar{f}(s)$ that the convolution integral is commutative.

Inversion Integral

Assume that the Laplace transform $\bar{f}(s)$ is an analytic function of the complex variable s except at isolated singular points. The inverse transform of $\bar{f}(s)$ is given by

$$f(t) = \mathscr{L}^{-1}\bar{f}(s) = \frac{1}{2\pi i}\int_{\gamma-i\infty}^{\gamma+i\infty} e^{st}\bar{f}(s)\, ds \tag{B.4}$$

where $i = (-1)^{1/2}$ and the line $\mathrm{Re}(s) = \gamma$ is to the right of all singularities of $\bar{f}(s)$. Thus, the inversion formula represents an integration along a line parallel to the imaginary axis in the complex s plane. The evaluation of this integral is usually accomplished through the use of residue theory, as is now considered in the following special case.

Inversion for Simple Poles

Let $\bar{f}(s)$ have the form

$$\bar{f}(s) = P(s)/Q(s) \tag{B.5}$$

and let all the singularities of $\bar{f}(s)$ be due to simple pole singularities from the zeros of $Q(s)$. The residues of $\bar{f}(s)\, e^{st}$ at its singularities, $s = a_j$, are given by

$$\frac{P(a_j)\, e^{a_j t}}{\lim_{s\to a_j}\,[Q(s)/(s-a_j)]}\,.$$

Now, as shown in Fig. B.1, the line integral in (B.4) is closed by adjoining the infinite semicircular arc of radius $R \to \infty$ to it. The integral around the entire contour is, from residue theory, given by $2\pi i$ times the sum of the residues within the contour. For $|\bar{f}(s)| < cR^{-k}$ along the arc $s = Re^{i\theta}$ where k and c are

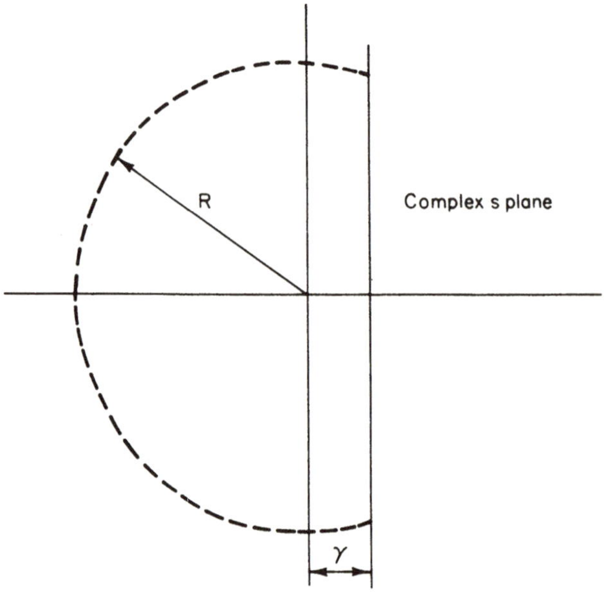

Fig. B.1. Contour integral.

constants with $k > 0$, the integral along the semicircular arc vanishes and the closed contour integration then corresponds to that in (B.4). Using residue theory gives (B.4) as

$$f(t) = \sum_{j=1}^{N} \frac{P(a_j)\, e^{a_j t}}{\lim_{s \to a_j} [Q(s)/(s - a_j)]} \tag{B.6}$$

where N is the number of simple poles. When $P(s)$ and $Q(s)$ are polynomials in s, (B.5) could be written in partial fraction form with the inversion following directly from that form.

Initial Value Theorem

The initial value $f(0)$ of $f(t)$ can be obtained from the transform $\bar{f}(s)$ through

$$\lim_{t \to 0} f(t) = \lim_{s \to \infty} s\bar{f}(s). \tag{B.7}$$

Actually, this result is easily generalized to obtain $f(t)$ for small values of time if $\bar{f}(s)$ can be expanded in a power series of terms involving $(1/s)^n$, $n \geqslant 1$. Then a term by term inversion applies.

Shifting Theorem

If the inverse transformation of $\bar{f}(s)$ is $f(t)$, the inverse transformation of $e^{-as}\bar{f}(s)$ is given by

$$\mathscr{L}^{-1}e^{-as}\bar{f}(s) = f(t - a)\, h(t - a) \tag{B.8}$$

where $h(t - a)$ is the unit step function defined in Appendix A.

REFERENCES

B.1. Churchill, R. V., *Operational Mathematics*, 2nd ed. McGraw-Hill, New York, 1958.

B.2. Carslaw, H. S., and J. C. Jaeger, *Operational Methods in Applied Mathematics*. Dover, New York, 1963.

B.3. Doetsch, G., *Handbuch der Laplace-Transformation*. Birkhäuser, Basel, Vol. 1, 1950; Vol. 2, 1955; Vol. 3, 1956.

B.4. Widder, D. V., *The Laplace Transform*. Princeton Univ. Press, Princeton, New Jersey, 1946.

Index

A

B

C

H

I

J

K

L

A CATALOG OF SELECTED

DOVER BOOKS

IN SCIENCE AND MATHEMATICS

A CATALOG OF SELECTED

DOVER BOOKS

IN SCIENCE AND MATHEMATICS

Astronomy

BURNHAM'S CELESTIAL HANDBOOK, Robert Burnham, Jr. Thorough guide to the stars beyond our solar system. Exhaustive treatment. Alphabetical by constellation: Andromeda to Cetus in Vol. 1; Chamaeleon to Orion in Vol. 2; and Pavo to Vulpecula in Vol. 3. Hundreds of illustrations. Index in Vol. 3. 2,000pp. 6⅛ x 9¼.
23567-X, 23568-8, 23673-0 Three-vol. set

THE EXTRATERRESTRIAL LIFE DEBATE, 1750–1900, Michael J. Crowe. First detailed, scholarly study in English of the many ideas that developed from 1750 to 1900 regarding the existence of intelligent extraterrestrial life. Examines ideas of Kant, Herschel, Voltaire, Percival Lowell, many other scientists and thinkers. 16 illustrations. 704pp. 5⅜ x 8½. 40675-X

A HISTORY OF ASTRONOMY, A. Pannekoek. Well-balanced, carefully reasoned study covers such topics as Ptolemaic theory, work of Copernicus, Kepler, Newton, Eddington's work on stars, much more. Illustrated. References. 521pp. 5⅜ x 8½.
65994-1

AMATEUR ASTRONOMER'S HANDBOOK, J. B. Sidgwick. Timeless, comprehensive coverage of telescopes, mirrors, lenses, mountings, telescope drives, micrometers, spectroscopes, more. 189 illustrations. 576pp. 5⅝ x 8¼. (Available in U.S. only.)
24034-7

STARS AND RELATIVITY, Ya. B. Zel'dovich and I. D. Novikov. Vol. 1 of *Relativistic Astrophysics* by famed Russian scientists. General relativity, properties of matter under astrophysical conditions, stars, and stellar systems. Deep physical insights, clear presentation. 1971 edition. References. 544pp. 5⅝ x 8¼. 69424-0

Chemistry

CHEMICAL MAGIC, Leonard A. Ford. Second Edition, Revised by E. Winston Grundmeier. Over 100 unusual stunts demonstrating cold fire, dust explosions, much more. Text explains scientific principles and stresses safety precautions. 128pp. 5⅜ x 8½. 67628-5

THE DEVELOPMENT OF MODERN CHEMISTRY, Aaron J. Ihde. Authoritative history of chemistry from ancient Greek theory to 20th-century innovation. Covers major chemists and their discoveries. 209 illustrations. 14 tables. Bibliographies. Indices. Appendices. 851pp. 5⅜ x 8½. 64235-6

CATALYSIS IN CHEMISTRY AND ENZYMOLOGY, William P. Jencks. Exceptionally clear coverage of mechanisms for catalysis, forces in aqueous solution, carbonyl- and acyl-group reactions, practical kinetics, more. 864pp. 5⅜ x 8½.
65460-5

THE HISTORICAL BACKGROUND OF CHEMISTRY, Henry M. Leicester. Evolution of ideas, not individual biography. Concentrates on formulation of a coherent set of chemical laws. 260pp. 5⅜ x 8½. 61053-5

A SHORT HISTORY OF CHEMISTRY, J. R. Partington. Classic exposition explores origins of chemistry, alchemy, early medical chemistry, nature of atmosphere, theory of valency, laws and structure of atomic theory, much more. 428pp. 5⅜ x 8½. (Available in U.S. only.) 65977-1

GENERAL CHEMISTRY, Linus Pauling. Revised 3rd edition of classic first-year text by Nobel laureate. Atomic and molecular structure, quantum mechanics, statistical mechanics, thermodynamics correlated with descriptive chemistry. Problems. 992pp. 5⅜ x 8½. 65622-5

Engineering

DE RE METALLICA, Georgius Agricola. The famous Hoover translation of greatest treatise on technological chemistry, engineering, geology, mining of early modern times (1556). All 289 original woodcuts. 638pp. 6¾ x 11. 60006-8

FUNDAMENTALS OF ASTRODYNAMICS, Roger Bate et al. Modern approach developed by U.S. Air Force Academy. Designed as a first course. Problems, exercises. Numerous illustrations. 455pp. 5⅜ x 8½. 60061-0

DYNAMICS OF FLUIDS IN POROUS MEDIA, Jacob Bear. For advanced students of ground water hydrology, soil mechanics and physics, drainage and irrigation engineering and more. 335 illustrations. Exercises, with answers. 784pp. 6⅛ x 9¼. 65675-6

ANALYTICAL MECHANICS OF GEARS, Earle Buckingham. Indispensable reference for modern gear manufacture covers conjugate gear-tooth action, gear-tooth profiles of various gears, many other topics. 263 figures. 102 tables. 546pp. 5⅜ x 8½. 65712-4

MECHANICS, J. P. Den Hartog. A classic introductory text or refresher. Hundreds of applications and design problems illuminate fundamentals of trusses, loaded beams and cables, etc. 334 answered problems. 462pp. 5⅜ x 8½. 60754-2

MECHANICAL VIBRATIONS, J. P. Den Hartog. Classic textbook offers lucid explanations and illustrative models, applying theories of vibrations to a variety of practical industrial engineering problems. Numerous figures. 233 problems, solutions. Appendix. Index. Preface. 436pp. 5⅜ x 8½. 64785-4

STRENGTH OF MATERIALS, J. P. Den Hartog. Full, clear treatment of basic material (tension, torsion, bending, etc.) plus advanced material on engineering methods, applications. 350 answered problems. 323pp. 5⅜ x 8½. 60755-0

A HISTORY OF MECHANICS, René Dugas. Monumental study of mechanical principles from antiquity to quantum mechanics. Contributions of ancient Greeks, Galileo, Leonardo, Kepler, Lagrange, many others. 671pp. 5⅜ x 8½. 65632-2

Physics

OPTICAL RESONANCE AND TWO-LEVEL ATOMS, L. Allen and J. H. Eberly. Clear, comprehensive introduction to basic principles behind all quantum optical resonance phenomena. 53 illustrations. Preface. Index. 256pp. 5⅜ x 8½. 65533-4

ULTRASONIC ABSORPTION: An Introduction to the Theory of Sound Absorption and Dispersion in Gases, Liquids and Solids, A. B. Bhatia. Standard reference in the field provides a clear, systematically organized introductory review of fundamental concepts for advanced graduate students, research workers. Numerous diagrams. Bibliography. 440pp. 5⅜ x 8½. 64917-2

QUANTUM THEORY, David Bohm. This advanced undergraduate-level text presents the quantum theory in terms of qualitative and imaginative concepts, followed by specific applications worked out in mathematical detail. Preface. Index. 655pp. 5⅜ x 8½. 65969-0

ATOMIC PHYSICS (8th edition), Max Born. Nobel laureate's lucid treatment of kinetic theory of gases, elementary particles, nuclear atom, wave-corpuscles, atomic structure and spectral lines, much more. Over 40 appendices, bibliography. 495pp. 5⅜ x 8½. 65984-4

AN INTRODUCTION TO HAMILTONIAN OPTICS, H. A. Buchdahl. Detailed account of the Hamiltonian treatment of aberration theory in geometrical optics. Many classes of optical systems defined in terms of the symmetries they possess. Problems with detailed solutions. 1970 edition. xv + 360pp. 5⅜ x 8½. 67597-1

THIRTY YEARS THAT SHOOK PHYSICS: The Story of Quantum Theory, George Gamow. Lucid, accessible introduction to influential theory of energy and matter. Careful explanations of Dirac's anti-particles, Bohr's model of the atom, much more. 12 plates. Numerous drawings. 240pp. 5⅜ x 8½. 24895-X

ELECTRONIC STRUCTURE AND THE PROPERTIES OF SOLIDS: The Physics of the Chemical Bond, Walter A. Harrison. Innovative text offers basic understanding of the electronic structure of covalent and ionic solids, simple metals, transition metals and their compounds. Problems. 1980 edition. 582pp. 6⅛ x 9¼. 66021-4

HYDRODYNAMIC AND HYDROMAGNETIC STABILITY, S. Chandrasekhar. Lucid examination of the Rayleigh-Benard problem; clear coverage of the theory of instabilities causing convection. 704pp. 5⅜ x 8¼. 64071-X

INVESTIGATIONS ON THE THEORY OF THE BROWNIAN MOVEMENT, Albert Einstein. Five papers (1905–8) investigating dynamics of Brownian motion and evolving elementary theory. Notes by R. Fürth. 122pp. 5⅜ x 8½. 60304-0

THE PHYSICS OF WAVES, William C. Elmore and Mark A. Heald. Unique overview of classical wave theory. Acoustics, optics, electromagnetic radiation, more. Ideal as classroom text or for self-study. Problems. 477pp. 5⅜ x 8½. 64926-1

CATALOG OF DOVER BOOKS

METHODS OF THERMODYNAMICS, Howard Reiss. Outstanding text focuses on physical technique of thermodynamics, typical problem areas of understanding, and significance and use of thermodynamic potential. 1965 edition. 238pp. 5⅜ x 8½. 69445-3

TENSOR ANALYSIS FOR PHYSICISTS, J. A. Schouten. Concise exposition of the mathematical basis of tensor analysis, integrated with well-chosen physical examples of the theory. Exercises. Index. Bibliography. 289pp. 5⅜ x 8½. 65582-2

RELATIVITY IN ILLUSTRATIONS, Jacob T. Schwartz. Clear nontechnical treatment makes relativity more accessible than ever before. Over 60 drawings illustrate concepts more clearly than text alone. Only high school geometry needed. Bibliography. 128pp. 6⅛ x 9¼. 25965-X

THE ELECTROMAGNETIC FIELD, Albert Shadowitz. Comprehensive undergraduate text covers basics of electric and magnetic fields, builds up to electromagnetic theory. Also related topics, including relativity. Over 900 problems. 768pp. 5⅝ x 8¼. 65660-8

GREAT EXPERIMENTS IN PHYSICS: Firsthand Accounts from Galileo to Einstein, edited by Morris H. Shamos. 25 crucial discoveries: Newton's laws of motion, Chadwick's study of the neutron, Hertz on electromagnetic waves, more. Original accounts clearly annotated. 370pp. 5⅜ x 8½. 25346-5

RELATIVITY, THERMODYNAMICS AND COSMOLOGY, Richard C. Tolman. Landmark study extends thermodynamics to special, general relativity; also applications of relativistic mechanics, thermodynamics to cosmological models. 501pp. 5⅜ x 8½. 65383-8

LIGHT SCATTERING BY SMALL PARTICLES, H. C. van de Hulst. Comprehensive treatment including full range of useful approximation methods for researchers in chemistry, meteorology and astronomy. 44 illustrations. 470pp. 5⅜ x 8½. 64228-3

STATISTICAL PHYSICS, Gregory H. Wannier. Classic text combines thermodynamics, statistical mechanics and kinetic theory in one unified presentation of thermal physics. Problems with solutions. Bibliography. 532pp. 5⅜ x 8½. 65401-X